CRITICAL APPROACHES TO SCIENCE AND RELIGION

CRITICAL APPROACHES TO SCIENCE AND RELIGION

EDITED BY MYRNA PEREZ SHELDON,
AHMED RAGAB, AND TERENCE KEEL

Columbia University Press
New York

Columbia University Press
Publishers Since 1893
New York   Chichester, West Sussex
cup.columbia.edu

Copyright © 2023 Columbia University Press
All rights reserved

Library of Congress Cataloging-in-Publication Data
Names: Sheldon, Myrna Perez, editor. | Ragab, Ahmed, editor. | Keel, Terence, editor.
Title: Critical approaches to science and religion / edited by Myrna Perez Sheldon, Ahmed Ragab, and Terence Keel.
Description: New York : Columbia University Press, 2023. | Includes index.
Identifiers: LCCN 2022030047 | ISBN 9780231206563 (hardback) | ISBN 9780231206570 (trade paperback) | ISBN 9780231556545 (ebook)
Subjects: LCSH: Religion and science.
Classification: LCC BL240.3 .C7475 2023 | DDC 201/.65—dc23/eng20221013
LC record available at https://lccn.loc.gov/2022030047

Columbia University Press books are printed on permanent and durable acid-free paper.
Printed in the United States of America

Cover design: Noah Arlow

We dedicate this book to bell hooks, Al Raboteau, and Howard Thurman for their courage, vision, and commitment to crafting a better world.

CONTENTS

Acknowledgments ix

Introduction 1
MYRNA PEREZ SHELDON, TERENCE KEEL, AND AHMED RAGAB

PART I: VALUES
TERENCE KEEL, AHMED RAGAB, AND MYRNA PEREZ SHELDON

1 Scripture of False Smiles: Lying with Erving Goffman 21
   KATHRYN LOFTON

2 Nihilism, Race, and the Critical Study of Science and Religion 41
   TERENCE KEEL

3 A Feminist Theology of Abortion 61
   MYRNA PEREZ SHELDON

4 Can Originalism Save Bioethics? 87
   OSAGIE K. OBASOGIE

PART II: BOUNDARIES
MYRNA PEREZ SHELDON, TERENCE KEEL, AND AHMED RAGAB

5 Spiriting the Johnstons: Producing Science and Religion Under Settler Colonial Rule 105
   TISA WENGER

6 Dark Gods in the Age of Light: The Lightbulb, the Japanese Deification of Thomas Edison, and the Entangled Constructions of Religion and Science 127
   JASON ĀNANDA JOSEPHSON STORM

7   Questioning the Sacred Cow: Science, Religion, and Race in
    the United States and India   159
    CASSIE ADCOCK

8   "And God Knows Best": Knowledge, Expertise, and
    Trust in the Postcolonial Web-Sphere   181
    AHMED RAGAB

PART III: NARRATIVES
AHMED RAGAB, TERENCE KEEL, AND MYRNA PEREZ SHELDON

9   Secular Grace in the Age of Environmentalism   207
    ERIKA LORRAINE MILAM

10 Performing Polygenism: Science, Religion, and Race in
    the Enlightenment   229
    SUMAN SETH

11 Out of Africa: Where Faith, Race, and Science Collide   255
    JOSEPH GRAVES JR.

PART IV: COHERENCE
AHMED RAGAB, TERENCE KEEL, AND MYRNA PEREZ SHELDON

12 Kānaka Maoli Voyaging Technology and Geography Beyond
    Colonial Difference   281
    ELI NELSON

13 Speculation Is Not a Metaphor: More than Varieties of
    Cryobiological Experience   304
    JOANNA RADIN

14 Maroon Science: Knowledge, Secrecy, and Crime in Jamaica   325
    KATHARINE GERBNER

15 Obeah Simplified? Scientism, Magic, and the Problem of Universals   348
    J. BRENT CROSSON

Conclusion   375
MYRNA PEREZ SHELDON, TERENCE KEEL, AND AHMED RAGAB

List of Contributors   381

Index   385

ACKNOWLEDGMENTS

The work for this volume began with support from a grant Critical Approaches to Science and Religion from the Templeton World Charity Foundation. We would like to thank the foundation, particularly Andrew Serazin, Peter Jordan, and Syman Stevens for their generous insights into the intellectual development of the project. The foundation supported a two-day symposium in 2019, research leave, and course development at Ohio University. Particular thanks go to Julie White and Cindy Anderson in the Women's, Gender and Sexuality Studies Program, as well as Bill Owens and Brian Collins in the Department of Classics and Religious Studies for their ongoing support in the administration and hosting of the grant and volume activities. Thank you also to Kristin Distel for her thoughtful work at many stages of the volume in copyediting and proofreading. We also want to thank Joseph Graves for introducing us to Eric Schwartz and facilitating the placement of the volume with Columbia University Press.

The inspiration for this book has come through many years of collaboration, conversation, and work. We would like to thank the Center for Brown, Black, and Queer Studies for hosting the archive of *Cosmologics* magazine, which was our space of creativity and experimentation for many years. We are grateful for the ongoing efforts of the center to make the vision of this volume into new pedagogical, academic, and political worlds.

CRITICAL APPROACHES TO SCIENCE AND RELIGION

INTRODUCTION

MYRNA PEREZ SHELDON, TERENCE KEEL, AND AHMED RAGAB

In *Critical Approaches to Science and Religion*, we offer a new method for studying science and religion that centers critical race theory, feminist and queer theory, and postcolonial and Indigenous analysis. Through this, we ask, How have histories of slavery, patriarchy, and colonialism shaped the emergence of science and religion in modernity? We are inspired by the sobering realization that those who inherit the unsolved problems of this generation will need resilient and creative tools to design a future that is more just and humane. A politically and ethically conscious understanding of science and religion should be a part of these future designs.

This critical orientation is a radical departure from the two paradigms that have shaped academic and popular thinking about science and religion over the past century. The first of these, the conflict thesis, emerged out of debates within Anglophone societies over science, evolution, and Christianity in the latter half of the nineteenth century.[1] It proposed that science and religion constitute timeless and irreconcilable ways of knowing the natural world. The conflict thesis has had an outsized influence beyond that immediate context, as it remains widely influential in global media, scholarly journals, political discourse, and K–12 education.[2] It has also persisted despite the efforts of historians to demonstrate that science and religion are categories bound to the specifics of time, place, and cultural context. In what would later be described as the complexity thesis, European and American scholars starting in the late 1960s sought to move beyond the use of military imagery and looked to emphasize the codependent relationship between science and religion.[3] They argued that the historically contingent nature of these two ways of knowing makes any one descriptive metaphor for their relationship an inadequate distortion. Scholars in the humanities have largely adopted the complexity thesis, even as the conflict position remains prominent throughout the life and social sciences.

Our critical approach draws on the historical methodology first advocated by the authors of the complexity thesis. Our own perspective centers histories of patriarchy, slavery, and colonialism in an analysis of science and religion in order to clarify the effects and lived experiences of both categories. This critical lens reveals how science and religion emerged alongside—and, indeed, constitute—modern power structures and identities, and, in turn, undergirded the emergence of race, gender, sexuality, class, and nationality in the modern world. Ultimately, defamiliarizing what is commonly understood about science and religion involves critique alongside new frameworks, methods, and writing practices to document the work these social creations have performed.

Nevertheless, we offer something more than a progression beyond conflict or even a simple addition to the lens of historical complexity. Our aim is not merely a new and lasting methodological validity. Rather, we seek something that these earlier paradigms were not designed to provide: the willingness to use historical insight to engage in contemporary political and social struggle. Recent global developments implicating both science and religion have helped make clear the need for frameworks that go beyond the political capacities of previous scholarly paradigms.

Consider, for example, that in the spring of 2015, hundreds of Kānaka Maoli gathered to block the only access road to Mauna Kea, a dormant volcanic peak of the island of Hawai'i, which was the proposed site of the new Thirty Meter Telescope (TMT). Because of its immense size and the optimal conditions on top of Mauna Kea, the TMT would enable astronomers to conduct research not possible with the existing instruments on the mountain. But Mauna Kea is deeply sacred to Kānaka Maoli. She is a relative.[4] When protesters blocked the road again in 2019 to prevent this violation of the āina, their actions energized the movement for Kānaka solidarity and self-determination, bringing the fact of U.S. colonial presence in Hawai'i sharply into focus.[5]

Or how in 2010, Turkey became the first country to ban its citizens from travel for the purposes of certain types of assisted reproductive technology (ART).[6] A secular country, Turkey nevertheless conforms to the Sunni Muslim fatwa issued by Al-Azhar University in 1980 stipulating that in vitro fertilization (IVF) treatment must be pursued in the context of heterosexual marriage.[7] It specifically prohibits the use of donor eggs and sperm, preventing queer and straight individuals who need third-party gametes from obtaining IVF treatment within Turkey. In response to this restriction, many Turkish citizens began traveling to nearby countries to receive donor-assisted IVF treatment. Turkey's 2010 prohibition of this travel was an extraordinary attempt to control the religio-reproductive lives of its citizens.

Or that currently in the United States, an increasing number of young Black women are leaving Christianity in favor of Black witchcraft and are forming digital covens.[8] Black witchcraft is just one expression of African diasporic religions that evolved in the response to the dehumanizing conditions of the Atlantic slave trade. Traditions such as Vodou in Haiti and Lucumí in Cuba kept African worldviews alive under the cover of enslavement and Christian conversion. They are finding renewed life in video conferencing technologies like Webex, which enable the transmission of copresence across temporal and geospatial realities throughout the Americas.[9] For Black witch covens, like Dawtas of the Moon, performing ancestral magic within a digital space has transformed the material and spiritual dimensions of Black diasporic identity, connecting practitioners to premodern rituals via late modern technologies.[10]

These three stories sit at the intersection of science and religion. Yet little is offered by analyzing whether science and religion are in conflict or harmony in each context. Nor are these situations adequately addressed by observing that they emerge from complex social and historical contexts. For the Kānaka Maoli, the threat to Mauna Kea comes from the steady march of American imperialism alongside astronomical science. A quasi-secular state is governing the reproductive lives of Turkish families and using IVF to police heteronormative values that predate this biotechnology. Contemporary Black witch covens can conjure and heal using much more than sage, root, and gemstones. Conflict, harmony, and complexity—the tried and true frameworks of scholarly analysis—fail to capture the political, bioethical, and materialist stakes of science and religion in these contemporary situations. It is this insufficiency in the current field and the desire to meaningfully address urgent political realities that prompt this volume.

Within a critical approach, a new set of analytical possibilities becomes available that allow us to recognize how science and religion alternatively function in both liberating and oppressive capacities. In the case of Mauna Kea, for instance, rendering the Kānaka Maoli worldview as a religion makes protection possible under U.S. freedom of religion statutes.[11] But it also exposes the community to denigration as backward or primitive in contrast to the universalizing claims of astronomic scientific research. Activists, artists, and Indigenous scholars are developing strategies of political resistance and social change grounded by the understanding that religious freedom and scientific research are predetermined by the colonial relationship between the United States and Hawaiʻi.[12] What is paramount in this case is the ability to design a future where there are mechanisms that value Indigenous worldviews and hold state and scientific actors accountable when religious belief is undermined in the name of apparent universal truth. At stake is developing knowledge for social change, not speculation about whether religion and science exist in an antagonistic or complementary relationship.

Importantly, for each of the contemporary issues recounted previously, critical theory scholars have already brought forward a wealth of insights. For instance, the field of feminist science studies has articulated the moral frontiers confronted by individuals who undertake IVF treatment.[13] And in the case of Black magic, critical scholars of religion have demonstrated how African diasporic spirituality emerges from racial and sexual assemblages in cooking and other daily rituals.[14] Beyond these examples, there is a wide and multidecade literature where the critical assessment of science and religion has grown among scholars who have avoided the intellectual traps of the conflict thesis while recognizing that the complexity thesis does not easily align with the progressive political and social sensibilities of scholars now working within the field. At its simplest and most fundamental, this volume offers many models and possibilities for what critical theories can offer to the study of science and religion.

Here we begin a conversation that is intended to be promising and provocative rather than exhaustive. We do believe that this effort should have the effect of expanding and transforming the existing contours of the field. In particular, a critical approach should call into question and reframe the field's persistent focus on Christianity and the intellectual history of Europe and the United States. But our intention is not solely, or even primarily, to correct existing scholarship by simply widening its gaze. Rather, we intend to reorient the ethical trajectory of this field to a more overtly political framework that centers the raced, gendered, sexualized, and colonized natures of science and religion. Ultimately, our hope is to use the study of science and religion as a venue for imagining alternative futures where we might live better with one another.

CRITICAL THEORIES

The term *critical approaches* is taken from our engagement with three areas of critical inquiry: critical race theory (CRT), feminist and queer theory, and postcolonial, decolonial, and Indigenous theories. Although each of these fields has distinctive histories and methodologies, together they emerged through the global transformations of the twentieth century, including the civil rights, women's, and decolonization movements. The 1950s and 1960s witnessed the rise of the New Left in the United States and Europe, as well as the slow decline of the nineteenth-century colonial enterprise. Connected to these politics, critical disciplines in the academy questioned the key legacies of Western modernity in the Americas, Europe, and Europe's former colonies. CRT developed in the United States within the context of legal studies as a theoretical orientation that exposed how racism is not an accidental feature but rather is integral to modern legal and political systems. Feminist and queer theory arose alongside international feminist and gay rights activism, focusing on the way that social power shapes gender relations and human sexuality. Postcolonial studies tackled debates around modernization, sovereignty, and nationalism in the newly independent states in order to grapple with the legacy of colonialism and the challenges of new world orders. While decolonial and Indigenous studies offered important space for reflecting on decolonization and on the continued lives of settler colonial regimes. Together these fields form the framework for our critical approach to the study of science and religion.

We rely on feminist studies in order to analyze the historical impact of patriarchy on the emergence of science and religion in modern times. Since the 1970s, feminist science studies has critiqued the models, modes of evidence, and rhetoric of modern science by calling attention to the gendered commitments in the modern study of life and the production of new technologies.[15] This work, in turn, was instrumental in the ongoing disciplinary development of the history of science and of science and technology studies (STS). Feminist approaches to science pushed these fields to abandon their fealty to the political neutrality of science and to instead grapple with how the knowledge regimes of modern science are bound up in views of gender, reproduction, and reproductive labor. For example, feminists highlighted the ways in which women were excluded from the institutions of science, even as scientific investigations were used to construct women as incapable of doing science.[16] The very definition of *science*—its empiricism, rationality, and objectivity—was tied to arguments that only masculine intellects were capable of science. This tautological definition of science excluded women from its practices and its category.

One result of this has been the feminization of religion as modern science's problematic other. Within this frame, contemporary science advocates portray religious belief as the cowardly purview of feminized minds. And this feminization was capacious enough to also signify, at one time or another, nearly all nonwhite European peoples who could not be emancipated from the hysteria and superstition of their primitive worldviews.[17] However, this gendering of religion has not meant that religious institutions and practices are liberatory for women or LGBTQ+ persons. On the contrary, feminist and queer scholars of religion have interrogated the deep history of patriarchy and heterosexism in various religious traditions while also working to imagine more progressive religious futures. Feminist and queer scholars of Islam, for example, have turned to inventive modes of the jurisprudential tradition of Muslim societies to advocate for social reform.[18] Key to these literatures are theoretical accounts of religion that imagine new possibilities, whether queer formations in Islam or feminist egalitarianism in evangelical Christianity. Thus, a feminist and queer approach to the study of science and religion foregrounds how these categories have been sexualized and gendered and how these dynamics have influenced their interplay in modern life.

We turn to postcolonial studies in order to examine the role of colonialism in shaping the epistemologies and practices of both science and religion. Since at least the nineteenth century, secularization narratives cast Eastern religions (such as Islam and Hinduism) and Indigenous cultures as inadequately evolved in relation to Western religion and, subsequently, to science. In this framework, Western Christianity (especially Protestantism) was framed as the inevitable precursor to scientific enlightenment. Western elites claimed that science was both universally true and the exclusive property of Euro-America. In particular, the glorification of European science grew out of a perceived contradiction between the "spirit of science" and religion. The former, often characterized as outgoing, inquisitive, and freethinking, was posited against the latter, labeled as backward, superstitious, and submissive to authority. During decolonization and the postcolonial period of the twentieth century, the struggle for progress in the Global South was often equated with a secularist narrative that borrowed from and reproduced the science and religion dynamic of the colonial era. In this context, the scholarly theses of conflict and conciliation are no longer abstract methods by which to understand how science and religion interact but are, in fact, paradigm-making concepts that attempt to define the postcolonial subjects and their relationship to Western power and its traditions. The failure to see, realize, or take sides in this perceived conflict/conciliation of science and religion was an even more damning sign of backwardness. As such, during the twentieth century, postcolonial states in the Global South had to contend with the complex, uncomfortable, and unsettled nature of the couple "science and religion," both as a problem that needs solving and as a necessary part of progress. Through this era, postcolonial scholars of religion and of medicine and science have critiqued these narratives, highlighting the racialized and sexualized legacies of these formations.

In recent decades, decolonial and Indigenous studies have further expanded this critique of science and religion as self-evident or coherent categories. Scholars in these areas have shown how native epistemologies have been subjected to various acts of violence, including forced schooling, the deprivation of legal status, and genocide. Nevertheless, the insufficiency of Euro-American thought in accounting for Indigenous and native epistemologies and worlds is obfuscated by the violent imposition of Euro-American ways of knowing. The already referenced debate over the proposed construction of the TMT is just one example of how American settler colonial legal systems, including those that define science and religion, fail to grasp Indigenous reality.

Ultimately, we intend for this project not only to bring critical theory into the study of science and religion but also to encourage scholars in critical studies fields to rethink their commitments to disciplinary formation. Our project puts questions that have been explored within science and religion—those of existential purpose, meaning making, and alternative imaginaries—at the center of critical reflection over race, political representation, material accumulation, and the epistemologies of science, medicine, and technology. With this volume, we encourage critical theorists to foreground issues of belonging, wholeness, and human flourishing in their analyses of identity categories such as race, gender, sexuality, class, region, and religion.

SCIENCE, RELIGION, AND WHITE SUPREMACY

Existing Euro-American accounts of science and religion, particularly in Anglophone settings, owe much of their shape to three narratives, each of which is critical to the cultural project of Western identity. These narratives give shape to the West, forming its epistemic intuitions, cultivating its moral virtues, and undergirding the logic of liberal governance. The first of these is the encounter between the West and Islam, in which the former is constructed as secular and rational and the latter as religious and illiberal. Here, the evident incommensurability of a directional adjective—*West*—and a religion—*Islam*—is not accidental. The West is almost never just a direction or a geography, and Islam is in many cases both an identity and a geography marking Muslims in the West as foreigners, suspects, or relics of the past.[19] The second of these is the ongoing colonial endeavors of Europe and the United States, in which Indigenous ontologies are depicted as nascent evolutionary forms on a progressive development trajectory from the primitive, to monotheism, to modern secularism. And the last of these is the flourishing of new religious and fundamentalist movements within Europe and the United States, which challenge the progressive march of reason and disrupt liberal governance that presumes shared, rational, and universal knowledge. Together these might be described as orientalism, colonialism, and creationism, and they have driven popular and scholarly accounts of science and religion since at least the Cold War period. And they have their own roots in nineteenth-century anthropological accounts of world religions and the development of modern science, which presumed Christianity as the necessary precursor to modern science.

Telling more diverse stories of science and religion—that is, heeding calls to examine contexts outside the intersection of Protestant Christianity and professional science—is a difficult task because it necessitates more than merely adding information to the existing scholarly paradigms. It requires us to recognize that the very frameworks of conflict, harmony, and complexity are themselves bound up in the ongoing project of Western identity. The ideas that we hold about the relationship between science and religion are a reflection of subject positions available to us within these frameworks that shape our understanding of truth. It is only through confronting this that we equip ourselves to recognize how both science and religion are deeply classed, raced, sexualized, and gendered—to see that individuals and cultures are "scientific" not simply because they follow a set of recognizable methods and practices but because their identities allow them to assume and claim ownership over a set of epistemic virtues.

Therefore, if we understand science and religion as concepts that induce and sustain a larger set of sensibilities, beliefs, and orientations that set limits on truth, meaning, and history, then we should have the courage and integrity to acknowledge that for the United States and former European colonies, science and religion have worked together to serve the interests of white supremacy. What we mean by white supremacy here is not reducible to colonial governments of the past, the identity politics of white nationalism, or the fringe, seemingly misguided, and distinctly modern extremists. White supremacy in the United States and Europe stems from cultural forces long in the making. As a historical project and conceptual paradigm, white supremacy could not be sustained if its destiny depended solely on an armed force to police and defend its territory. White supremacy is engendered and sustained often through passive consent and the uncritical belief that the values and ideas of the West produce unassailable truths for the future of our species. Combining postcolonial theory and new shifts in CRT scholarship, we recognize white supremacy as a universalizable worldview, a paradigm to end all paradigms, a historical destination for orienting all knowledge, spiritual formations, and material conditions of life. Race, gender, political orientation, and economic idealism present only temporary barriers to assimilation under the project of white supremacy as long as one is willing to accept that the end of history has been settled—that there is no global future outside of the universal humanism of European traditions with their cultural, scientific, and political orientations.

Where did white supremacy derive this power to set the course for all of the world's history? Surely modern racism did not emerge ex nihilo.[20] One answer now available to us implicates the religious traditions of the West. Scholars who have embraced the conceptual breakthroughs of CRT have argued that the anti-Jewish beliefs and practices of the early Christian church created the conditions of possibility for anti-Native, anti-Muslim, and anti-Black sentiments within early modern Europe and then later in its colonial territories during the modern era.[21] Yet not all scholars of science and religion have been willing to believe that the social practices and color symbolism of the ancient world have any connection to present-day racism. Indeed, there is a debate across the humanities and social sciences about the premodern origins of racism and its ties to Christianity and the cultural traditions of Europe.[22] Many who oppose the notion that contemporary racism has premodern antecedents have a habit of misrepresenting the arguments of scholars on the other side; there is the belief that those with more critical sensibilities around race are advocating for a constant and unbroken chain of racist ideas and practices that persist from antiquity to the present.[23] Or that revealing the presence of racism within the premodern context by default implies that racism was a central or the only organizing principle for premodern societies. A more careful assessment, however, would reveal that many scholars interested in racism argue that medieval and early modern thinkers drew on the racial concepts of the ancient world and the Abrahamic faith traditions to articulate and sustain modern forms of racism. This claim should not be controversial given the wealth of historical evidence to support it—some of which is discussed in this volume. Additionally, and this is particularly crucial, it is easy to mistake historical work on the formation of race in the premodern world with scientific ideas *about* race that, in fact, do argue contemporary humans are descendants of ancient people, constituting a largely unbroken chain of genetic traits and dispositions that distinguish them from other races.

We raise these issues to point out that fears about critical race scholarship have prevented historians of science and scholars of the ancient world alike from seriously considering how racism has shaped their subject matter and the very intellectual frameworks used to make claims about science and religion. We believe it is not unreasonable to consider racism/white supremacy as one of the many possible outcomes of the religious and cultural traditions of the West–in fact, no more unreasonable than it is to claim objective reason, democracy, or science as one of the natural outcomes produced by the Occident. Yet to deny honest and robust debate over the possibility that racism has ancient roots with contemporary implications is akin to the arguments currently in circulation among American political and cultural conservatives who reject the work of the 1619 Project on the grounds that they believe the Constitution and the jurisprudence of the early republic created a benevolent, race-neutral, and inherently inclusive social arrangement that was occasionally and sporadically manipulated by a few white men with racist agendas.[24]

Moreover the shadow cast by the racism of the Anglophone world has contemporary implications. For instance, when considering why racial minorities are underrepresented in the fields of science, technology, engineering, and mathematics in North America and Europe, we must recall the use of scientific knowledge itself to argue that persons of African, Asian, and Native American descent—who had been feminized as irrational, superstitious, and hysterical—were incapable of scientific work and thus exclude them. Moreover, modern medicine in the United States has prioritized white health over that of marginalized communities of color, producing and sustaining an unequal distribution of illness and death that all but guarantees shorter life expectancies for Black, Native, and Latinx people when compared to whites, regardless of social and economic status. This inequality might appear discordant with recent developments within biomedicine, where explicit scientific racism appears to have fallen out of favor since the end of World War II with the Nuremberg trials and the influence of liberalism within evolutionary biology.[25] Yet agents of white supremacy need not be militiamen or card-carrying nationalists. The spirit of white supremacy lives through the research of geneticists who attribute genetic causes to socially produced diseases that disproportionately affect nonwhite populations.[26] White supremacy is at play when the state-appointed medical examiners of Hennepin County declared that the biological mechanism responsible for George Floyd's death was his preexisting heart condition, not the lethal physical restraint used by police.[27] It is at work when border patrol agents, the U.S. Department of Agriculture, white American cattle herders, and environmental conservationists intentionally force African, Caribbean, and Muslim migrants crossing the Mexico border to travel across terrains that have known lethal levels of sun and heat exposure.[28] Its presence was felt by Nigerian and Black South African immigrants denied entry into Poland during Russia's aggressive assault on Ukraine, while scores of white refugees were given safe haven.[29]

We could continue. The point here is that given Europe and America's long history of valuing white populations over all others, we should not be surprised by persistent health disparities in the United States or in former European colonies, structural racism within biomedical research, the weaponization of terrain to kill immigrants, or anti-Blackness in Europe. White supremacy promises to recruit us all through designs that appear race neutral, objective, and reasonable. Antiracist work involves opposing these creations and revealing them for what they are. We believe scholars of religion and science can no longer afford to avoid this antiracist work and must be involved in conceiving more humane futures.

VOLUME ORGANIZATION

The contributions to this volume range across geography and time period, and they include both archival histories and anthropological studies. They are not a comprehensive survey of a critical approach to science and religion. They are a conversation of interconnected examples that make this approach tangible and reorient the animating themes of this scholarship.

The volume is organized into four parts, each of which centers around an animating question. In "Values," Kathryn Lofton, Terence Keel, Myrna Perez Sheldon, and Osagie K. Obasogie ask, How ought scholarship on science and religion shape our moral and political lives? These chapters move beyond historical and discourse analysis into the normative work of living well together in the contemporary world. In "Boundaries," Tisa Wenger, Jason Ānanda Josephson Storm, Cassie Adcock, and Ahmed Ragab respond to the question, How does a critical approach to science and religion reframe any attempt to define or distinguish these categories? Their histories demonstrate that science and religion are entangled in the power dynamics of patriarchy, colonialism, and slavery. In "Narratives," Erika Milam, Suman Seth, and Joseph Graves ask, What ethical and political work has been done with narratives about science and religion? In their chapters, they examine instances when a distinction made between science and religion animates different ethical and political agendas. And in "Coherence," Eli Nelson, Joanna Radin, Katharine Gerbner, and J. Brent Crosson explore the question, Should the study of science and religion seek coherence? These chapters confront the complexity of writing from a subaltern position and ask whether translation across a power divide is possible, necessary, or even desirable.

Each part begins with a brief introduction that further explains and situates its theme, in addition to discussing the ways that the chapters speak to one another or take the animating question in unexpected directions. This structure provides grounding for scholars and students who are interested in this critical approach while also leaving space for new possibilities in the field. It is in this interplay between a tangible specificity and further possibility that we see the most potential for exciting new work in the study of science and religion.

IMAGINING AN ALTERNATIVE FUTURE FOR SCIENCE AND RELIGION SCHOLARSHIP

Many of the scholars who contributed to this volume noted that the framework of this project asked them to work beyond the conventional boundaries of their disciplinary (or even interdisciplinary) fields. Because of this, we are deeply grateful for the intellectual generosity and creativity of these contributors—and especially their willingness to stretch beyond the limits that usually circumscribe their field. But it is worth pausing for a moment to ask, Why would this be? We suggest that these experiences have little to do with individual limitations but rather reflect constraints of existing disciplinary and interdisciplinary formations.

The most fundamental of these is the lack of support for scholars in STS or the history of science to theorize religion or for scholars of religion to theorize science. Within these separate domains, often with the critical theories we have outlined, scholars work with an expansive understanding of their categorical object of study. That is, STS scholars and historians of science do not limit themselves to analyses of science that are delimited by "what scientists do." Instead, they go beyond the laboratory and field to study the bureaucracies, institutions, epistemic postures, rhetorics, and narratives of science. Similarly, scholars of religion do not confine themselves to the activities of sectarian groups—religion is not defined solely by the walls of a synagogue, mosque, or church. Rather, religion is understood to manifest in cultural rituals, sacralizing practices, and myth-making exercises.

In each of these fields, it is not that science or religion is anything and everything and therefore also nothing. Rather, religion—or science—might be many things not immediately assumed or recognized. One of the volume contributors, Kathryn Lofton, argues in previous work that when references to devotion, worship, scripture, or priests happen outside of the confines of sectarian religion—when they are applied to Kardashian reality television or corporate culture—we should take this seriously if we want to understand the forms of religion in contemporary life.[30] Similarly, claims of accuracy and objectivity and appeals to experimentation that happen outside of the sanctioned spaces of science (for example, laboratories, universities, and hospitals)—as in the case of political polling, masking and unmasking in dealing with the COVID-19 pandemic, market speculation, and sports statistics—need to be taken seriously as part of the infrastructure that defines the attributes of science.

But the training to analyze religion or science both capaciously and specifically is rarely encouraged for both categories. And the result is that historical or contemporary accounts of science and religion are often framed at literal intersections—for example, the influence of sectarian religion on scientific practice, the impact of scientific discoveries on religious communities, or the individual religious commitments of working scientists. The result is that the field of science and religion does not take up many of the themes that occupy critical work, such as how to theorize materiality, the limitations of liberalism, or the reproductive logics of ethnonationalism. It is in a new scholarly intuition that we hope to train ourselves, and to inspire similar action by other scholars, so that the academic study of science and religion will no longer be defined in narrowly literal terms. Instead, a critical approach to science and religion reveals how the workings of these epistemic postures and institutional arrangements have an outsized influence on modern life.

But even as we push toward new disciplinary formations, we recognize the tensions and limitations in our own project. There is no single vision of an alternative future in this volume. There is no consistent rejection of the sensibilities of scientific liberalism that continue to shape much of our scholarship, even as we may seek to undermine it. Rather, we offer this series of provocations in the hope that they might embolden others to imagine more freely how the study of science and religion might bring about justice in the world.

NOTES

1. John Draper, *History of the Conflict Between Religion and Science* (New York: Appleton, 1874); Andrew Dixon White, *A History of the Warfare of Science with Theology in Christendom* (New York: Appleton, 1896).
2. For variations of the conflict thesis within more scholarly work in the twentieth and twenty-first centuries, see Emile Durkheim, *The Elementary Forms of Religious Life: A Study in Religious Sociology*, trans. J. W. Swain (London: Allen & Unwin, 1915); Edward Grant, *The Foundations of Modern Science in the Middle Ages: Their Religious, Institutional, and Intellectual Contexts* (Cambridge: Cambridge University Press, 1996); and Stephen J. Gould, "Nonoverlapping Magisteria," in *Intelligent Design Creationism and Its Critics*, ed. Robert T. Pennock (Cambridge, MA: MIT Press, 2001), 737–749.
3. Ian Barbour, *Issues in Science and Religion* (New York: Vantage, 1968); Thomas Torrance, *Theological Science* (London: Oxford University Press, 1969); Pierre Teilhard de Chardin, "Christology and Evolution," in *Christianity and Evolution*, trans. R. Hague (San Diego, CA: Harcourt, 1971), 76–95; John Hedley Brooke, *Science and Religion: Some Historical Perspectives* (Cambridge: Cambridge University Press, 1991); Peter Harrison, *The Territories of Science and Religion* (Cambridge: Cambridge University Press, 2015).
4. Cassie Ordonio, "Kanaka Maoli Views of Maunakea and TMT," *Ka Leo*, July 29, 2019, http://www.manoanow.org/kaleo/features/kanaka-maoli-views-of-maunakea-and-tmt /article_ f97a7272-b1a5-11e9-a227-1be3df6b2521.html.
5. Maile Renee Arvin, *Possessing Polynesians: The Science of Settler Colonial Whiteness in Hawai'i and Oceania* (Durham, NC: Duke University Press, 2019), 224–228.
6. Zeynep B. Gurtin, "Banning Reproductive Travel: Turkey's ART Legislation and Third-Party Assisted Reproduction," *Reproductive BioMedicine Online* 23, no. 5 (November 2011): 555–564; "Dawtas of the Moon," accessed August 1, 2020, https://www.dawtasofthemoon

.org/; Kecia Ali, *Sexual Ethics and Islam: Feminist Reflections on Qur'an, Hadith, and Jurisprudence* (Oxford: Oneworld, 2006), 193–198.
7. Marcia C. Inhorn, *Local Babies, Global Science: Gender, Religion and In Vitro Fertilization in Egypt* (New York: Routledge, 2003), 96.
8. Sigal Samuel, "The Witches of Baltimore," *The Atlantic*, November 5, 2018, https://www.theatlantic.com/international/archive/2018/11/black-millennials-african-witchcraft-christianity/574393/.
9. Aisha M. Beliso-de Jesus, *Electric Santería: Racial and Sexual Assemblages of Transnational Religion* (New York: Columbia University Press, 2015).
10. "Dawtas of the Moon."
11. Tisa Wenger, *Religious Freedom: The Contested History of an American Ideal* (Chapel Hill: University of North Carolina Press, 2017); Marisa Peryer, "Native Hawaiians on Coverage of Mauna Kea Resistance," *Columbia Journalism Review*, July 29, 2019, https://www.cjr.org/opinion/mauna-kea-telescope-protest-hawaii.php.
12. Arvin, *Possessing Polynesians*, 195–223.
13. Sarah Franklin, *Biological Relatives: IVF, Stem Cells, and the Future of Kinship* (Durham, NC: Duke University Press, 2013), 103–111.
14. Elizabeth Pérez, *Religion in the Kitchen: Cooking, Talking, and the Making of Black Atlantic Traditions* (New York: New York University Press, 2016).
15. Genes and Gender Collective, "Genes and Gender: A Symposium," January 29, 1977, box 1, folder 1, Records of the Genes and Gender Collective, Schlesinger Library, Radcliffe Institute, Harvard University, Cambridge, MA; Dorothy Burnham, "Biology and Gender," in *Genes and Gender: On Hereditarianism and Women*, ed. Ethel Tobach and Betty Rosoff (New York: Gordian, 1978); Marian Lowe, "Sociobiology and Sex Differences," *Signs* 4, no. 1 (October 1, 1978): 118–125; Helen E. Longino, "Can There Be a Feminist Science?," *Hypatia* 2, no. 3 (1987): 51–64; Londa Schiebinger, *Nature's Body* (Boston: Beacon, 1993); Londa Schiebinger, *Has Feminism Changed Science?* (Cambridge, MA: Harvard University Press, 2001).
16. Cynthia Eagle Russett, *Sexual Science: The Victorian Construction of Womanhood* (Cambridge, MA: Harvard University Press, 1989); Kimberly A. Hamlin, *From Eve to Evolution: Darwin, Science, and Women's Rights in Gilded Age America* (Chicago: University of Chicago Press, 2014).
17. These were crucial issues raised by Sander Gilman's *The Jew's Body* (New York: Routledge, 1991), 60–103, and also by Nancy Lee Stepan's "Race and Gender: The Role of Analogy in Science," in *The Racial Economy of Science: Toward a Democratic Future*, ed. Sandra Harding (Bloomington: Indiana University Press, 1993), 261–277.
18. Examples include the foundational works of Amina Wadud, Fatema Mernissi, and Zahra Eshraghi, among others, who developed important tools and methods to interpret the Quran from a feminist perspective. Ali, *Sexual Ethics and Islam*, 193–198.
19. See, for example, Zareena Grewal, *Islam Is a Foreign Country: American Muslims and the Global Crisis of Authority* (New York: New York University Press, 2014).
20. Benjamin Isaac argues that the social prejudices of the ancient Greek and Roman world were, in fact, racist and constituted early antecedents for modern racism. See Benjamin Isaac, *The Invention of Racism in Classical Antiquity* (Princeton, NJ: Princeton University Press, 2004). More recently, Terence Keel argues that conceptual formations born

within premodern Christianity created enduring frameworks used by modern thinkers to develop racist ideas with the life sciences that were novel in their content but ancient in their form. See Terence Keel, *Divine Variations: How Christian Thought Became Racial Science* (Stanford, CA: Stanford University Press, 2018), 14–17.

21. For key texts involved in the recent racial turn in early church and Christian studies more generally, see Denise Kimber Buell, *Why This New Race: Ethnic Reasoning in Early Christianity* (New York: Columbia University Press, 2005); J. Kameron Carter, *Race: A Theological Account* (Oxford: Oxford University Press, 2008); Willie Jennings, *The Christian Imagination: Theology and the Origins of Race* (New Haven, CT: Yale University Press, 2011); Susannah Heschel, "The Slippery Yet Tenacious Nature of Racism: New Developments in Critical Race Theory and Their Implications for the Study of Religion and Ethics," *Journal of the Society of Christian Ethics* 35, no. 1 (2015): 3–27; Keel, *Divine Variations*; and Katherine Gerbner, *Christian Slavery: Conversion and Race in the Protestant Atlantic World* (Philadelphia: University of Pennsylvania Press, 2019).

22. For an example of scholarship sympathetic to the linking of modern racism with the premodern world, see Thomas Hahn, ed., "Race and Ethnicity in the Middle Ages," special issue, *Journal of Medieval and Early Modern Studies* 31, no. 1 (Winter 2001).

23. See, for example, Venita Seth's representation of this debate in "The Origins of Racism: A Critique of the History of Ideas," *History and Theory* 59, no. 3 (September 2020): 343–368.

24. Nikole Hannah-Jones et al., "The 1619 Project," *New York Times Magazine*, August 18, 2019.

25. See Maurizio Meloni's discussion of what he calls the "four pillars of democratic biology" that emerged after World War II in his *Political Biology: Science and Social Values in Human Heredity from Eugenics to Epigenetics* (New York: Palgrave Macmillan, 2016), 159–187.

26. See Keel's critical discussion of the Slim Initiative in Genomic Medicine for the Americas (SIGMA) and its discovery that Neanderthal genes put Mexican and other Latin American populations at risk of developing type 2 diabetes, which is an illness widely recognized by public health experts as being produced by social, political, and environmental conditions. Terence Keel, "Race on Both Sides of the Razor," in *Symposium on Race and Science*, 5, no. 1 (Spring 2018): 12–15.

27. Scott Neuman, "Medical Examiner's Autopsy Reveals George Floyd Had Positive Test for Coronavirus," NPR News, June 4, 2020, https://www.npr.org/sections/live-updates-protests-for-racial-justice/2020/06/04/869278494/medical-examiners-autopsy-reveals-george-floyd-had-positive-test-for-coronavirus.

28. See Mary E. Mendoza, "Treacherous Terrain: Racial Exclusion and Environmental Control at the U.S.-Mexico Border," *Environmental History* 23, no. 1 (January 2018): 117–126. Scientists with antiracist commitments working in the fields of evolutionary biology, microscale adaptation, and environmental niche modeling have now begun to study the biological effects of these weaponized terrains on undocumented migrants. See Shane Campbell-Staton et al., "Physiological Costs of Undocumented Human Migration Across the Southern United States Border," *Science* 374, no. 6574 (December 17, 2021): 1496–1500.

29. Mehdi Chebil, " 'Pushed Back Because We're Black': Africans Stranded at the Ukraine-Poland Border," France 24, February 28, 2022, https://www.france24.com/en/europe/20220228-pushed-back-because-we-re-black-africans-stranded-at-ukraine-poland-border.

30. Kathryn Lofton, *Consuming Religion* (Chicago: University of Chicago Press, 2017).

PART I

VALUES

TERENCE KEEL, AHMED RAGAB, AND MYRNA PEREZ SHELDON

We believe that science and religion must be understood through the historical framing of slavery, patriarchy, and colonialism. These points of reference allow us to better represent the political realities of our contemporary world and reveal the roles science and religion play within our everyday lives in light of the long reach of history. Our focus on these framings also reflects value commitments that shape our scholarship: we believe we need new forms of human empathy, flourishing, and liberation. Scholars of science and religion have an opportunity to contribute to new concepts and systems of belief that mend and repair the violence of the past.

Because of this, we lead the volume with a part on values. In it, the contributors ask, How do we live well with one another? By raising this question, they challenge existing expectations of what histories of science and religion are meant to do. We are inspired by the willingness of scholars like Ronald Numbers, David Livingstone, Janet Browne, and Peter Harrison to think beyond philosophy and polemics in their accounting of these categories. But our desire is not for mere characterization; nor are we content with telling histories determined by the values of liberal tolerance and social pluralism. In other words, we are committing ourselves to proposals for what we might do, what our political lives must be, and what our collective values might look like.

Each of these chapters takes on an urgent issue and, in the process, makes a claim about value commitments that ought to shape the thinking of religion and science scholars. In "Scripture of False Smiles: Scholarship and Lying with Erving Goffman," Kathryn Lofton reflects on the ethics and purpose of lying in academic contexts. She begins with the question of whether scholarship can be scripture. Her question assumes that scholarship and activism are part of the same work in feminist, critical race, and postcolonial theory. But it is significant that Lofton's question is not whether academic work can be scholarship and be political or whether it can be scholarship and also activism. Her question is "Can it be scripture?" And in this, Lofton uses an almost scandalous word, one that evokes belief, faith, and metaphysical commitment. Whatever we might be doing when we study religion and science, surely writing scripture isn't one of the options. But in her meditation on lying—the reasons for lying, the twentieth-century sociologist Erving Goffman's fascination with lying, and the lying that is part of academic life—she exposes the mimetic relationship between theology and religious studies and also between the study of science and its practice. Because isn't religious studies or science studies itself a kind of lie? A lie born out of the assertion that religion (or science) is best studied objectively, that the men who created the discipline could study the things that they loved without fear of losing their respectability, lest anyone accuse them of actually believing anything. And so the question, "Can scholarship be scripture?" is a transgression and a claim that academic work might, and possibly should, be oriented toward something that enables us to live well together.

Terence Keel's chapter, "Nihilism, Race, and the Critical Study of Science and Religion," proposes a radical reorientation of the study of science and religion toward a communal ethic of "faith in the designs of others." Keel contends that the Christian value commitments that underwrote European colonization produced a form of nihilism that alienated us from one another and constrains our ethics to the life-denying goals of white supremacy. He provides a novel reading of Nietzsche and the development of scientific racism to explain our connection to the nihilistic values behind Dylann Roof's mass murder at the Emanuel African Methodist Episcopal Church in South Carolina on June 17, 2015. Keel argues that the white settler ontologies embedded in the religious beliefs and scientific practices of the early republic have nihilistic effects on those of us who inherit and make homes within Euro-American worldviews. He believes this has been hidden from our awareness, leaving us incapable of seeing how science and religion have contributed to racism.

In her chapter "A Feminist Theology of Abortion," Myrna Perez Sheldon considers the role of morality in public health policy by arguing for a new ethics of abortion that traverses the dichotomy of the pro-choice versus pro-life conflict. Drawing on a history of the legal and cultural contexts of the twentieth- and twenty-first-century United States, she argues that the legalization of abortion has been a tool of scientific population control, while the restriction of abortion rights has emerged from the politics of white Christian nationalism. Furthermore, she argues that we are able to understand the contours of contemporary abortion debates more clearly if we map them onto this historical interplay between scientific and religious authority. However, she moves beyond historical analysis and uses a textual practice of deep listening in order to advocate a radical trust in women. In this, she develops a scholarly practice that sets aside desire for epistemological control and moral righteousness. As a historian of science and feminist scholar, Sheldon advocates for a postsecular theology that does not rely on metaphysical assertion or doctrinal authority and is not delimited by liberal forms of political recognition. She proposes instead something more profound: making trust in women sacred in contemporary debates over abortion policy.

Finally, Osagie K. Obasogie, in his chapter "Can Originalism Save Bioethics?," examines the origins of contemporary bioethics. He reminds us how the field was conceived as a corrective to the long history of Christian Europe's anti-Semitism and modern scientific racism, which were largely responsible for the Holocaust. Importantly, he makes both historical and normative arguments in this chapter. Historically, he contends that bioethics has moved away from its initial provenance and has instead become an entirely procedural field that shies away from morality. But he moves beyond descriptive characterization to propose that the principle of originalism (borrowed from legal theory) can aid bioethics in returning to its former ethical strength. Rather than consigning morality to religion, Obasogie insists that bioethics should be the very field in which medicine directly confronts issues of value and meaning rather than solely biological definitions of life and death. Through a renewed understanding of the original principles of bioethics, Obasogie believes medicine can once again ask such questions as these: What is a good life? What is a valuable life? What does care look like?

Together these chapters reshape existing expectations for the form and content of scholarship on science and religion. They are informed by the recognition that we need to design and preserve value systems that redress and move beyond the histories of slavery, Western colonialism, and patriarchy that continue to haunt and hold captive our scholarly gaze. Recognizing the importance of this task changes the scopes, content, and geographic focus of a critical approach to scholarship on science and religion.

CHAPTER 1

SCRIPTURE OF FALSE SMILES

Lying with Erving Goffman

KATHRYN LOFTON

Can a liar tell truths? Can scholarship be scripture? These two questions animate this chapter. I think of them together because a particular scholar, a scholar whose work I have treated like scripture, led me to think a lot about lying in scholarship and in life.

Most definitions of scripture agree that, whatever scripture is, it is not science. Scripture may include theories of the earth and its emergence and records of societies and their reproduction, as well as accounts of miracles and interpretations of their meaning. Readers of scripture may hypothesize what scientific practice, if any, scripture encourages. But as concepts—*science* and *scripture*—each comprises the alternative to the other. Science is a procedure: the systematic study of the physical and natural world through observation and experiment. Scripture is a result: the revered texts of the world, understood as scripture because those devoted to them regard their words as sacred. Students of science and students of scripture agree that there is enormous variety within each category—a wide variety of things studied under the rubric of experiment or understood by communities as sacred. Work in science studies undermines any easy recourse to assuming science as universally systematic; work in religious studies undermines any neat assumptions about scripture's sacrality.

This chapter emerges from a sense that critical approaches to science and scripture have not accounted for the conjoined relationship between the social need for scripture and the social prescriptions for science. In his famous volume on scripture, the Islamic studies scholar and theorist of comparative religions Wilfred Cantwell Smith argued that we need a new common scripture, since society lacks "a shared vocabulary to enable it corporately to live well, or even to talk and think about doing so, or its members to encourage each other to aim so."[1] Here Smith, a scholar of religious studies—that is, a scientist of religion—is doing more than studying scripture. He is recommending its development for the formation of common good.

When I ask if a work of scholarship could be scripture, I do not mean to ask if it is right about everything or if it has its facts straight. I mean to ask if that work can be something that helps enable humans corporately to live well. Can scholarly work ever do that? Can it do that *and be scholarship*? Can scholarship be both scripture and science?

The person who made me think about these things, sociologist Erving Goffman (1922–1982), is not coincidentally a scholar with a complicated relationship to scholarship. People cannot stop reading him. He shows up on "best of" and "most significant" lists. Yet there is a little scientific nervousness about his scholarly legacy: you may nod as you read what he wrote, but scholars find his conclusions hard to replicate scientifically and wonder about the smoothness with which everything is narrated. Do readers often nod along because Goffman was a brilliant human interpreter or because he was an effective con? Is the smoothness because he was good with words or because he worked to hide his research flaws? Through a review of his writings, I return to the trampled bifurcation between science and religion, a division raised with fresh insight by the editors of this volume, to think about whether there is a relationship between being a genius and being a deceiver in the crafts of science, scholarship, and religion. His work offers a way to think about the interrelation of science and religion because his methodological slippage and his pile of well-thumbed books press us to think about the practice of both: about what it means to be methodologically careful and what it means to be sacred in word, and whether a person needs the former to achieve the latter.

My purpose here is not to convince you that Goffman authored scripture or to offer a new verdict on his career. Rather, it is to ask whether, on their way to improving scholarly science, intellectuals neglect the opportunity to forge the common good. The scholarly fight for a certain idea of right science may keep scholars from providing the scripture society needs.

We could measure by the ton the available social scientific literature on lying. Here are some relevant articles that appeared in the first months of 2019:

- "Carving Pinocchio: Longitudinal Examination of Children's Lying for Different Goals"
- "Lying in Bed: An Analysis of Deceptive Affectionate Messages During Sexual Activity in Young Adults' Romantic Relationships"
- "It's Alright, It's Just a Bluff: Why Do Corporate Codes Reduce Lying, but Not Bluffing?"
- "The Impact of Lying About a Traumatic Virtual Reality Experience on Memory"
- "Lying About God (and Love?) to Get Laid: The Case Study of Criminalizing Sex Under Religious False Pretense in Hong Kong"[2]

What unites these works—published in journals representing a variety of academic disciplines—is a commitment to understanding the social contexts and consequences of lying. As these research projects imply, humans mislead others to achieve many different goals: to stay out of trouble with their parents, to succeed professionally, to keep their heads straight, and to get themselves laid. Social scientists have decided that, in all verbal interactions, people negotiate their desire to tell the truth, their desire to benefit themselves, and their desire to please others. When these forces come into conflict, one way to resolve the tension—the tension between their desire to tell the truth and their sense that telling the truth won't benefit themselves or please others—is to lie. Research suggests lying is very common. Reports from diary studies suggest that, on average, people lie in one out of every three to five interactions.[3]

Scholars have worked to see if they can help humans better detect the lies of other humans and if science can help people lie less. It turns out, for example, that trying to make people cheat less is hard because cheating *feels good* to the cheaters.[4] Likewise, many criminological studies prove that the ability to detect lies is unreliable. Lie detection accuracy for laypersons in 54 percent, and most law enforcement groups do not perform much better. Despite this inability, people—especially the police—think they can tell when someone is lying. Studies show that law enforcement groups consistently overestimate their lie detection accuracy and express unwarranted confidence in their ability to detect lies.[5]

A great deal of sociological work has focused on the different cultural and historical norms about lying. Lying is a learned behavior. Psychologists observe that the deceptions by young children can be hard to detect because their behavior may not conform to what we think of as deceptive action. Young children are not as familiar with display rules that establish the appropriateness of behaviors in various contexts.[6] Those various contexts are culturally determined. For example, consider Chinese and U.S. medicine. Telling the truth to competent patients is widely affirmed as a biomedical obligation in contemporary medical practice in the United States. This was not always the case. For a long time, medical practitioners in the West took lying for granted. Plato argued against lying in every position save that of the physician because he thought that lying could be a useful form of treatment. Chinese medical ethics agrees and remains to this day committed to hiding the truth as well as to lying when necessary to achieve the best interests of the patient as determined by patient's family (a practice depicted in the 2019 Lulu Wang film *The Farewell*). The Chinese physician's first responsibility is to cooperate with the broader social need—not to tell a patient the truth about their condition.[7]

Lying is a subject studied in qualitative sociology, quantitative sociology, psychology, economics, and political science. It is a subject about which scientists write and think confidently, even as they also expose the confusions of finding the truth scientifically. Researchers proceed in their work on lying, exposing how lying happens, without ever worrying that their research subjects may be lying in ways that the researchers' own perceptions cannot yet grasp. And this is without even entering the murkier waters of the researchers themselves lying—that is, smudging or tilting findings to serve social need. The science of lying scrutinizes everything but itself.

In 1982, the Canadian sociologist Erving Goffman died of stomach cancer in Philadelphia. The photos accompanying his obituaries depict a middle-aged white man, seated and unsmiling, with unremarkable physical features. "Smiling, it can be argued, often functions as a ritualistic mollifier, signaling that nothing agonistic is intended or invited," he wrote in 1976.[8]

Trust me, then, when I tell you that Goffman would want people to ask: Why *isn't* this man smiling? Social psychologists have explained that scientific research shows people like photographs of smiling people better than those of unsmiling people. Science says that people normally judge a smiling person less harshly. They usually think smiling people are better in-law material and are less likely to maul other people on the subway or cause much trouble.[9]

Goffman is the reason science credits smiles with so much. Most of contemporary social science traces back to him. Nudge economics, behavioral modification, image management—anytime you see a headline telling you how small actions, gestures, or facial expressions affect your life (and how changing them could *improve* your life), you have run into Goffman's way of looking at the world. Even more, the entirety of performance studies—as well as the world of queer and feminist theorizing, with which performance studies overlaps—has him as one of its signifying progenitors. Few syllabi on performance studies begin without a passage from Goffman; reading his *Gender Advertisements* (1976), one discovers a direct preface to Judith Butler's seminal text *Gender Trouble* (1990).[10] Goffman is the grandfather of a vast amount of social science and humanistic research that observes how specific human gestures articulate contradictory social meaning.

So why isn't Goffman smiling in the photo that accompanied his obituary? The easiest answer is that smiling just isn't what serious men do in photographs if they want to be taken seriously as men. Waitresses and stewardesses smile, a gendered fact Goffman's brilliant student Arlie Russell Hochschild would examine with wry deftness in her 1983 book, *The Managed Heart: Commercialization of Human Feeling*. Professors, lawyers, and accountants stare grimly at the camera. Their unsmiling stare is their scientific neutrality and an assertion of masculinity.

Another reason Goffman didn't smile in the photographs is that he didn't smile much in his everyday life. He was born an outsider, a Jew on the Canadian frontier, and he spent much of his life beating against the walls of a social establishment that did not warm to his irritable replies to their social pleasantries. Colleagues said he could be kind to people without power, to people who were sick or poor or (as Goffman would describe them) *stigmatized*. To carriers of any kind of social power, though, he was impossible: railing against petty bureaucracies, dismissing hierarchy, and disrespecting anyone whose mind he deemed unworthy of his time. This included students and staff, family members, and people who thought they were his friends. An unsmiling Erving was a familiar sight to those with whom he worked at the National Institute of Mental Health at the University of California, Berkeley, and at the University of Pennsylvania;

to those with whom he lived as husband, father, and friend; and to those with whom he studied in Winnipeg, Toronto, and Chicago.[11]

Goffman's obituary photograph could be a performance to align him with other serious men; it could reflect his annoyance at everyone around him, including the intruding photographer. You may think there is something very different between the two: in one, his lack of a smile is a performance; in the other, his lack of a smile is an archive of his reality.

His gift to thought was to argue powerfully that there is no real difference. We are what we perform.

Published in 1952, "On Cooling the Mark Out: Some Aspects of Adaptation to Failure" is one of Goffman's earliest articles, and it sets the terms for everything he authored subsequently. In it, he captured from a marginal part of lived experience something big about how human beings handle one another and therefore define themselves. As a starting point, he described how cons operate their rackets. The con allows the person they target, the "mark," to win a little money in a rigged game. The con then convinces the mark, whose earlier success emboldens them, to invest a larger amount. Suddenly, there is some "accident," the mark is left penniless, and the operators of the con disappear quickly.

This is familiar ground for anyone who has ever watched a heist movie. Yet Goffman observed that the con had a further strategy to deal with a mark who was not willing to keep quiet about the embarrassing affair and "squawked" or "beefed" to the police. When an angry mark confronted the con, an additional step called "cooling the mark out" was added to the game: "one of the operators stays with the mark and makes an effort to keep the anger of the mark within manageable and sensible proportions . . . and exercises upon the mark the art of consolation." This "cooler" tries to define the situation for the mark "in a way that makes it easy for him to accept the inevitable and quietly go home. The mark is given instruction in the philosophy of taking a loss."[12]

Goffman abstracted the "cooling out" concept from the setting of the confidence game and applied it to human interactions more generally. He knew that con games rarely occur in everyday life. But he argued there are plenty of persons who need cooling out. Cooling out is needed when someone feels wronged, involuntarily deprived of success in a circumstance that suggests their own lack of ability or appeal contributed to their failure. There is, for example, the person who considers themselves a "lover" but is involuntarily relegated to the status of a "friend." There is also the long-serving employee who believes themself entitled to a promotion but who is passed over by management. If the institution does not provide a means to console the humiliated person in such a situation, the victim may make a scene, become violent, or sue. Cooling out is something someone with power over the mark does to make sure the mark doesn't create more trouble in their loss of power—and maybe also to keep the mark from exposing just how much power the con has and giving away the game to others.

Goffman is very funny in this article, at one point referring to the psychotherapist as "society's cooler."[13] For scholars of religion, it is hard not to think of religious rituals and religious leaders as potential fits to his model. Is religious ritual a cooling-out process? Are imams, pastors, priestesses, and rabbis serving as society's coolers?

Whatever humor Goffman deploys in this piece is balanced by the devastation of his overall thesis. He points out that people who *don't* get cooled out can—and often do—lose even more than the con took. Sandra Bland (1987–2015), the Black Texan who refused to stay calm when stopped by a traffic cop, refused to be cooled. She later took her own life in a jail cell. Goffman asked, chillingly, "What happens if the mark refuses to be cooled out?" There is a norm in Western society, he explained, that tries to persuade people to keep their chins up and make the best of it. What if people didn't? As I revise this chapter at the end of a summer of international protest about racialized and murderous policing, we have new assuredness that cooling out is a suppressive tool. It can also be a tool of the collective resistance when marks choreograph dissent in ways that resist the con's ability to individuate the refusal to be cooled.

To keep themselves safe from harm and to maintain their social place, humans tell a lot of little lies that limit personal risk and ensure individual standing. I'm smiling at you so that you don't think I'm trouble; I'm telling a joke to cover that awkward comment and try to keep everyone feeling good. Over the course of his thirty-year career, Goffman argued that humans do this work to present the self in a certain light during the various interactions that make up daily experience: how a person greets their waiter, how they shake the hand of a new colleague, whether they speak in a whisper or a shout. He defined human beings as performers whose main business is managing the impression they make.

Goffman observed that all people—from saints to sociopaths, from elder white men to young queer radicals—perform theatrically in their everyday interactions. Everyone has emotions they suppress and emotions they express; everyone has norms they think they follow and norms they flout. Humans convey multiple different kinds of things to survive in the multiple contexts of their survival. Goffman explained why humans do this using language drawn from religion. He wrote that people tell lies to manage impressions because they are "ritually delicate objects," because they are objects of "sacred value."[14] Humans require familiar practices of care to tend to their fragility. This care indicates something Goffman understood as having no other analogy but religious ritual. The origin of human care for human beings as a sacred practice is found in religion.

Realizing how everyday life is filled with acts of performance is less difficult for some people to recognize than for others. For marginalized people, it is often easier to realize the world's a stage. Minoritized persons learn the hard way how to figure out who in the room has power and how to anticipate the violence they might use to keep it. Such people drew Goffman's attention because he liked how they, the victims of society's violence, exposed the truth behind society's false fronts of politeness and order. He went to asylums and hung out in gambling halls because there he saw more clearly what civilized society likes to obscure—namely, how much we're all always performing to survive insult, impoverishment, and ignominy.

What goes on in a psychiatric hospital or prison in Goffman's work is a "ritual game of having a self," where the staff holds most of the face cards and all the trumps. A tête-à-tête, a jury deliberation, "a task jointly pursued by persons physically close to one another," and a couple dancing, debating, or boxing—indeed, all face-to-face encounters—are games in which, "as every psychotic and comic ought to know, any accurately improper move can poke through the thin sleeve of immediate reality."[15] Rules drive human behavior, but no individual simply submits to rules. They manage, finesse, interpret, and alter them to their benefit. People try to get little rewards; they try to avoid embarrassment; they are specific about how they differentiate themselves from the rest of the herd. People put in public what they can tolerate to expose.

And this is not a bad thing. As the anthropologist Victor Turner would explain, when people perform, they are "making, not faking."[16] People work to construct the interpersonal reactions they want. They explain themselves to other people.

Making the authentic self is not unveiling a secret. It is an imminent realization of representation and its reactions.

In the world of *The Colbert Report* (2005–2014) and *RuPaul's Drag Race* (2009–), everybody knows that people are performers. Still, though, individuals hold that there is something real behind the performance, that behind the scenes is the truth and onstage is the fiction. Goffman worked assiduously to show how false such a belief in true selves is. Performing does not mean someone is faking. It is about being more intentionally something a person wants to be. A drag queen isn't a fake woman; a drag queen is a person who wants the world to see her as a woman. Goffman helped us understand this work as something everyone does. Even more, he suggested people ought to think about what they can affect once they understand the universal fact of impression management.

Resistance to norms is often imagined as something requiring great energy or power. Goffman would point to how life is the most dramatically disrupted in the intimacy of regular interaction (during a toast at a graduation party, when walking into a library, while waiting at a traffic stop) and how the most powerful protests engage at this intimate level of everydayness. Consider the fact the mid-twentieth-century civil rights protests in the U.S. South saw individuals sitting in different spots on public buses or swimming in segregated pools. Nothing was more frightening to those in power than the shifting of (what Goffman would call) the interaction order. By looking at how human beings keep the world as it is through tiny acts of submission, Goffman taught how to resist through collective acts of choreographed dissent. Human beings determine social life by how they engage each other. Changing the world can also mean agreeing to disagree with the normal that humans collectively make, every day, through how they speak, with whom they speak, and what they do when they walk into a room.

When I tell people I am thinking a lot lately with Erving Goffman, I get often get one of two responses: "Who is that?" *or* "Oh, wow, I *love* him." The person who says the latter is the respondent that interests me, since people often point to a specific work by him to say how much it meant to them, how they read it at a particular moment, or how much he described the life they have had.

There is something of self-help in the way people refer to Goffman. (I'll leave it there for now—at the word *self-help*—even as I'm reaching, too, in how

I write about him to speak about it as scripture.) Many contemporary social scientists worry self-help talk does not help us as much as it keeps us the same. In *Stand Firm: Resisting the Self-Improvement Craze* (2017), the psychologist Svend Brinkmann argues that if people want a better world, it is better for them to stop helping themselves and instead help others. Goffman would agree and disagree with Brinkmann. He would agree that self-help does not do much for society. He would disagree that you can order people to serve the public good. Goffman was clear: You cannot convince people to stop helping themselves. You can only get them to realize how much they are doing so, helping themselves brace and survive through tiny efforts of trumping the con.

A lot of self-help writers have liked Goffman, thinking that his work gives ways to help people get what they want. Goffman claimed he specifically *wasn't* seeking to write handbooks on how to behave; he just saw himself observing how people survived their inequality. But his readers have not agreed. Among management and business self-help consultants especially, he is a go-to guru. Many "how to" guides rely on key concepts from his work. They include books on how to give a speech, how to deal with your in-laws, and how to live with a person who struggle with their mental health, as well as several more on how to transition to a new work environment. There are books advising leaders how to manage an organization in crisis and books on how to develop a *Second Life* avatar, how to give the impression of quality service, how to testify in court, how to be a bicyclist who shares a road with automobiles, how to be a coach, how to be a better conversationalist, how to understand why social encounters make you so unhappy, and how to perform gender.

I highlight these "how to" subjects because I think these works—popular and academic alike—show how Goffman's work is a resource for making social life work. His writings read like guides to what society is. But he didn't want us to stop there. As he wrote in his 1974 work, *Frame Analysis*, he studied the world of micro-interactions because he hoped there would be "greater resources to draw upon for intentionally unhinging the frame of ordinary events."[17] He didn't do all the work he did to teach a manager how to give a successful speech that would get her a promotion as much as he did it to get people to think about how they tend to get jobs and make speeches *in certain ways* and these certain ways maintain society. On the other side of all this observation about how human beings make society is a manual on how to unmake society, interaction by interaction.

When I began researching Goffman, I found the academic world divided about how to think about him. Everyone agrees he introduced a new vocabulary for describing public behavior, including concepts like frontstage, backstage, in-group, out-group, multiple selves, social roles, subordinated persons, total institutions, institutionalization, gatherings, situations, social occasions, stigma, framing, strategic interaction, interaction order, social order, social frame, impression management, role distance, face engagements, remedial work, and the performance of self. Scholars agree, too, that he made possible an increased appreciation for the role of affect and emotion (alienation, embarrassment, fun, shame, sympathy, approval, exoneration, understanding, amusement, coolness) in the strategic management of everyday social life.

To what end? For some people, like Geoffrey Nunberg, a Stanford linguist who reviewed Goffman's final book, *Forms of Talk*, "Mr. Goffman's moral is very sad" because in his evidence, Goffman found only the "management of impressions." "There is rarely an intimation that anything animates our performances beyond the terrible fear of being caught out," Nunberg wrote.[18] Nunberg is right: Goffman described in great detail how much work people put into staying within a frame of social understanding to avoid being "caught out" (to borrow from his language), to avoid being shamed or uncool. Nunberg is also wrong: this doesn't have to be understood as very sad. Indeed, as shown by subsequent decades of theorists thinking about race and sexuality and gender, being conscientious about performance can be enlivening and inciting. There is something powerful in the recognition that social life is not an intractable fact but something made through specific acts of interaction. I could walk out of my house this morning singing a Doris Day song and wearing a salmon pink shift dress; I could walk out of my house this morning wearing a military uniform and murmuring Nitty Scott lyrics. Social life isn't an assignment. It is a thing I get to coproduce.

This is a moment where there is a lot of complaining about snowflake college students and their dramatic reactions to things the old guard used to "take in stride," like being harassed on the street by construction workers or getting misgendered by the cashier. This is a moment when things that used to be products are increasingly now services (hence, the label *service economy*). In other words, this is a time when some older people are mad at younger people for being so sensitive about everything just as the main jobs those older people have given to younger people (in financial services, hospitality, retail, health, human services, information technology, and education) require those younger folks to circumscribe their feelings. The contemporary service economy is—as inheritors of Goffman have explained—also a *smile economy*, in which one's facial movements, gestures, and word choices could lead to a better or worse tip, a thumb's up or thumb's down rating, or a good or bad Yelp review. More and more, economic survival depends on how workers control their feelings to serve the customer's demands. More and more, an older guard of observers seems disgusted by all the feelings the precarious youth flout on social media.

Goffman helps us think about a world defined by service and discomfort with its realities. Today's economic livelihoods are more and more about humans working as social people who sell and salve and soothe other people (customers, students, tourists) to keep them buying, learning, and coming back for their specific service. Goffman mediates between the two sides—the members of the establishment who are uncomfortable with a culture of feelings and the workers whose work requires their feeling—showing how everyone shows feeling, even when they force smiles or keep quiet; he also shows how some people are more conscious of this than others. He describes those who abuse, those who call out the abuse, and the cycle that makes victims and losers of all.

Goffman's impact on social theory has been simultaneously "great and modest."[19] Several scholarly works consider the particularity of his significance and the role he played in the mid-twentieth-century establishment of the social sciences.[20] He is among the most influential sociologists—revered and cited but never experimentally reproduced—both daunting to those who try to write like him and worrisome to those who look closely at his irregular citational practices. Perhaps unsurprisingly, he chafed at the limits of academe. His work flouted scholarly rules: he conducted unsystematic ethnography, he wrote with few footnotes, he could not care less about statistical samples, and he cherry-picked illustrations from far-flung sources (bleak modernist novels, cheery hostess etiquette guides, *Playboy* interviews) to exhibit the truth of his theories.[21] Despite his widespread readership and influence, remarkably few scholars continue his work; there has been no distinct Goffman school.

If academics couldn't quite get over Goffman's dismissal of their devotion to transparent methods, regular readers were frustrated by the fact that he never told them what to do. If you wanted exact thoughts on how to behave in the modern world, he could be cagey. He told people how they behaved with each other but never said exactly how people might do otherwise. On the final page of his most famous work, *The Presentation of Self in Everyday Life* (1959), he explained that his key metaphors—"the language and mask of the stage"—had served their purpose over the course of the book and the reader could now drop them. "Scaffolds, after all, are to build other things with, and should be erected with an eye to taking them down."[22]

Goffman used thick descriptions to show how, precisely, all humans avoid risk and calculate their dealings to acquire a bit of dignity in the humdrum grind. He argued that this was a neutral observation. Even if *neutral* is the right word to apply to such a description of humanity, it remains a powerful gauntlet thrown. In the "Introduction" to his 1974 book *Frame Analysis*, Goffman included a response to readers who thought his work should evaluate social disadvantage and reveal the "true" reality behind performed appearances. "I think that is true," he wrote, observing that he does not do those things. "I can only suggest that he who would combat false consciousness and awaken people to their true interests has much to do, because the sleep is very deep. And I do not intend here to provide a lullaby but merely to sneak in and watch the way people snore."[23] He stared at the depth of the sleep and, in doing so, saw the pathways through which people might wake others up.

Queens and starlets, lawyers and bus drivers—everybody interacts with other people, and as they do, they make themselves. "Personal identity, then, has to do with the assumption that the individual can be differentiated from all others and that around this means of differentiation a single continuous record of social facts can be attached, entangled, like candy flows, becoming then the sticky substance to which still other biographical facts can be attached," Goffman wrote in *Stigma* (1963).[24] An individual's sense of being one of a kind is not opposed to their social identity but is dependent on social experience to appraise. There is no *I* without some *us*, some *we*, some *them*. This is painful for many, as society and social life do not extend a simple warm embrace. It is also powerful, underlining that society is something human beings contribute to create.

Religion offered a significant repository of metaphor for Goffman. In his work, he often used the category of ritual as an inevitable secular form. Consider this passage, drawn from a 1956 essay:

> In this paper I have suggested that Durkheimian notions about primitive religion can be translated into concepts of deference and demeanor and that these concepts help us to grasp some aspects of urban secular living. The implication is that in one sense this secular world is not so irreligious as we might think. Many gods have been done away with, but the individual himself stubbornly remains as a deity of considerable importance. He walks with some dignity and is the recipient of many little offerings. He is jealous of the worship due him, yet, approached in the right spirit, he is ready to forgive those who may have offended him. Because of their status relative to his, some persons will find him contaminating while others will find they contaminate him, in either case finding that they must treat him with ritual care. Perhaps the individual is so viable a god because he can actually understand the ceremonial significance of the way he is treated, and quite on his own can respond dramatically to what is proffered him.[25]

Any student of Émile Durkheim is unsurprised by Goffman's easy move from religious rites to secular rites. Like Durkheim, Goffman thought individual identity was secondary to social identity. Underlying all social rituals is what he called the "expressive order." This is most clearly on display in moments of disruption, such as when someone makes a "mistake" in the processing of a religious ritual by dropping a chalice or fumbling to close the ark. He helped to describe how the clergy manages mistakes to maintain the expressive order of the liturgy.[26]

Thinking with Goffman as scholars of religion might mean that we focus more on instances of fraud and forgery as instances in which the expressive order of religion is most clearly established. *Fraud* is not a term commonly used in religious studies—it does not appear in most dictionaries and encyclopedias of religion—even though designating frauds (e.g., false idols) is very much a part of religion's vocabulary and, from a cynical view, religion's history.[27] For example, of the twenty-seven writings that make up the Christian New Testament, at least ten are almost certainly forgeries—that is, writings whose authors falsely claim to be someone else. In his writing on these forgeries, Bart Ehrman argued that early Christian authors lied to advance the larger cause of truth. They lied about who authored what they wrote so that those who read the text would be more likely to believe the value of what was said. Reviewing Ehrman's work, David Brakke asks, leaning into Goffman's work, how these early Christian forgeries indicate the "performances of self and other that forgery facilitated." He suggests that this literary deceit wasn't only about missionary plausibility but also allowed the authors to play the part of the Christian author, to be seen as the thing they wanted most to be.[28] The study of religion is not new to thinking about religious identity as a performance. Goffman underlined this performance as strategic labor necessary for social survival *and* social resistance.[29] The fraud isn't just a lie; it facilitates a new social reality.

Lately I have come to realize that I will be spending the second half of my life explaining to people the lies I told in the first half. In the first half of my life, I lied constantly: about how I was feeling, about what I thought of your book, about whether I was having fun or liked you or agreed with your outrage. If I piled up all the lies that I told, it would be more words than there are in every scripture ever written.

If I told you the specific individual lies that I told, you might try to comfort me, saying that those are harmless or trivial lies. You might even use the phrase *white lie* to describe them, suggesting that I told these lies to avoid hurting someone's feelings. (Nobody alive in the twenty-first century thinks it's an inconsequential coincidence that the lies we excuse quickly we call *white*.) And this is true. I told most of these lies because I assumed that if I didn't tell them, the person to whom I was speaking would be hurt. Or worse—*I'd* be hurt.

Then I watched Dr. Christine Blasey Ford testify about her sexual assault by the U.S. Supreme Court nominee Brett Kavanaugh during a Senate Judiciary Committee confirmation hearing. I saw how nice she was, inviting the members of the inquisition tribunal who were pummeling her with questions to come over to her house and making light of her caffeine addiction just so everyone felt relaxed, so her interrogators might feel relaxed because they might perceive in her casual reference that she was relaxed. And I thought of that line attributed to Zora Neale Hurston: "If you are silent about your pain, they'll kill you and say you enjoyed it." And I knew that the problem wasn't just the white Republican men asking Dr. Ford ignorant questions. I knew that the problem was also how much people have lied about what they feel.

The social scientific scholarship on lying captures the frequency of lying but doesn't imagine that lying happens even in the frame of research, in the relations that produce research findings. This is a subject about which Goffman was an expert, but it is hard to incorporate into how scholars today live out their lives as researchers and into the practice of scholarship as a relation that presupposes something known as truth. Scholars don't talk about how much lying there is on the way to success, how many insincere smiles are offered, how many cheerful yeses are uttered while inside thinking "Please, no."[30] Scholars don't talk about what they hide in order to seem like they want to seem. Or, rather, they do talk about it, but they don't say it's lying. They say it's just good form.

Several years before the Immigration Act of 1967, long before the elections of Barack Obama and Donald Trump, and many years before *diversity* entered the academic lexicon, Goffman explained the nature of privilege:

> In an important sense, there is only one complete unblushing male in America: a young, married, white, urban, northern, heterosexual Protestant father of college education, fully employed, of good complexion, weight, and height, and a recent record in sports. Every American male tends to look out upon the world from this perspective, this constituting one sense in which one can speak of a common value system in America.[31]

During my time working in the academy, I have had to figure out how to talk about that common value system and the way it forces everyone—and that is to say, *everyone*—who doesn't look out on the world from that perspective into a lying posture: that is, if they want to succeed; if they want to win; if they want to be, as Senate Judiciary Committee member Orrin Hatch said of Dr. Ford, *an attractive and pleasing person*.[32]

Erving Goffman was the person who taught me that lying was whatever everyone did to avoid being more miserable and to experience some victory in the struggle of life. This was a revelation to me like those about which I have read many times in relation to converts to various religions. When a person recognizes a scripture as one that they think could be theirs, it isn't because they perceive in it something fantastical or fake. They recognize it as a truth. Imams, pastors, priestesses, and rabbis serve as society's coolers in a simple sense, offering counsel to distressed persons. They are also coolers in the deepest sense, insofar as they don't just say, "There, there"; they also work to return upset persons with their distressed questions—who am I? how can I live in a world like this?—to texts and rituals that encode answers that offer them truth. Science is a procedure toward reasoned answer; religion is too. The end for both is right answers for the big and small things that ail.

Goffman became scripture to some who read him because he seemed to see not only their struggle but also a way forward. "We find ourselves with one central obligation," he wrote in his final essay, "to render our behavior understandably relevant to what the other can come to perceive is going on."[33] This idea could describe conservative actions; it could require submission or deceit. Or it could describe how people alter perception and shift what is going on. "You have to act as if it were possible to radically transform the world," Angela Y. Davis wrote. "And you have to do it all the time."[34] I don't know if reading Goffman will have the same result for you. I do know that reading him and understanding how lying becomes integral to avoiding hurt might help realize the universal fact that people play themselves in everyday life. This may not be mere social science anymore. Maybe it is mission. What exactly is the difference?

NOTES

1. Wilfred Cantwell Smith, *What Is Scripture?* (Minneapolis: Fortress, 1993), 239.
2. Victoria Talwara, Jennifer Lavoie, and Angela M. Crossman, "Carving Pinocchio: Longitudinal Examination of Children's Lying for Different Goals," *Journal of Experimental Child Psychology* 181 (May 2019): 34–55; Margaret Bennett and Amanda Denes, "Lying in Bed: An Analysis of Deceptive Affectionate Messages During Sexual Activity in Young Adults' Romantic Relationships," *Communication Quarterly* 67, no. 2 (2019): 140–157; Jörg R. Rottenburger, Craig R. Carter, and Lutz Kaufmann, "It's Alright, It's Just a Bluff: Why Do Corporate Codes Reduce Lying, but Not Bluffing?," *Journal of Purchasing and Supply Management* 25, no. 1 (January 2019): 30–39; Tameka Romeo, Henry Otgaar, Tom Smeets, Sara Landstrom, and Didi Boerboom, "The Impact of Lying About a Traumatic Virtual Reality Experience on Memory," *Memory and Cognition* 47, no. 3 (April 2019): 485–495; Jianlin Chen, "Lying About God (and Love?) to Get Laid: The Case Study of Criminalizing Sex Under Religious False Pretense in Hong Kong," *Cornell International Law Journal* 51, no. 3 (2019): 553–607.
3. Heather Mann, Ximena Garcia-Rada, Daniel Houser, and Dan Ariely, "Everybody Else Is Doing It: Exploring Social Transmission of Lying Behavior," *PLoS ONE* 9, no. 10 (October 2014).

4. Neil E. Garrett, Stephanie C. Lazzaro, Dan Ariely, and Tali Sharot, "The Brain Adapts to Dishonesty," *Nature Neuroscience* 19 (2016): 1727–1732.
5. Amy-May Leach, R. C. L. Lindsay, Rachel Koehler, Jennifer L. Beaudry, Nicholas C. Bala, Kang Lee, and Victoria Talwar, "The Reliability of Lie Detection Performance," *Law and Human Behavior* 33, no. 1 (February 2009): 96–109.
6. Carolyn Saarni and Maria von Salisch, "The Socialization of Emotional Dissemblance," *Lying and Deception in Everyday Life*, ed. Michael Lewis and Carolyn Saarni (New York: Guilford, 1993), 106–125.
7. Ruiping Fan and Benfu Li, "Truth Telling in Medicine: The Confucian View," *Journal of Medicine and Philosophy* 29, no. 2 (2004): 179–193.
8. Erving Goffman, *Gender Advertisements* (Washington, DC: Society for the Anthropology of Visual Communication, 1976), 48.
9. For a review of the scholarship on smiling, see Eric Jaffe, "The Psychological Study of Smiling," *Observer* 23 (December 2010), https://www.psychologicalscience.org/observer/the-psychological-study-of-smiling
10. Judith Butler, *Gender Trouble: Feminism and the Subversion of Identity* (New York: Routledge, 1999). For two indicating examples, observe the prominent documentary place Goffman has in Henry Bial and Sara Brady, eds., *The Performance Studies Reader* (New York: Routledge, 2004), and the significant theoretical and historical role for Goffman in Richard Schechner, *Performance Studies: An Introduction* (New York: Routledge, 2002).
11. From Dmitri N. Shalin, in "Interfacing Biography, Theory and History: The Case of Erving Goffman," *Symbolic Interaction* 37, no. 1 (2013): 2–40, we learn that Goffman would not want his personal connected to his professional. As Shalin writes, "Everything we know about Erving Goffman indicates that he was averse to self-disclosure. He forbade his lectures to be tape-recorded, did not allow his picture to be taken, gave only two known interviews for the record, and sealed his archives before he died with the explanation that he wished to be judged on the basis of his publications. More than that, Goffman specifically disavowed research where scholars turn their attention to themselves" ((2). Shalin founded the Erving Goffman Archives, an online project that collects documents and memoirs illuminating Goffman's life and work. Most of the contributors are Goffman's students or scholars who knew him well and who blend their analysis with personal recollections. Shalin discusses the ambition and purpose of this archive at length in "Interfacing Biography, Theory and History."
12. Erving Goffman, "On Cooling the Mark Out: Some Aspects of Adaptation to Failure," *Psychiatry* 15, no. 4 (1952): 452.
13. Goffman, "On Cooling the Mark Out," 461.
14. Erving Goffman, *Interaction Ritual: Essays on Face-to-Face Behavior* (New York: Pantheon, 1967), 31, 33.
15. Goffman, *Interaction Ritual*, 91; Erving Goffman, *Encounter: Two Studies in the Sociology of Interaction* (Indianapolis: Bobbs-Merrill, 1961), 81.
16. Victor Turner, *From Ritual to Theatre: The Human Seriousness of Play* (New York: PAJ Publications, 1982), 93.
17. Erving Goffman, *Frame Analysis: An Essay on the Organization of Experience* (Boston: Northeastern University Press, 1974), 495.

18. Geoffrey Nunberg, "The Theatricality of Everyday Life," *New York Times*, May 10, 1981.
19. Gary Alan Fine and Philip Manning, "Erving Goffman," *The Blackwell Companion to Major Contemporary Social Theorists*, ed. George Ritzer (Malden, MA: Blackwell, 2003), 56. See also Ann Branaman, "Erving Goffman," *Profiles in Contemporary Social Theory*, ed. Anthony Elliott and Bryan S. Turner (Thousand Oaks, CA: SAGE, 2001), 94–106.
20. Philip Manning, *Erving Goffman and Modern Sociology* (Palo Alto, CA: Stanford University Press, 1992), and "Ethnographic Coats and Tents," in *Goffman and Social Organization: Studies of a Sociological Legacy*, ed. Greg Smith (London: Routledge, 1999); F. G. Bailey, *The Saving Lie: Truth and Method in the Social Sciences* (Philadelphia: University of Pennsylvania Press, 2003).
21. There is an essay to be written in which the mistakes of the father are viewed through the scholarly rise and fall of his daughter, Alice Goffman. Such an essay might focus on how Erving Goffman evaded in his lifetime the methodological and ethical indictments to which Alice Goffman was subject precisely through his elisions, selection of subjects, and occupation of a different moment of scholarly scientism. On Alice Goffman, see Christina Sharpe, "Black Life, Annotated," *New Inquiry*, August 6, 2014, https://thenewinquiry.com/black-life-annotated/; Paul F. Campos, "Alice Goffman's Implausible Ethnography," *Chronicle of Higher Education*, August 21, 2015, https://www.chronicle.com/article/alice-goffmans-implausible-ethnography/; Philip Manning, Sarah Jammal, and Blake Shimola, "Ethnography on Trial," *Society* 53 (2016): 444–452; and Lawrence Ralph, "The Limitations of a 'Dirty' World," *Du Bois Review* 12, no. 2 (Fall 2015): 441–451.
22. Erving Goffman, *The Presentation of Self in Everyday Life* (New York: Doubleday, 1959), 254.
23. Goffman, *Frame Analysis*, 14.
24. Erving Goffman, *Stigma: Notes on the Management of Spoiled Identity* (New York: Simon & Schuster, 1963), 57.
25. Erving Goffman, "The Nature of Deference and Demeanor," *American Anthropologist* 58, no. 3 (June 1956): 499.
26. Christopher M. Donnelly and Bradley R. E. Wright, "Goffman Goes to Church: Face-Saving and the Maintenance of Collective Order in Religious Services," *Sociological Research Online* 18, no. 1 (2013).
27. A. J. Droge, "'The Lying Pen of the Scribes': Of Holy Books and Pious Frauds," *Method and Theory in the Study of Religion* 15 (2003): 117. For emerging work on fraud in the study of religion and the United States, see David Walker, "The Humbug in American Religion: Ritual Theories of Nineteenth-Century Spiritualism," *Religion and American Culture* 23, no. 1 (Winter 2013): 30–74; Emily Ogden, *Credulity: A Cultural History of U.S. Mesmerism* (Chicago: University of Chicago Press, 2018); and Charlie McCrary, *Sincerely Held: American Secularism and Its Believers* (Chicago: University of Chicago Press, 2022).
28. David Brakke, "Early Christian Lies and the Lying Liars Who Wrote Them: Bart Ehrman's Forgery and Counterforgery," *Journal of Religion* 96, no. 3 (2016): 378–390.
29. Sumita Raghuram, "Identities on Call: Impact of Impression Management on Indian Call Center Agents," *Human Relations* 66, no. 11 (2013): 1471–1496.
30. Someone reading this may say "I never do that! I have only ever been perfectly forthright about my feelings and experiences!" I have always thought that someone who thinks this either has internalized their lying as the truth or is draped in privilege.

31. Goffman, *Stigma*, 128.
32. "Sen. Hatch Calls Ford 'Attractive,' 'Pleasing,'" PBS NewsHour, September 27, 2018, https://www.pbs.org/newshour/politics/sen-hatch-calls-ford-attractive-pleasing.
33. Erving Goffman, "Felicity's Condition," *American Journal of Sociology* 89, no. 1 (July 1983): 51.
34. Angela Y. Davis, "Black History Month 2014: Civil Rights in America," Southern Illinois University, Carbondale, February 13, 2014, https://www.youtube.com/watch?v=6s8QCucFADc (last accessed July 7, 2022).

CHAPTER 2

NIHILISM, RACE, AND THE CRITICAL STUDY OF SCIENCE AND RELIGION

TERENCE KEEL

The time is coming when we shall have to pay for having been Christians for two thousand years.

—Friedrich Nietzsche, *The Will to Power*

In this chapter, I identify nihilism as an important problem for scholars who want to study science and religion while taking seriously the effects of racism within societies built on Christian settler colonialism. I explain how life under Christian colonial designs produces a debilitating form of nihilism that alienates us from the social sources of our value commitments and threatens to limit our worldview to the intellectual horizon of white supremacy.[1] My understanding of settler colonialism is informed by the thought of critical Indigenous studies scholar Kim TallBear. What I mean by Christian nihilism draws on Friedrich Nietzsche and Albert Camus. For Nietzsche, nihilism has its origins in the Christian rejection of the world made by man for the eschatological kingdom of God—the true power governing life and human history. Taken in its fullest sense, Christian nihilism cannot simply mean the denial of God's existence—Christians were, after all, God-believing people. Christian nihilism, I argue, is the refusal to believe in the power and consequences of human designs, most especially the designs of nonwhite others. I explain how scientific racism is a paradigmatic expression of Christian nihilism; it involves the belief that the world is designed in such a way that the behavior, appearance, and life chances of human groups are determined by omnipresent biological forces inherited from past ancestors. We express our nihilism every time we pursue genetic explanations for human differences that assault our belief in the power of human conventions (e.g., discriminatory policies, exploitative economic relations, and state violence) to determine our life chances. Scientific racism has left us doubting the power of our own discriminatory creations and the liberatory designs of non-European others.

Scholarship within the field of science and religion was born outside of, and largely remains situated against, the designs of nonwhite others: the critical orientations of Indigenous studies, critical race theory, and postcolonial theory. We continue to produce stories about the history of science that remain unaffected by these (critical) designs. Such denial is a symptom of nihilism hidden within our own intellectual and spiritual gaze. It also leaves us ill prepared to document the continued violence and discrimination inspired by modern anthropology and perpetuated across the life sciences and the field of medicine. I show in this chapter how a critical assessment of scientific racism and its ties to the violence of white supremacy offers one model for integrating the critical designs of nonwhite others into the narratives we produce about science and religion.

THE ONTOLOGY OF COLONIAL DESIGNS

On November 3, 2018, in Seattle, Washington, the social scientist Kim TallBear, citizen of the Sisseton-Wahpeton Oyate, became the first Indigenous person to address the History of Science Society (HSS) in its annual Distinguished Lecture series. The theme of that year's meeting was Telling the Stories of Science, and I recall a palpable sense of anticipation and anxiety among HSS conference goers before her lecture. What story would TallBear share with a title like "Science v. Spirituality: A Dead-End Settler Ontology and Then What?"[2] This was a significant departure from the origins of the HSS Distinguished Lecture series, whose inaugural address in 1981 featured the historian Charles C. Gillispie and his account of how the Montgolfier brothers created the world's first piloted hot air balloon.[3] Gillispie, one of the "founding fathers" of the field, taught some of the first history of science courses in the United States during the 1950s while at Princeton.[4] Like many in his generation, he was of the view that science moved steadily toward an objective, value-neutral, and quantifiable worldview that produced knowledge that was transcultural in its significance and applicability.[5] TallBear and Gillispie could not be further apart.

TallBear clarified the danger of modern biology's assumed march toward value-free knowledge that attempted to narrate all of human global history. This intellectual campaign displaced Indigenous people from the material locations that inform their sense of being. She explained to us:

> Old school narratives of vanishing Natives, noble exploration, and "knowledge for the good of all" forged in the fires of American colonization and early Anthropology continue to animate 21st century science and the stories it tells about human history and nature. Indigenous critics of human genome diversity research know all of this. They know that the disciplines broadly have been key to disrupting Indigenous relations with place because the disciplines have justified the appropriation of so-called natural resources for nation building.[6]

Evolutionary science has helped make the disorder produced by Europe's stealing lands and bodies appear inevitable and permanent with claims that Indigenous people have been absorbed through genetic admixture or that they are on the verge of being lost to history. TallBear explained that these evidence-based claims turn European settler colonialism into a natural fact that "de-animates us, understands us as less alive." "We are produced of particular bodies of land and water that continue to be necessary to sustain us AS peoples."[7] She argued that Indigenous temporal and geographic orientations have endured and evolved despite the workings of settler colonialism.[8]

Modern anthropology in the West is not possible without settler colonialism. Land appropriation, European nation-states warring over natural resources, grave robbing, the beheading of deceased natives, enslaved Africans, south Asian subjects, the forced conversion of those same peoples to the Christian religion, and the denial of Indigenous sovereignty were the material and social conditions that made it possible for Johann Blumenbach, Samuel Morton, Josiah Nott, and so many others to study human diversity. This is now an undeniable truth in the field. Yet the colonialism that underwrote "the science of man," TallBear insisted, was an expression of a very specific Western *religious* worldview. The ontological distinctions between human/nonhuman and animate/deanimate, which we now take for granted, were not organic to Indigenous ways of knowing; they reflect Western faith traditions that declare God, "man," races, and nonhumans as distinct entities.[9] Western thinkers operating under these cultural assumptions integrate these ontological divisions into the frameworks used to study and manipulate human biology and solidify our domination over the natural world. Life viewed under these terms, TallBear argued, was an intellectual dead-end. So then she asked, "What alternative forms of science and religion are you willing to imagine for the future?"

I heard TallBear's address to contain a warning about the collateral effects of making an intellectual and spiritual home within Western colonial designs. She conjured for me Frantz Fanon, who reminds us in *Black Skin, White Masks* that the anti-Blackness of colonialization deforms the subjectivity of both the target of this violence and its (white) perpetuators.[10] In my judgment, settler colonial ontologies that distinguish spirituality from science, the living from the nonliving, and God from creation not only displace Indigenous notions of time, place, and being—they also *alienate* and *constrain* the moral conscience of those who inherit, inhabit, and defend this Euro-American ontology.

By *alienation*, I mean someone who finds value and purpose almost exclusively within the history and science that anchors whiteness. Our understanding of science and history remains at a distance from other value systems, most especially nonwhite creations—we don't center the history of America and its science from the vantage point of Indigenous nations or the descendants of enslaved Africans. There is a Eurocentric canon, and nonwhite actors are intelligible to the degree that their understanding of science and history accommodates this vision. The distance between ourselves and nonwhite others has been so long in the making and so effective that it no longer appears as a social convention; we have forgotten that the space was produced by defending provincial Eurocentric stories about human origins in the face of alternative non-Western creations.[11] With the proper distance secured, we inheritors of these stories are freed to experience the truth of history and science as though it were transhistorical and ultimately beyond human control. Uncritically ensnared in the experience of our own truths, we lose the need to remember that our science reflects the social history of the West. In the wake of this loss stand commitments to truths without accountability for the violence and discrimination they produce.

Tragically, the critical orientations of nonwhite others could be curative if only we could overcome our skepticism, distrust, and scientific chauvinism. Distance from the designs of nonwhite others then is a trap of our own creation. It becomes a type of alienation from the social and historical ground—namely, white supremacy—that produced and animated the values of modern anthropology and the study of race in the life sciences. Sylvia Wynter bridged the distance of this estrangement by exposing the history and values that established what she called "the overrepresentation" of the white European as the universal "Man."[12] But without closing the gap, the Christian colonial-settler values responsible for the scientific orientations of the West remain obscured, leaving us tethered to values we can't fully understand without the critical insights of nonwhite others who have suffered as a result of these orientations or who have maintained alternative designs. You see now the ethical dilemma: we Westerners almost exclusively use science to tell the story of human origins without the burden of remembering the racial violence and colonialism that provided its evidence base. Remind us of this history—like TallBear in her address to the HSS—and we double down on our alienation and insist that despite the work of a few bad-faith actors, the core of our scientific traditions and our practice as historians remains virtuous.

Yet the mere survival of the Indigenous, Blacks, and the other nonwhites for which the American republic was not designed means we must continue defending the experience of our scientific truths. Despite being occasionally confronted by the lives taken by those truths, we so desperately want to remain within their reassuring limits; within them, we've made identities, careers, and even kinships more meaningful than familial bonds. Asking us to bridge the space between ourselves and nonwhite or non-Western "others" as equals or on their own terms is too much to bear. From this vantage point, the moral compass of those who occupy the intellectual systems of settler colonialism is not entirely diminished. It is instead *constrained* to discovering and preserving through science and history, where the truth of white supremacy gives purpose and meaning to all of human life. When I use the term *white supremacy*, I am not speaking merely of card-carrying extremists or the identity politics of white nationalists. As George Lipsitz has argued in *The Possessive Investment in Whiteness* and Eddie Glaude in *Democracy in Black*, white supremacy is sustained and promoted through value systems that produce habits of mind, socioeconomic beliefs, and institutional arrangements that can appear race neutral and thus mask their connections to designs that reinforce the sovereignty and moral universe of whiteness.[13] The values and intellectual orientations that make the project of white supremacy a compelling destination for all human history are the problem. White supremacy can reign over our lives while being inclusive of nonwhite people—so long as they do not threaten these values. Herein lies the distinction between integration and abolition, where the former intends to create a home under the values of white supremacy (most often out of the will to survive) and the latter seeks to expose and remove such spiritual and intellectual commitments. Both positions are surely entangled with western values, even though the political stakes of integration and abolition are not the same. TallBear's address to the unwelcomed guests of a Sheraton Hotel, occupying the land of six Indigenous nations, left us with a troubling question: What are the costs to one's moral and social conscious when we uncritically inhabit colonial ontologies designed for white supremacy?

I believe one answer to this question can be found in the thoughts of Nietzsche. Although he did not theorize whiteness as such, I believe his discussion of nihilism offers tools for diagnosing the intellectual and spiritual effects of existence under life-denying beliefs. In its hostility to non-European ways of being, white supremacy is a rather obvious life-denying orientation. Nietzsche wrote at the end of the nineteenth century that Christian values guided European thought to seek explanations for life beyond the one where we find ourselves. This produced what he called Christian nihilism: a sustained distaste for life as it is given, substituting the world as it appears for an imagined one. Drawing on Nietzsche, I argue that life under Christian colonial designs manifests as a debilitating form of nihilism. Reflecting on the emergence of scientific racism from Christian thought, I expand Nietzsche's moral indictment of the West to argue that Christian nihilism appears within the contemporary life sciences and that scientific racism—derived and sustained by settler colonial ontologies—is a paradigmatic expression of this form of Christian nihilism. I demonstrate this by discussing the connections between the history of scientific racism and Dylann Roof's 2015 massacre of a Bible study group at the Emanuel African Methodist Episcopal Church in South Carolina. The (Black) life-denying actions of Roof bear the symptoms of a type of nihilism not limited to his extremism but also present within our historical and scientific imagination.

CHRISTIAN NIHILISM

Nietzsche left fragments of his thoughts about nihilism in published and unpublished works edited by his Nazi-sympathizing sister, Elisabeth Förster-Nietzsche.[14] The curation of these incomplete texts raises some doubts about what we can definitively say. Reproducing Nietzsche's view of nihilism has its challenges because he was forming that view near the end of his life and because there is evidence to suggest he never intended for nihilism to be a complete intellectual destination, a method of life, or a consistent worldview.[15]

Nietzsche's discussion of nihilism is often misunderstood as celebrating the elimination of all values and the destruction of accountability within society. Many of his liberal and conservative critics, recently joined by those among the alt-right in the United States, view Nietzsche as making a case for how the skepticism of modern science, the bureaucratic rationality of liberal state institutions that attempt to create equity across the body politic, and the destabilizing movement of modern capitalism have dissolved the social ties necessary to sustain the belief that humans are part of a grand cosmological unity created by God.[16] Having lost our sense of the divine reality that binds us as a species, we lose the ability to accept that life has any meaning at all. This permits all forms of behavior, as morality now appears to be nothing more than the tools of oppression deployed by those in power. I believe this to be a misreading of Nietzsche even though there is evidence suggesting he called for the suspension of our value commitments to morality and modern structures like liberal democracy.[17]

This interpretation overlooks the fact that Nietzsche understood nihilism to be rooted within the emergence of Christianity itself, not modernity. He writes that

> it is an error to consider "social distress" or "physiological degeneration" or, worse, corruption, as the cause of nihilism. . . . Distress, whether of the soul, body, or intellect, cannot of itself give birth to nihilism (i.e., the radical repudiation of value, meaning, and desirability). Rather: it is in one particular interpretation, the Christian-moral one, that nihilism is rooted.[18]

Economic inequality, opposition to liberal democracy, sociopathic behavior, and civil unrest are the symptoms of nihilism, not its cause. These symptoms, he argues, derive from designs that predate modernity and can be found within the constellation of value judgments that sustain a Christian worldview. He suggests there is something profoundly inhumane about the world-renouncing orientations of Christian morality.

Susannah Heschel's critique of Christian theology as a colonialist enterprise is helpful here.[19] She reminds us that after appropriating the God and eschatology of the ancient Israelites—and integrating it with Platonism—the early church formed a worldview that denied the validity of life as it appeared on earth.

Rejected were the truth claims, social systems, and beliefs of non-Christians occupying alternative cosmologies and forms of governance. This, of course, included the beliefs of Jews themselves, which had been superseded by God's new convent with the Christian followers of Jesus.

Yet Christian morality involved more than a renunciation of the ancient Israelites. It also exhibited a general hostility toward human conventions—what Nietzsche calls humanity's *will to power*. Nietzsche writes that

> it is the experience of being powerless against men, not against nature, that generates the most desperate embitterment against existence. Morality treated the violent despots, the doers of violence, the "masters" in general as the enemies against whom the common man must be protected, which means first of all encouraged and strengthened. Morality consequently taught men to hate and despise most profoundly what is the basic character trait of those who rule: their will to power.[20]

In my reading of Nietzsche, human conventions (e.g., legal regulations, religious beliefs, scientific practices, historical narrative, etc.) are the vehicles that move our will to power about the world, reshaping society in the image of its architect. Nietzsche suggests that the moral orientations of the West—as a result of its Christian heritage—were designed to be hostile toward human willing through social conventions designed to rule over the weak, the just, and the virtuous. We can find this moral hostility toward the political power of human conventions in John's gospel. Jesus, anticipating his death on the cross, is reassured by God through the voice of an angel about the universal significance of his sacrifice. Jesus says, "This voice came not because of me, but for your sakes. Now is the judgment of this world: now shall the prince of this world be cast out. And I, if I be lifted up from the earth, will draw men unto me."[21] In this Christian formulation, true power does not lie with the social systems of humans that govern the world—with Christ, the prince is cast out. Power lies with God, whose kingdom appears among believers who are drawn into what Denise Buell calls a "new race" that constitutes the body of Christ.[22] This holy kingdom has a history but is not limited by it—this nation of people sits within God, the unmoved mover beyond this world. Moral hostility toward the world of human convention is what Nietzsche meant by Christian nihilism. I believe the clearest expression of this nihilism is characterized by the conviction that our lives are ultimately governed by forces outside of human control.²³ Nietzsche allows us to see how religious tools designed for liberation from social and political tyranny can paradoxically create forms of disbelief and resentment that produce unfreedom.

Nietzsche believed that holding space for a truth that remains beyond human limits and outside social convention results in the "devaluing of values," where the world shaped by the human hand is set up against "an artificially built 'true, valuable' one."²⁴ The world of agency and social convention is a false world because it is limited by those qualities that make us human: fallibility, perspectivism, passion, and bias. According to our Christian culture, God, and thus also the truth we seek, do not (and should not) have these mortal qualities. Camus understood what was at stake here for Nietzsche, writing in *The Rebel* that

> Christianity believes that it is fighting against nihilism because it gives the world a sense of direction, while it is really nihilist itself in so far as, by imposing an imaginary meaning on life, it prevents the discovery of its real meaning: "Every Church is a stone rolled onto the tomb of the man-god; it tries to prevent the resurrection, by force." Nietzsche's paradoxical but significant conclusion is that God has been killed by Christianity, in that Christianity has secularized the sacred.²⁵

The stone rolled on the tomb of Christ is a dual metaphor. In one sense, it represents how Christian morality inhibits a type of intellectual and spiritual freedom that, if allowed to flourish, would mean the undoing of the Christian worldview itself.

In another sense, the stone—made of the earth and moved by "man" in the service of the church—represents how the secular in the West has been configured to sustain Christian morality and its life-denying nihilism. *Life-denying* in this context means a repudiation of alternative ways of being and the preservation of violent social conventions. The stone rolled over the tomb of the god-man represents Christianity's use of secular means to keep at bay supersession of its own worldview. This is a problem of its own making: Christianity's colonial appropriation of Judaism has oriented the faith toward supersessive commitments that, if allowed to flourish unrestrained, threaten Christianity's own truth and worldview. Built within Christianity is an effort to stop its replacement by ceding the life-denying power of its morality to the control of secular forces.²⁶ This strategy works as long as we maintain a civic Christian culture whose values and orientations can coexist within secular power. I am describing something akin to a type of civil religion that doesn't seek the domestication of Christianity in the service of state power but is closer to an unconscious theocracy that puts the state in the service of Christian values while evading recognition.[27]

To continue with Camus's metaphor of the stone rolled over the tomb of Christ, our theocratic culture can avoid detection and replacement through an agreement with state power. What could Christianity offer this bearer of a stone whose sense of modern freedom is shaped by rational empiricism, a hostility toward religious fanaticism, and anthropocentrism?[28] The history of scientific racism tells us that the answer is divinized whiteness, anointed by the church, secured by the natural sciences, and encoded into law. In this covenant with whiteness, our laws, politics, and scientific commitments become vehicles for the life-denying power of what Nietzsche called a "Christian moral interpretation of the world"; the nihilism of this arrangement is concealed by rendering Eurocentric orientations natural, genetic, and politically and biologically necessary. This explains the entitlement of American white supremacists who in August 2017 gathered in Charlottesville, Virginia, during the Unite the Right rally to chant: "You will not replace us! Jews will not replace us!" This covenant with Christian morality, whiteness, and secular power also explains the ease with which conservative Republicans declared that the insurrection of white supremacists at the U.S. Capitol four years later was a legitimate political protest. Contemporary white supremacists in the United States are radical agents tasked with pressing secular power to uphold the Christian settler values that established the republic.

From this vantage point, science in the service of whiteness is one of many (secular) life-denying forces that keep at bay the supersession of the Christian moral interpretation of the world—scientific racism is the metaphorical stone commissioned by the church to deny resurrection of the nonwhite, non-Christian "man-god." This same stone also suppresses our ability to believe in the power of human social conventions. Through science, white supremacy has sustained psychological and spiritual attachments to life-denying orientations.

LIFE-DENYING ORIENTATIONS

On February 26, 2012, twenty-one-year-old Dylann Roof came to the realization that white Americans were on the verge of becoming a vanishing people, a population that would be superseded biologically and politically by Black Americans and the Latinx. This was the day Trayvon Martin was murdered by self-appointed vigilante George Zimmerman while walking back from a convenience store in Sanford, Florida. Roof would write in his five-page manifesto that the killing of Trayvon Martin brought him to see that Zimmerman was bringing balance to a three-hundred-year race war that white Americans were losing. He had this to say in his journal, written before the shootings:

> The event that truly awakened me was the Trayvon Martin case. I kept hearing and seeing his name, and eventually I decided to look him up. I read the Wikipedia article and right away I was unable to understand what the big deal was. It was obvious that Zimmerman was in the right. But more importantly this prompted me to type the words "Black on White crime" into Google, and I was never the same since. The first website I came to was the Council of Conservative Citizens. There were pages upon pages of these brutal, disgusting black on White murders. I was in disbelief. At this moment I realized that something was very wrong. How could the news be blowing up this Trayvon Martin case while hundreds of these black on White murders got no airtime?[29]

Convinced that whiteness was under assault, Roof, like Zimmerman, took it upon himself to restore the social order against Black life. On the evening of June 17, 2015, he shot and killed nine African Americans who had gathered for Bible study at the Emanuel African Methodist Episcopal Church in Charleston, South Carolina. The location was carefully chosen. The church had been founded in 1816 and was under constant assault by whites who feared the political visions of Black folk with the courage to design spiritual and intellectual spaces beyond the horizon of white supremacy.[30] Denmark Vesey, a free Black man, preacher, and abolitionist, was one of the church's founders.[31] He and his congregation paid dearly for their abolitionist designs. Denmark was executed on July 2, 1822, in Charleston for plotting a slave revolt.[32] Several iterations of the church he helped create were burned to the ground by Americans defending whiteness and gripped by the fear that they might be replaced by Black life.

The structure Dylann walked into on June 17, 2015, was a sacred representation of what Robin Kelly would call Black freedom dreams.[33] This iteration of the building was completed in 1892 and endured the racial redemption of the South by white America, the systematic implementation of Jim Crow segregation, the assassination of Dr. Martin Luther King Jr., the starts and stops to the Black power movement, and the election of Barack Obama. Hidden in the church is a brass diorama honoring Vesey and its other founders. Roof walked into the fellowship hall of the Emanuel Church during the evening. He sat for almost forty minutes, which was long enough to take in Reverend Clementa Pinckney's reading of the Gospel of Mark's account of the Parable of the Sower before he started shooting.[34] None of the Black freedom designs of this historic space made an impression on Roof: their worldview, their truth telling, and their openness to him—the only white person in church that day according to the cell phone footage we have—were irrelevant and frankly outside of the designs of white supremacy he was there to protect and defend.

After his arrest and during trial, Roof's convictions were apparent and his remorse absent. His intention was to inspire an active commitment by whites to the long race war against Blacks in America. Surely his hatred for Black America was a reflection of his living in South Carolina, where monuments to Confederates still reign, myths of the "Lost Cause" are pervasive in the culture, and neoliberalism has unbraided belief in democratic practices needed to foster a progressive political culture.[35]

As the political theorist Wendy Brown has argued, white poor and working-class men—along with their female counterparts—have been drawn by the designs of neoliberalism to the destructive forms of freedom enabled by the undermining of society as a space of recognition, justice, and accountability.[36] If society is the enemy of freedom and social justice is taken to be an imposition on that freedom, then Black life cannot matter in any substantive way; in fact, Black life can be only the enemy of this form of white self-determination. Roof's act of taking the lives of nine Black members of a church founded by an abolitionist was an assault on the social designs challenging the spiritual and intellectual horizon of whiteness.

There is another dimension to the problem that Roof poses: there is a science behind his opposition to Black life, to accountability, to "the social." Scientific racism—along with the settler ontologies that draw distinctions between white and native, living and inanimate, and the world as it appears and the biological laws set into motion by some unmoved mover—has given us the language to imagine race as the product of forces outside the control of social conventions. Racism within the biological and medical sciences is the belief that the health, behavior, and physiological traits of an individual are shared by other people who are thought to belong to the same population.[37] Collectively, people in this race inherit who they are, what they look like, and their likelihood of illness and health from their family and their ancestors. Scientific and medical racism creates the impression that the biological characteristics or the genes we receive ultimately determine the course of our lives, not the society that sustains or discriminates against those lives. It also creates the belief that races can be augmented only indirectly—namely, through eugenic elimination or curation. According to this view God/Nature create race, not humans. The conservation or elimination of these races by scientists or white nationalists is an act that remains within the moral universe of this firstly Christian and now scientific interpretation of the world.

During his trial, Roof dismissed his legal counsel and represented himself because he believed his lawyers were his religious and biological enemies. He, in fact, saw his elimination of Black life as a faith conviction. In his motion to remove his appointed counsel, he penned the following:

> My two currently appointed attorneys, Alexandra Yates and Sapna Mirchandani, are Jewish and Indian respectively. It is therefore quite literally impossible that they and I could have the same interests relating to my case.
>
> Trust is a vital component in an attorney client relationship, and is important to the effectiveness of the defense. Because of my political views, which are arguably religious, it will be impossible for me to trust two attorneys that are my political and biological enemies. The difficulties during my trial are evidence of this.[38]

Roof's understating of race is an expression of the nihilistic worldview that has haunted the scientific study of race. Roof and his nineteenth-century scientific counterparts were convinced they were stewards of human history set into motion by God/Nature and thus also had authority over others concerning the preservation (or elimination) of the races. Roof could say that his legal counsel were his religious and biological enemies because he feared his natural birth right from God as a member of the chosen white race was being lost to history. He, like all of us under colonial designs, inherited a Christian moral interpretation of the world in which races are produced outside of human conventions. This was the same worldview championed by anthropologists during European colonialization and then by eugenicists at the turn of the twentieth century who looked to evolution, medicine, and public health to wage a biological war against the unfit and nonwhites. Like Roof, they worked to defend racial purity, which was threatened by the designs of a pluralist democracy freed from the horizon of white supremacy and the Divine right to power over others.

Roof can be an extremist only if we extract him from the historical and social conditions that produced him. If neoliberalism, as Brown argues, has made comprehension of society and its social effects optional, then we have to take Roof's white supremacist designs seriously or else lose the lucidity we need to navigate the social creations we've made that have the power to take lives.[39] If we don't take these designs seriously, we maroon Roof within our consciousness—leaving no reasonable (or believable) links to be made between him and the displacement of Indigenous people, the racism of American anthropology, and the eugenic spirit of modern genetics and public health. Severing the ties between Roof and our broader value commitments mirrors the effort of so many scientists who are convinced that their truth has no ties to religion and supersedes non-Western knowledge systems. Dismissing Roof as merely an extremist or as psychologically ill denies the banal intellectual structures that have been the handmaiden to white supremacy throughout the history of Europe and its former colony in North America. Not taking our connection to Roof seriously is itself a symptom of nihilism. Again reflecting on Nietzsche, Camus writes:

> If nihilism is the inability to believe, then its most serious symptom is not found in atheism, but in the inability to believe in what is, to see what is happening, and to live life as it is offered. This infirmity is at the root of all idealism. Morality has no faith in the world. For Nietzsche, real morality cannot be separated from lucidity.[40]

Casting Roof as simply an extremist—to say that he is not an expression of the United States, its religion, or the history of its science—would turn him into a white supremacist without a society accountable for his formation. Paradoxically, this would divinize Roof, as his actions would be the result of forces beyond social conventions and thus beyond social control. This is what the loss of comprehension looks like under the effects of Christian nihilism.

CONCLUSION: SCIENCE AND RELIGION AFTER NIHILISM

Nihilism is not simply the idea that there is no God. Nietzsche reminds us that nihilism is far more nefarious: it involves living in a society where, after we have lost confidence in social conventions and the designs of others, our pursuit of truth cannot be satisfied within this world. This has left us in a state of paralysis, in which we continue to feel psychological and emotional needs for God and thus construct explanations for life that hide the historical and social ground that has given purpose to such debilitating needs. This is a trap. As we continue to sustain faith in a world beyond this one, it erodes our belief in the power of human convention and, most importantly, the designs of others oriented toward different types of constructions and alternative value systems concerning the formation and maintenance of human and nonhuman life and what counts as its proper history.

Older scholarship on science and religion, by lacking the empathy and courage to detail the links between the colonial designs of white supremacy and modern science, has exacerbated the effects of Christian nihilism on our field. For too long, scholars working at the intersection of science and religion have been oriented toward internal debates and problems far from the social designs of nonwhite and non-Christian others. Within the intellectual horizon of the field as currently configured, there is no single actor or historical institution that can be held responsible for this state of affairs.

Some time ago Cornel West identified a similar problem embedded within contemporary postmodern American philosophy, arguing that while it may be conceptually demanding, it remained a "culturally lifeless rhetoric mirroring a culture (or civilization) permeated by the scientific ethos, regulated by racist, patriarchal, capitalist norms and pervaded by debris of decay."[41] There is analogy here for the study of science and religion where Christian nihilism has captured our imagination. Living under colonial designs has meant carrying on exacting work about science and religion while remaining blind to racism and the legacy of white settler colonialism that made use of Christianity and science to accomplish its life-denying ends. Seeing these connections is a skill that starts with restoring faith in the conventions of others—which really means restoring faith in the world we create and the critics who remind us that we can do better.

As scholars of science and religion, we can ignore the value of this skill only at our own peril—tragically, there will be more Dylann Roofs until American intellectuals commit to abolishing our nation's colonial legacy and join efforts already underway to design a more humane society.

NOTES

1. My use of design language to speak of colonialism and structural racism has been greatly influenced the work of Ruha Benjamin and Safiya Noble. While working separately, they have helped us see racism within technology as a product of specific human design and intention that reflects larger structural relationships supporting white supremacy. Benjamin writes in *Race After Technology* that technical designs "often hide, speed up, and even deepen discrimination, while appearing to be neutral or benevolent when compared to the racism of a previous era. This set of practices that I call the New Jim Code encompasses a range of discriminatory designs" (8). Noble in *Algorithms of Oppression* coined the term *technological redlining* to capture how unequal social conditions appear on the internet and everyday technologies where discrimination is "embedded in computer code and, increasingly, in artificial intelligence technologies that we are reliant on, by choice or not" (1). Design within the context of my discussion here complements what we know about technoscience from Benjamin and Noble by drawing attention to the social conventions and discriminatory logics "encoded" in religion and the life sciences. It also signals social designs that correct racial oppression within religion and science and thus are what Benjamin calls "abolitionist" in their creation and orientation. For more, see Ruha Benjamin, "Discriminatory Design, Liberatory Imagination," in *Captivating Technology: Race, Carceral Technoscience, and Liberatory Imagination in Everyday Life*, ed. Ruha Benjamin (Durham, NC: Duke University Press, 2019), 1–24, and *Race After Technology: Abolitionist Tools for the New Jim Code* (Cambridge: Polity, 2019); and Safiya Umoji Noble, *Algorithms of Oppression: How Search Engines Reinforce Racism* (New York: New York University Press, 2018).
2. The following citations from Kim TallBear are from an unpublished address shared with the author: "Science v. Spirituality: A Dead-End Settler Ontology and Then What?," Distinguished Lecture: History of Science Society, Seattle, WA, November 3, 2018. Please do not cite or copy.
3. A list of all previous HSS distinguished lectures can be found at https://hssonline.org/about/honors/history-of-science-society-distinguished-lecture/.
4. Obituary of Charles C. Gillispie, Mather Hodge Funeral Home, accessed November 9, 2020, https://matherhodge.com/tribute/details/713/Charles-Gillispie/obituary.html.
5. See Charles C. Gillispie's *Edge of Objectivity: An Essay in the History of Ideas* (Princeton, NJ: Princeton University Press, 1960). His insights and observations on the development of science were based on the first courses on the history of science taught in the American academy, offered at Princeton during the late 1950s.
6. TallBear, "Science v. Spirituality."

7. TallBear, "Science v. Spirituality."
8. For more on the persistence of native temporal and geographic orientations within and beyond settler colonialism, see Mark Rifkin, *Beyond Settler Time: Indigenous Self-Determination* (Durham, NC: Duke University Press, 2017).
9. TallBear, "Science v. Spirituality."
10. Frantz Fanon, *Black Skin, White Masks*, trans. Charles Lam Markmann (1952; repr., London: Pluto, 1986). For a recent engagement with Fanon on the problem of modern liberal forms of political recognition within the context of critical Indigenous scholarship, see Glen Sean Coulthard, *Red Skins, White Masks: Rejecting the Colonial Politics of Recognition* (Minneapolis: University of Minnesota Press, 2014).
11. There are surely rival traditions in the West that were also confronted and absorbed in route to the development of the postmodern synthesis of evolutionary biology, life science, and empirical medicine that reigns in the present. One very clear example from the study of human origins would be the resolution of the controversy created by polygenism (a theory on the separate origins of the human races) in the seventeenth century by the French theologian Isaac La Peyrère, in the nineteenth century by American polygenists, and in the twentieth century by Milford Wolpoff and his students, who championed the multicentric origins of anatomically modern humans. My focus is not on these internal rival traditions that share moral and conceptual ground while also never challenging the objectivity of Western truth or the unique privilege of European stories about the origins of human life.
12. Sylvia Wynter, "Unsettling the Coloniality of Being/Power/Truth/Freedom: Towards the Human, After Man, Its Overrepresentation—An Argument," *CR: The New Centennial Review* 3, no. 3 (Fall 2003): 257–337, https://doi.org/10.1353/ncr.2004.0015.
13. George Lipsitz, *The Possessive Investment in Whiteness: How White People Profit from Identity Politics* (Philadelphia: Temple University Press, 1996); Eddie Glaude, *Democracy in Black: How Race Still Enslaves the American Soul* (New York: Broadway Books, 2016).
14. See Carol Diethe, *Nietzsche's Sister and the Will to Power: A Biography of Elisabeth F. Förster-Nietzsche* (Champaign: University of Illinois Press, 2007).
15. On this point about Nietzsche not seeing nihilism as a final intellectual destination or a consistent worldview, I would start with his thinking in "On the Uses and Disadvantages of History for Life," in *Untimely Meditations*, trans. R. J. Hollingdale (New York: Cambridge University Press, 1997), 57–124. Also see Ruth Burch, "On Nietzsche's Concept of 'European Nihilism,'" *European Review* 22, no. 2 (May 2014): 196–208; Nitzan Lebovic, "The History of Nihilism and the Limits of Political Critique," *Rethinking History* 19, no. 1 (2015): 1–17; and Steven Michels, "Nietzsche, Nihilism, and the Virtue of Nature," *Dogma: Revue de Philosophie et de Sciences Humaines* (2004), http://www.dogma.lu/txt/SM-Nietzsche.htm.
16. For recent treatment of how Nietzsche and Heidegger have laid the philosophical groundwork for the resurgent Right and its dissatisfaction with modern political liberalism, see Ronald Beiner, *Dangerous Minds: Nietzsche, Heidegger, and the Return of the Far Right* (Philadelphia: University of Pennsylvania Press, 2018).
17. Nietzsche does say in *The Will to Power*, "Skepticism regarding morality is what is decisive. The end of the moral interpretation of the world, which no longer has any sanction after

it has tried to escape into some beyond, leads to nihilism" (7). We might read Nietzsche here to be suggesting nihilism abounds when we no longer believe morality has a place in the world—that we are no longer obligated to act as an ethically responsible liberal subject committed to democracy, as the values that might guide such behavior are not anchored in anything solid or necessary. With this reading, it is easy to conclude that Nietzsche is arguing that, under modernity, God, morality and the liberal state are the trappings of those in power looking to solidify their sovereignty and domination. See Friedrich Nietzsche, *The Will to Power*, trans. W. Kaufmann and R. J. Hollingdale (New York: Vintage, 1968).

18. Nietzsche, *Will to Power*, 7.
19. Susannah Heschel, "Theology as a Vision for Colonialism: From Supersessionism to Dejudaization in German Protestantism," in *Germany's Colonial Pasts: An Anthology in Memory of Susanne Zantop*, ed. Eric Ames, Marcia Klotz, and Lora Wildenthal (Lincoln: University of Nebraska Press, 2005), 148–164.
20. Nietzsche, *Will to Power*, 36–37.
21. John 12:30–32 (King James Version).
22. Denise Buell, *Why This New Race: Ethnic Reasoning in Early Christianity* (New York: Columbia University Press, 2005).
23. Nietzsche's specificity about Christian morality being the source of a unique form of nihilism is generative and shows how he believed something was to be gained through reconsidering our value commitments to God and other trappings of our Christian culture. On this point, Ruth Burch draws attention to the distinction Nietzsche makes between passive and active nihilism. Burch argues that for Nietzsche, passive nihilists take on a life of self-destruction and spiritual weakness. They remain immobilized by the destructive power of nihilism and refuse to establish new values and goals beyond those of the Christian moral interpretation of the world. Following Burch, it seems to me passive nihilists continue to be shaped by values that orient them to a world beyond where we all live. This orientation maintains the psychological and spiritual dependence on an eschatological power that subverts human control and thus becomes the inspiration for assaulting the designs of those not in step with this otherworldly orientation. Activist nihilists, on the other hand, are marked by an enhanced spiritual power and use the resolve of deconstruction to clear space for designing new values and orientations. See Ruth Burch, "On Nietzsche's Concept of 'European Nihilism,'" *European Review* 22, no. 2 (May 2014): 202. Moreover, if we follow Burch's reading, nihilism potentially has a transformative role for the future, and discussion of its various stages by Nietzsche signals this. Nietzsche writes in *The Will to Power*: "For why has the advent of nihilism become necessary? Because the values we have had hitherto thus draw their final consequence; because nihilism represents the ultimate logical conclusion of our great values and ideals—because we must experience nihilism before we can find out what value these 'values' really had" (4). Albert Camus, in *The Rebel*, also interprets Nietzsche in this way, as he writes: "According to Nietzsche, he who wants to be a creator of good or of evil must first of all destroy all values" (66). See Albert Camus, *The Rebel* (New York: Vintage, 1991). Active nihilists, in their deconstruction of the world beyond, restore faith in the power of human conventions and our ability to imminently shape society.

What else does it mean to create new values other than to renew faith in our power to create social conventions that inform and direct thought and behavior? Nietzsche's intentions are very specific: he was looking for an alternative to the life-denying values and orientations of the West, held captive by a Christian culture that insists there can be only one horizon for humanity.

24. Nietzsche, *Will to Power*, 24.
25. Camus, *The Rebel*, 69.
26. Noah Feldman argued not long after the United States declared its "War on Terror" "that politics and religion are, in fact, themselves better conceptualized as kinds of technology, and subject to kinds of questions that we regularly consider in the space of conceptual design." See Noah Feldman, "Politics and Religion Are Technologies" (Ted Talk, 2003), https://www.ted.com/talks/noah_feldman_politics_and_religion_are_technologies?language=en.
27. I benefited greatly from Ronald Beiner's incredibly lucid discussion of what he identifies as the four traditions of civil religion within modern Western political philosophy in *Civil Religion: A Dialogue in the History of Political Philosophy* (Cambridge: Cambridge University Press, 2010).
28. See Charles Taylor's *A Secular Age* (Cambridge, MA: Harvard University Press, 2007) and his discussion of the growing intolerance in Europe for religious fanaticism following the English Civil War. This hostility toward religious fanaticism accompanied what he calls the emergence of "Providential Deism"—an intermediate position between a biblical worldview where God is an agent in history and an exclusive humanism where God is not necessary for social life. Taylor argues that under Providential Deism, God relates to Europeans "by establishing a certain order of things, whose moral shape we can easily grasp, if we are not misled by false and superstitious notions. We obey God in following the demands of this order" (234). In my judgment, the laws of natural science that established white Europeans as a superior racial group allowed for a relationship with God to persist without violating their burgeoning secular sensitivities. I argue that within what Taylor calls "exclusive humanism," underwritten by anthropocentrism, white Europeans replace God as the prime historical agent, but whiteness does not supplant the psychological need for European superiority to stand on "a certain order" given by Christian theology. The alliance between white supremacy and the modern life sciences reveals this order and in doing so betrays the purity of exclusive humanism that Taylor argues is a hallmark of our age.
29. "Dylann Roof's Journal," *Post and Courier* (Charleston, SC), December 6, 2014, https://www.postandcourier.com/dylann-roofs-journal/pdf_c5f6550c-be72-11e6-b869-7bdf860326f5.html.
30. David Robertson, *Denmark Vesey: The Buried Story of America's Largest Slave Rebellion and the Man Who Led It* (New York: Vintage, 2000), 27–40.
31. Robertson, *Denmark Vesey*, 41–56.
32. Robertson, *Denmark Vesey*, 88–99.
33. Robin Kelly, Freedom Dreams: The Black Radical Imagination (Boston: Beacon Press, 2003).
34. Rachel Kaadzi Ghansah, "A Most American Terrorist: The Making of Dylann Roof," *GQ*, August 21, 2017, https://www.gq.com/story/dylann-roof-making-of-an-american-terrorist.

35. Wendy Brown writes that in the late twentieth century, neoliberalism transformed American notions of society, the individual, and the family through a set of reforms designed to dismantle our trust in society as a space for mediating justice and equality. See Wendy Brown, *In the Ruins of Neoliberalism: The Rise of Antidemocratic Politics in the West* (New York: Columbia University Press, 2019), 38. This has had the effect of repurposing the family as a "state actor" responsible for the welfare of the young, the sick, the unemployed, and those in debt (39). These neoliberal transformations have had a remarkable effect on the relationship between the social body and the state—and ultimately on our conceptions of society. As Brown explains: "The epistemological, political, economic, and cultural dismantling of mass society into human capital and moral-economic familial units, along with the resulting recuperation of both the individual and the family at the very moment of their seeming extinction, are among neoliberalism's most impressive achievements" (39). Neoliberalism has enervated our ability to think and speak meaningfully about what she calls "the social," which is "where subjections, abjections, and exclusions are lived, identified, protested, and potentially rectified" (40). The neoliberal assault on society as both a concept and a space for shared accountability and justice has given rise to destructive forms of freedom that are dangerously antisocial and antidemocratic.
36. See Lorrie Frasure's excellent study of white women voters who supported Trump during the 2016 presidential election in "Choosing the Velvet Glove: Women Votes, Ambivalent Sexism, and Vote Choice in 2016," *Journal of Race, Ethnicity, and Politics* 3, special issue no. 1 (March 2018): 3–25.
37. For a recent comprehensive account of scientific racism, see Joseph Graves Jr. and Alan Goodman's *Racism, Not Race: Answers to Frequently Asked Questions* (New York: Columbia University Press, 2021).
38. Pro Se Motion to Remove and Replace Appointed Counsel, United States of America v. Dylann S. Roof, No. 17–3 (4th Cir. Sept. 18, 2017).
39. Brown, *In the Ruins of Neoliberalism*, 28.
40. Camus, *The Rebel*, 67.
41. Cornel West, "Nietzsche's Prefiguration of Postmodern American Philosophy," in *The Cornel West Reader* (New York: Basic Books, 1999), 210.

CHAPTER 3

A FEMINIST THEOLOGY OF ABORTION

MYRNA PEREZ SHELDON

What happens to the tissues, cells, membranes, and blood of an aborted or miscarried fetus? In most parts of the United States, fetal material is collected by a medical hazard company along with other human bodily waste that results from testing or surgery and then incinerated. If the woman is at home for a miscarriage or a medicinally induced selective abortion, the fetal material is handled by the woman or her community as either sees fit—whether dropped into a toilet or a trashcan or buried. Treating fetal material up to the point of viability as bodily waste means that a fetus is not issued a death certificate. It does not need to be handled as human remains. This practice embodies the U.S. Supreme Court's decision in *Roe v. Wade* that a fetus is not legally a person, and, thus, fetal material is handled as if it were part of a body, not a dead human.[1] But in 2016, the Indiana state legislature challenged this fundamental premise with a law that requires abortion clinics to "provide for the final disposition of the aborted fetus."[2] By mandating that either women or abortion clinics treat fetuses as dead, the law enforces a view of the fetus as a grievable person rather than as bodily waste.

When the U.S. Supreme Court upheld the legality of this law in early 2019, it implicitly granted the premise of the statute—the grievable personhood of a terminated pregnancy. It was an extraordinary threat to the security of abortion rights in the United States. However, it was not the decision by the court but a second provision of the state law that created a media stir. The Indiana state legislature also attempted to prevent women from seeking abortions "solely because of the race, color, national origin or ancestry," "sex," or "potential diagnosis of . . . disability."[3] This sex-selective and disability abortion ban was declared an unconstitutional violation of a woman's right to privacy by the Seventh Circuit

Court of Appeals, a decision effectively affirmed by the Supreme Court when it refused to grant a hearing on this aspect of the case. But it was Justice Clarence Thomas's decision to write a twenty-page opinion tying abortion to eugenics that prompted several American historians to write fiery opinions condemning his interpretation of history and his claim that abortion has the potential to "becom[e] a tool of modern-day eugenics."[4]

In the course of his opinion, Thomas cited the work of several historians to argue that abortion is enmeshed in the legacy of American eugenics. Many of these historians publicly condemned his views, arguing that his "rendition of history is incorrect" and calling his decision to link "abortion to eugenics . . . as inaccurate as it is dangerous."[5] The various op-eds centered on two points in their criticism of Thomas's opinion. First, Thomas was said to have misrepresented the historical relationship between abortion and eugenics. One historian argued that "the most prominent American eugenicists did not support abortion" and that although Thomas's discussion of the role of eugenics in forced sterilization and race-based immigration law was "real history," it was "not particularly relevant to abortion."[6] Another historian contended, "I've been studying this stuff for 40 years, and I've never been able to find a leader of the eugenics movement that came out and said they supported abortion."[7] Second, many of these scholars claimed that because abortion in the United States is about individual choice rather than state-controlled breeding, it cannot be a technique of eugenics. As Adam Cohen argued in an *Atlantic* op-ed, "A woman in Indiana who has an abortion because the child will be born with a severe disability is not acting eugenically—she is not trying to uplift the human race."[8] A *Washington Post* op-ed on the topic made a similar claim, contending that "abortion—a personal choice by an individual—differed significantly from the state-mandated programs foisted involuntarily on others by eugenicists."[9] Karen Weingarten perhaps best summarized the indignation of these scholars by declaring in a post for the blog *Nursing Clio*, "The truth is that abortion was never, and will never be, a tool to prevent women from reproducing."[10]

The response by these historians is manifestly understandable; it must be unnerving to have one's scholarship—particularly histories done to address the injustices of eugenics—taken up by the conservative cause to limit women's reproductive self-determination. Fearing the power of the conservative court, these scholars have used American history as a proof text in their battle with Thomas.

But this fear inhibits us. It prevents us from clearly understanding the parallel violences done to women by racial biopolitics and patriarchal Christianity. It is a fear that takes from us the power of critiquing the racialized legacies of abortion policy in the United States and around the world. And it robs us of the imagination it must take to envision a world outside of this bind.

Justice-minded academics have struggled to respond to religiously inflected right-wing movements that criticize the public authority of science, public health, and rights-based arguments for reproductive freedoms. Since the 1970s, the critical left in the academy has critiqued the racial and patriarchal violences of Western medicine and science and has questioned the foundational assumptions of Western liberalism. But when conservative voices such as Thomas's seem to echo these criticisms in their attacks on abortion rights, the science of vaccines, or climate change, for example, many scholars have turned to a defense of liberal rights-based values and the sanctity of scientific truth. This is a broad crisis of ethics in the academy and the work of public scholarship, one that will take courage and empathy to address.

To that end, I begin this chapter by sketching one specific node of this crisis, the politics of abortion in the United States, in order to argue that clarity can be found by mapping this historical and contemporary debate as a struggle between scientific and religious authority. Since the early twentieth century, the U.S. debate over abortion has been caught between the racial biopolitics of public health and the patriarchal claims of Christian nationalism. The two provisions of the 2016 Indiana law reflect this parallel violence. Recognizing this helps to clarify the ongoing terms of the debate within broader struggles over secularism, national identity, and racialized sexuality in the United States. It also makes tangible what a critical approach to the study of science and religion might look like. Rather than focusing on abstract debates over divinity or cosmogony, an analysis of this nature moves the study of science and religion into the embodied and feminized realities of abortion politics.

However, my aim in this chapter extends beyond even these methodological clarities. In the second half of the chapter, I listen to two sets of voices—an abortion doctor who is a mother and reproductive justice activists in South Dakota and Georgia—in order to imagine a feminist theology of abortion. In this, I ask what it might look like to relentlessly and radically trust in the experiences of those who are, or can be, or might be, or wish not to be pregnant—whatever their gender.[11] I hope for a feminist theology that does not simply reproduce a new taxonomy of certainty and righteousness but that sets aside the quest for epistemological certainty and moral purity that has characterized much of the public and scholarly debate over abortion.

I claim this hope as a theology because I desire a new sacred.[12] The biopolitics of the secular state views the health of the nation—conceived in racialized and ableist terms—as sacred. Lives are valued for their contribution to the productivity and racial identity of the nation-state. The patriarchy of Christian nationalism holds the life of the fetus as sacred. The fetus is valued as a sign of both the nation's obedience to God and the future promise of a Christian America. These forms of the sacred are doing ongoing violence in American society—whether in the forced sterilizations at U.S. immigrant detention centers or in the deputization of citizenry to enforce an abortion ban in Texas.[13]

In the face of this violence, I imagine a feminist theology of abortion in order to believe in the sacredness of women's experiences. When women choose, their choices create and end life. When women choose, their choices make and unmake nations. And we must trust them to do this. We must trust not women *and their doctors* or women *and their pastors* or women *and their communities*—we must trust just women.

ABORTION AND CHRISTIAN NATIONALISM

In the summer of 1979, a few short months before the Republican primaries for president that would eventually usher Ronald Reagan into office, Francis and Frank Schaeffer embarked on a film tour that would help to transform the evangelical approach to abortion. The father and son team brought their film series *Whatever Happened to the Human Race?* to crowds in churches, town halls, and gymnasiums of fundamentalist colleges all across the nation. The first episode of the series, *The Abortion of the Human Race*, scandalized and shocked many in the audience—many churches refused to show the film at all. But it also galvanized a generation of evangelicals in support of the antiabortion cause and helped cement abortion as one of the key political issues in the newly forged alliance between the Republican Party and the Christian Right.[14]

The opening scene of *The Abortion of the Human Race* depicts two surgical teams.[15] One operates on a newborn, struggling to survive. The other lays out ghastly looking obstetric instruments on an operating table, preparing for an as-yet-unknown procedure. As the first team works, it becomes clear that they are saving the baby's life; triumphantly soft music soon starts to play as the baby is transferred to a recovery room. Meanwhile, the other team prepares for a different outcome; although an abortion is not explicitly depicted, the implication that the pregnancy is to be terminated is clear. A fully rounded pregnant abdomen appears in the frame; prepped for surgery, it is momentarily draped with a sign with the words "remove and discard." The juxtaposition between the premature baby and the aborted fetus in the film would be echoed in arguments Schaeffer made in his 1982 book coauthored with C. Everett Koop, the U.S. surgeon general under Ronald Reagan. In that book, also titled *Whatever Happened to the Human Race?*, Schaeffer and Koop argued that modern society had become "schizophrenic" because doctors were resuscitating infants at the same stage of development as it was legal to terminate a pregnancy.[16] Together the film and the book intended to convey one unassailable truth—that the United States was authorizing the murder of babies.

Although these kinds of arguments do not feel new in the early twenty-first century, at the time they represented a novel approach to the ethics of abortion. The cornerstone of Schaeffer's new theology was the argument that the practice of abortion was a sign of the decline of Christian morality and the rise of secular humanism in Western civilization. In *Whatever Happened to the Human Race?*, he asked, "Why has our society changed? The answer is clear. The consensus of our society no longer rests upon a Christian basis but a humanistic one."[17] Moreover, Schaeffer specifically blamed new evolutionary theories for the advancement of secularism. He derisively described the selfish-gene logic of the new sociobiological account of evolution to his readers: "Regardless of what you think your reasons are for unselfishness, say the sociobiologists, in reality you are only doing what your genes know is best to keep your gene configuration alive and flourishing into the future. This happens because evolution has produced organisms that automatically follow a mathematical logic."[18] And it was this materialist conviction, the view that humans were nothing more than their genes, that Schaeffer blamed for the unfeeling and cruel nature of humanism. Touting the writings of E. O. Wilson, Harvard zoologist and author of *Sociobiology: The New Synthesis* (1975), as an example of this depraved moral perspective, Schaeffer relayed Wilson's call for morality to be "biologized" and removed from "the hands of the philosophers."[19] Schaeffer was appalled by the view that "ethics and behavior" were now to be "put into the realm of the purely mechanical, where ethics reflect only genes fighting for survival."[20] He blamed evolutionary theory as the foundation of this humanism, arguing that "if man is not made in the image of God, nothing then stands in the way of inhumanity. . . . Human life is cheapened."[21]

This cheapening, he believed, was responsible for the totalitarian disregard for life exhibited by the Third Reich in the Final Solution. And it was the same moral philosophy he now believed was governing America's legalization of abortion.

Although there were piecemeal objections to abortion policy by Evangelical Protestants before this period, for most of the twentieth century the antiabortion movement had been a largely Catholic cause. The Catholic hierarchy objected to abortion on the same grounds it used to disapprove contraceptives: a belief in the procreative nature of sex and a conviction that the refusal of procreation was synonymous with sexual sin. That is, Catholic authorities believed there should be no need to prevent or terminate pregnancies if all sexual activity was confined within heterosexual marriage. Coupled with this heteronormative and patriarchal view of sexuality was another set of objections to abortion, which ran along the same lines as the Catholic opposition to eugenic policies in the United States, including forced sterilization. Since Catholic doctrine values the individual as the image bearer of God, Catholic religious leaders developed an anti-eugenic vocabulary that called on communities to care for "unfit" individuals rather than asking these individuals to sacrifice their right to reproduce for the good of society.

Catholics were the largest and best-organized opposition to eugenic ideology in the United States before the 1970s.[22] However, amongst Protestants, attitudes toward eugenics tended to fall along modernist/fundamentalist lines.[23] Because modernist Protestant groups were more open to doctrinal accommodations to science, they were similarly more eager to embrace eugenics. During the 1920s, in fact, the American Eugenics Society held a series of sermon contests in which pastors were invited to preach sermons on the question "Religion and Eugenics: Does the church have any responsibility for improving the human stock?"[24] Liberal ministers submitted entries that embraced the role of science in the work of the Christian church, arguing that "the old faith is not destroyed, for science is giving us a true technique of righteousness."[25] These pastors believed that the science of eugenics would allow the church to win "in its fight against poverty. Charity is no remedy. Poverty is not so much a state as a condition, inside people."[26]

Thus, during the early twentieth century, Catholics approached antiabortion and anti-eugenic activism as part of the same moral cause and drew on the championship of individuals as image bearers of God in their arguments against both. Protestants, on the other hand, were not consistently opposed to eugenics nor were they leaders in antiabortion activism. Because of this, recent histories have emphasized that the antiabortion movement in the pre–*Roe v. Wade* era was configured differently than the pro-life movement of the 1970s and 1980s. Daniel K. Williams, for instance, has argued that this early activism owed much to the championship of minority rights that animated anti-eugenics opposition.[27] He further contends that those who participated in this activism in the early part of the twentieth century often saw it as part of the same kind of liberal stance that supported the civil rights movement. When Evangelical Protestants began to be involved consistently in organized antiabortion politics in the late 1970s, this changed the political identity of the group, as well as of the movement itself. Schaeffer's theological arguments were a key factor in these transformations.[28] His argument that abortion was a sign of the rise of secular humanism and the collapse of Christian authority was transformative for many of the key figures of the Christian Right.

One of these men was Jerry Falwell, whose apocalyptic vision of American history shaped evangelical politics for two generations. He was a different brand of religious leader than the Schaeffers; he had begun his ministry empire by founding a Baptist church in the 1950s, and it soon grew to include Liberty University and a Christian television network. Through the 1960s, he was a leading evangelical segregationist, campaigning against the integration of public school systems and directly criticizing Civil Rights activists, including Martin Luther King Jr. By the late 1970s, Falwell's attention had turned to the women's and gay rights movements as the key threats to white Christian nationalism. Drawing on a Calvinist tradition that viewed the United States as being in a unique covenant with God, he analyzed history, politics, and culture for signs that America had now fallen out of God's favor. In his 1980 sermon series *Listen, America!*, he called out the forces turning America away from its godly path; the legalization of abortion topped the list, followed by "homosexuality, pornography, humanism, and the fractured family."[29] Through his political action committee, the Moral Majority, Falwell launched a legislative offensive that transformed Schaeffer's theological vision into a political juggernaut.

Antiabortion activism was not only at the heart of the Christian Right; it was also integral in the late twentieth-century political contests between creationism and evolution. One of the founders of the Moral Majority, Timothy LaHaye, was also a founder of the most important creationism organization of the era, the Institute for Creation Research (ICR). From its beginning in 1972, the ICR led the political agenda of creationism, supporting bills to mandate the teaching of creation science in public schools, publishing a creation research journal titled

Acts and Facts, and providing churches with a Bible study curriculum to help their members articulate an antievolutionary apologetic.[30] The ICR was part of an evangelical campus in San Diego County in Southern California that included a four-year liberal arts college and the megachurch Shadow Mountain. These entities, in turn, had deep formal and informal ties to such conservative think tanks as the Heritage Foundation and other politically influential megachurches in the region. Importantly, the moral perspective of the evangelical antiabortion movement was not merely associated with creationism by institutional proximity; it was one feature of a cohesive approach to national and global politics that saw secular science and progressive politics as twin threats to white Christian nationalism. The ICR leadership, particularly Timothy and his wife, Beverly, were key leaders in the conservative campaigns against the Equal Rights Amendment and in pro-life activism during this period. Beverly LaHaye founded the Concerned Women for America (CWA) in 1979 after watching Betty Friedan being interviewed by Barbara Walters on national television.[31] Incensed that the "humanist" morality of Friedan could take over America, through the CWA LaHaye helped lead the most politically influential conservative women's movement of a generation.

During the 1980s the emergent pro-life movement was intimately connected to American creationism. Recognizing this situates evangelical antiabortion activism within the broader political worldview of white Christian nationalism. Thus, contests between evolution and creationism, segregation and civil rights, pro-choice and pro-life, LGBTQ rights and religious freedom, or immigrant rights and nativism are not isolated political matters. Rather, they are a set of interconnected debates that have the question of Christian authority in American culture at their core. This point is further illustrated by continuing the story of Southern California evangelicalism.

Just down the road from the LaHayes was another megachurch, Skyline Wesleyan, founded by James Garlow in the early 1990s. It was part of the orbit of the ICR, utilizing creation-science materials for its adult Bible studies and Sunday school curriculums.[32] Garlow was also a key leader behind Proposition 8, a 2008 bill that attempted to amend the California state constitution to define marriage in solely heterosexual terms.[33] Through this period, he preached sermons that framed the fight against gay rights as a matter of religious liberty for American Christians. And during the 2016 presidential election cycle, he was one of the leading voices pressing evangelical support for the election of

Donald Trump, and he later sat on Trump's Presidential Spiritual Advisory Council.[34] Beginning in 2016, Garlow held weekly Bible studies for members of Congress based on his book *Well Versed: Biblical Answers for Today's Tough Questions*. In it, he weaved together a worldview that champions creationism and pro-life activism while attacking immigration from Muslim-majority countries. Denouncements of "evolutionary untruths" sit alongside arguments that immigrants must "understand and respect our national foundations as being based on eternal Judeo-Christian values" and that abortion in the case of rape or incest is "not biblical."[35] Through writing, teaching, and preaching, Republican evangelical pastors like Garlow helped to firmly root antiabortion activism as a keystone of Christian ethnonationalism.

These institutional developments were essential to the emergence of pro-life politics in the several decades after the summer of 1979 when the Schaeffers first shared *Whatever Happened to the Human Race?* with evangelical communities. Without the organizing energy and financial capacity of new parachurch organizations like CWA and the Heritage Foundation, the Schaeffers' moral vision would have done little to reshape the evangelical approach to abortion politics. But I began this short history of the late twentieth-century pro-life movement with the Schaeffers for an important reason. When Francis Schaeffer declared that the legalization of abortion was a sign that the West had abandoned its Christian foundation, his argument resonated with the broader current of apocalypticism running through twentieth-century evangelical theology. It was no longer just one of many cultural issues on which evangelicals might take a variety of stances. It was also not merely a cultural identity marker. *Whatever Happened to the Human Race?* helped make the unborn fetus into a sacred object within evangelicalism and thereby transforming abortion into a cause that Christians ought to—and did—vote on, protest about, create new spaces and rituals about (e.g., crisis pregnancy centers), and even kill over.

In other words, they helped to transform antiabortion activism into a sacred cause for evangelical Christians. This assertion may feel unintuitive for the reason that we tend to assume that moral values in religious communities originate from their sacred texts or traditional rituals and that if they do emerge in more contemporary contexts, they are a cover for some other kind of social ethic (e.g., nationalism or sexism). But religious studies scholars who have critically examined the role of the sacred in modern life can help us think through this point in a more nuanced way. Gordon Lynch, for instance, has argued that we should take seriously the stance of social theorists such as Émile Durkheim that it is not "economic structures, demographic shifts or technological change that were the primary drivers of social life but the sacred meanings that societies gave to these things."[36] For Lynch, "the sacred" is far from a synonym for sectarian religion; rather, it is a "way of communicating about what people take to be absolute realities that exert a profound moral claim over their lives."[37] The sacred is that which is almost unspeakably important: what people do not merely value but are willing to kill over or die for. And as Lynch argues, although it is possible through critical reflection to see the contingent nature of the sacred, the profoundly charged emotional dimension of this form of social life means that it becomes "the fixed point to which we orientate our lives."[38]

A feminist theology of abortion is a hope for a different fixed point than the politics of Christian ethnonationalism. But before we can turn to this, we must first examine the other thread of American public life that has created a sacred around abortion—the biopolitics of the secular nation-state.

ABORTION AS BIOPOLITICS

Clarence Thomas's 2019 opinion highlights the eugenic and racist history of feminist reproductive rights movements in the United States. And he concludes that this history should compel the state to prevent abortion from being used in the present as a tool of eugenics—which is why he supports the Indiana statute that would make it illegal for a provider to perform an abortion if they know the mother is seeking it because of the "race, sex, diagnosis of Down syndrome, [or] disability" of the fetus.[39] In their efforts to combat Thomas's judicial authority and political conclusions, academic historians focused on refuting Thomas's claim that campaigns for abortion rights emerged out of eugenic motivations. But we must be able to reject Thomas's attack on abortion access without denying the eugenic and racist foundations of the American reproductive rights movements. The very framework of critical biopolitical analysis compels us to recognize that what is most sacred to the modern nation-state is the scientific management of its population. Beginning with Michel Foucault, scholars have described the means by which nation-states create themselves through public health campaigns, immigration quotas, and, most critically, the management of sexuality.[40] Reproduction in the modern nation-state is always and already about the racial composition of the state—something that was as true globally as it was in the United States throughout the twentieth century.

In fact, in the decade before American evangelicals built the pro-life movement, campaigns for the legalization of abortion adopted new public health discourses that emphasized both race and ability. For instance, in her 2010 book *Dangerous Pregnancies: Mothers, Disabilities, and Abortion in Modern America*, Leslie Reagan unfolds the now largely forgotten episode of the rubella epidemic that began in the United States in 1964. When contracted during pregnancy, rubella often causes miscarriage or stillbirth. It can also affect cardiac, cerebral, auditory, and ophthalmic development, particularly if infection occurs during the first trimester of pregnancy. Reagan argues it was the connection between rubella and birth defects that fueled national anxiety over what she terms "dangerous pregnancies." This anxiety, in turn, prompted the development of a rubella vaccine and the transformation of the abortion rights debate. The respectability of the white middle-class women who sought abortions after contracting a disease "through no fault of their own" changed the cultural imaginary around abortion. Fear of disability made abortion seem reasonable, proper, and even necessary. And when these women lobbied Congress for access to legal abortions, it changed the face of the movement.[41] As Reagan concludes, "The uncomfortable truth is that the fear of disabilities opened up the first respectful public conversation in the United States that listened to women telling why they believed they needed an abortion."[42]

Alongside themes of disease and disability, activism for abortion legalization in the United States during the 1960s was laced with the discourse of population control. In a decade that was capped by the publication of Paul Ehrlich's *The Population Bomb* in 1968, the era's fear of unchecked population growth shaped U.S. approaches to domestic poverty, management of Native American nations, and soft power colonialism, particularly in Latin America.[43] The end of national origin quotas in immigration policy in 1965 meant that by 1980, for the first time in U.S. history, immigration from Asia and Latin America outpaced that from Europe.[44] This shift in the racial demographics of immigrants, coupled with the new policies allowing for nonwhite naturalization, began a generational crisis of citizenship that has yet to be resolved. Through the end of the century, the discourses on "welfare queens" in Black neighborhoods and "anchor babies" in Latino communities coupled population growth anxiety with racism. This broader context of demographic change is essential for understanding abortion policy conversations in this period.

One figure in the 1960s who explicitly tied together advocacy for abortion rights, population control, and eugenics was Alan Guttmacher. He was president of Planned Parenthood from 1962 to 1968, during which time he also served as vice president of the American Eugenics Society. Before his tenure at Planned Parenthood, he was director of obstetrics and gynecology at Mount Sinai Hospital in New York for more than a decade. Throughout his career, he wrote widely about abortion from both medical and legal standpoints. In his many academic articles, op-eds, and manuals on contraceptives and pregnancy, he intertwined medical advice with policy recommendations on abortion, and he often framed these recommendations in population control and eugenic terms.

For instance, in his 1961 book *Birth Control and Love*, Guttmacher referenced rubella as "one of the principle eugenic reasons for abortion."[45] The other "eugenic indication for abortion is created by x-ray therapy" unknowingly exposed to a uterus, causing congenital changes.[46] He also discussed his work with racial minorities at Mount Sinai Hospital, calling up images of overcrowded slums as the impetus for birth control and sterilization.[47] In a chapter titled "The Place of Sterilization," published in a 1964 edited volume titled *The Population Crisis and the Use of World Resources*, he described his recommendations for when therapeutic abortions are necessary. In the list, he included "eugenic: mental deficiency in the parents or the likelihood that a congenital disease or malformation will be present in the child."[48] In a chapter titled "The Problems of Abortion and the Personal Population Explosion" in the same volume, Jerome M. Kumer linked abortion and overpopulation even more explicitly, arguing that the United States would

> do well to concern themselves with the impending avalanche of population explosion. The direct bearing that abortion has on population figures is all too apparent. In the United States alone, if we were to take the estimated one million abortions per year and calculate population curves predicated on these abortions not having been accomplished, the contrast to present curves would indeed be vivid.[49]

Kumer went on in the piece to speak admiringly of Japan, whose reproductive policies allow for the "limited use of sterilization and abortion for eugenic reasons along with an extensive use of contraception for family planning."[50] These anxieties about population growth prompted medical and scientific experts to offer solutions in the form of sterilization, abortion, and contraceptive advice.

Indeed, population and public health rhetoric shaped campaigns for the legalization of abortion throughout the middle decades of the twentieth century. Johanna Schoen, in her history of the North Carolina Eugenics Board, contends that "in their fight for abortion reform, health and welfare officials across the country turned to the same financial and eugenic arguments that justified eugenic sterilization policies that stood behind the 1960s push for family planning programs."[51] Broadening the picture from the United States to a transnational frame only emphasizes the relationship between abortion and eugenics. After all, eugenics was far more than a narrow set of state-sponsored policies in the United States, Britain, and Germany in the 1920s and 1930s. It was, in fact, a global ideology with different scientific, cultural, and institutional instantiations that were often contradictory or discordant in their aims. What threaded through these disparate movements was the confidence that scientific knowledge was the means by which humanity could engineer its own progress. Through the science of genetics or reproductive medicine, nations believed that they could control their racial futures. And abortion policy was a key part of this confidence. As Alison Bashford has argued in her transnational survey of eugenics during the twentieth century, "Arguably the most overlooked trajectory of eugenics lies in its connection to the liberalization of abortion law."[52] She points specifically to the legalization of abortion in Japan, which was done for explicitly eugenic reasons, but also highlights the policies in the former Soviet Union, China, India, and the United States during the century.[53] Even though many American eugenicists, including Margaret Sanger, were explicitly against the legalization of abortion, this broader historical context demonstrates the manifold connections between abortion policy and eugenic ideologies.

To describe abortion as "eugenic" is to call to mind the physical or legal enforcement of abortion by a state in the manner of forced sterilization. It is this imagined evil that makes it difficult to understand how and why abortion has been or could be eugenic. Perhaps if we describe the set of biopolitical logics that governs abortion policies—the logics of health, normalcy, safety, sexuality, gender, class, ability, and race—then it becomes possible to see how the imperatives of the nation-state impose themselves continuously through the management of reproduction. We need not call this eugenic to see the complex role that technoscience plays in relationship to the intimate experiences of pregnancy, conception, contraception, termination, miscarriage, and maternal health. And as countless scholars have argued, the role of reproductive technoscience is not always, or even usually, liberatory for individuals and communities.

We are not saved from reactionary political attacks on reproductive freedoms by an uncritical championship of scientific authority and public health. And we cannot dismiss accusations of racism in the history of abortion policy, even when they come from those who would dismantle reproductive rights. We have to look somewhere else to free ourselves from the bind between Christian ethnonationalism and secular biopolitics.

DEEP LISTENING

The U.S. Supreme Court's opinion in the landmark 1973 case of *Roe v. Wade* consistently assumes that any doctor providing an abortion is a man. Throughout the text of the bench opinion, physicians in the abstract are referred to as "he" and "him." For instance, in addressing a potential hypothetical decision to perform an abortion in the early stages of pregnancy, the court refers to the reasoning of "a physician and *his pregnant patient*."[54] Throughout the opinion, the decision-making authority for an abortion flows more readily between the expertise of the (male) physician and the state than it does to the pregnant woman. This view that ultimate authority lies outside of the pregnant woman is seen again, for instance, in the Supreme Court's opinion in the 1992 case of *Planned Parenthood v. Casey*, where the court averred, "In some broad sense it might be said that a woman who fails to act before viability has consented to the State's intervention on behalf of the developing child."[55]

Certainly, there are extraordinary moments in these cases that grapple with the liberty of a pregnant woman. In *Casey*, for instance, the court makes this liberatory claim: "The destiny of the woman must be shaped to a large extent on her own conception of her spiritual imperatives and her place in society."[56] But the persistent assumption of the masculine identity of the decision-making physician in *Roe* and other abortion court cases reflects more than just historical sexism in the medical profession. It is a more essential feature of the biopolitics of the American nation-state, a deeply held belief that the knowledge to care for the health of the nation rests in certain bodies. Indeed, the intuition that epistemological authority rests within a masculine purview is threaded throughout the text of *Roe v. Wade*, even at the point when the court seemingly cedes a right to certainty: "We need not resolve the difficult question of when life begins. When those trained in the respective disciplines of medicine, philosophy and theology are unable to arrive at any consensus, the judiciary, at this point *in the development of man's knowledge*, is not in a position to speculate as to the answer."⁵⁷ Although in this statement the court seemingly gives up authority, it, in fact, suggests that there will be a point when "man's knowledge" will be enough to master the domain of fetal life and personhood.

What would it look like, instead, to have an approach to the biopolitical work of reproductive medicine that foregrounds the material experiences of pregnancy? Would an insistence on intellectual mastery be set aside? What is abortion when the performing doctor is a woman and, moreover, is pregnant?

In a beautiful and spare *New York Times* editorial from the spring of 2019 titled "When an Abortion Doctor Becomes a Mother," physician Christine Henneberg describes the entanglements between her medical practice and her motherhood.⁵⁸ She weaves together the physicality of her pregnancy with the work of providing abortions. Throughout she refuses boundaries—between herself and her patients, between fetus and child, between mother and doctor. She recounts her first ultrasound, "performed by a colleague at the abortion clinic," where the "first flicker of a heartbeat" calls forth a visceral fear of miscarriage. She recalls her experiences of witnessing miscarriages at all stages of pregnancy—"the clotted blood a woman thinks is just her late period" or "horrible preterm births in the hospital, fluid everywhere, the fetus—perfect and translucent and too tiny to live—slipping between her legs." Henneberg pushes the reader to confront this messiness, the slipperiness between miscarriage and abortion, between wanted and unwanted, between products of conception and child. She layers descriptions of her work with discussions of her daughter, admitting that "the fetus in the dish, the perfect curl of its fingers and toes. Sometimes it reminds me of my daughter—how could it not?" Perhaps most strikingly, Henneberg explains her inability to separate herself from her pregnancy: "As a doctor, I can draw a distinction, a boundary between a fetus and a baby. When I became a mother, I learned that there are no boundaries, really. The moment you become a mother, the moment another heartbeat flickers inside of you, all boundaries fall away."

Henneberg's rhetorical choices are challenging. Twice she references the heartbeat of an early pregnancy—language that must call to mind so-called heartbeat bills that prohibit abortions after a heartbeat can be detected through a vaginal ultrasound. Since a heartbeat is potentially detectable before many women know about a pregnancy, abortion rights advocates have described these measures as "forced pregnancy bills" and a "healthcare ban."[59] (By early 2022, the U.S. Supreme Court had declined to intervene in the implementation of a Texas state law that empowers private citizens to enforce an abortion ban after the detection of fetal cardiac activity.[60]) Advocates have pointed out that the use of "heartbeat" is deliberating humanizing, unlike the more medically accurate phrase "fetal pole cardiac activity."[61] This political context makes Henneberg's reference to "the first flicker of a heartbeat" a particularly charged decision. As a pregnant woman, she suggests that this flicker represents the first moments of tangible life, of a desired child, of a baby. But as a doctor, it is a moment that does not change the fact that "in reality, everything is messy. The world of doctors, the work of mothers . . . and yet somebody has to do the work." Ultimately, it is her courage to do this work, despite not being able to keep "things tidy and acceptable," that I find compelling.[62]

In this short editorial, Henneberg opens up one of the central issues for feminist and queer theorists on this topic: the relationship between "the morality of abortion and the status of the fetus."[63] For several decades, feminist scholars have investigated the ways in which the social meanings of fetal life emerge from the intersection of technology and cultural narratives. Feminist perspectives in science and technology studies have narrated the ways in which sonograms, fetal genetic screening, in vitro fertilization, and early pregnancy tests have constructed our individual and collective understanding of fetal life.[64]

An important perspective in this literature is the work of Karen Barad, whose leading work in new materialism pushes feminist and other critical theorists to engage with how "matter comes to matter."[65] Barad's 2007 book, *Meeting the Universe Halfway*, directly addresses the question of the social and biological status of a fetus and emphasizes that there is not some discoverable-for-all-time fetus. Instead, the fetus comes into being within and through the social and technological apparatuses that surround, touch, perceive, image, and sonify it. In the example of ultrasound technology, Barad highlights the role of technological apparatus in the emergence of a fetus as scientific object: "In obstetric ultrasonography, the piezoelectric transducer is the interface between the objectification of the fetus and subjectivation of the technician, physician, engineer, and scientist."[66] Her perspective is that the fetus comes into being through technological apparatuses, social practices, and medical knowledge. It is not an unmoving surface on which social meanings are transcribed, nor is it a transcendent static entity that conveys meaning in a timeless fashion.

Technology is not the only space out of which the meanings of fetal life emerge. Judith Butler's theorization of a "grievable life" is a key concept in the feminist understanding of abortion; she explores this concept in *Precarious Life* (2004) and in *Frames of War* (2009), both written in response to the militarized and warring culture of the post–September 11 era.[67] She asks how a life is recognized as grievable and questions how it is possible for more kinds of life to be worthy of grief. Her aim is to expand the boundaries of such lives without reifying a transcendent definition of *person* or *human* into which new lives are recognized and incorporated.[68] Importantly, she addresses the possibility that this concept can be used for a pro-life stance but ultimately rejects this, arguing that "the question is not whether a given being is living or not, nor whether the being in question has the status of a 'person'; it is whether the social conditions of persistence and flourishing are or are not possible."[69] Butler's suggestion, though not further developed, is that we must set aside the quest for a transcendent outside-of-time definition of life or personhood. As with her work on gender identity, she foregrounds the power of discursive processes, seeming to suggest in the instance of abortion that social meaning is the only ground for decision-making.

Ultimately, Henneberg's editorial points us to a bold but compassionate challenge to the pro-choice ideology that champions bodily autonomy and individual choice. It defies the clean difference between *fetus* and *baby*, calling into question the possibility of a biologically based separation between the two. The piece is haunted by the material entanglements between fetus and pregnant person, entanglements that refuse the ontological certainties that frame abortion as an individualized rights-based action that is premised on certainty about when life begins—or not. By centering her own embodied experiences of pregnancy, Henneberg frames the medical act of abortion as part of the love of a child and the care for a broader community of individuals with different values, views, and experiences. As she ends, "the work must be done."[70]

Work is being done in the bold embodiment and imaginative theorizing of reproductive justice activism. Rather than erasing the history of sexism and racism in reproductive politics, groups such as SisterSong and Spark Reproductive Justice Now have shown the way forward through critiques of those discourses. In 2010, for instance, SisterSong launched the national #TrustBlackWomen campaign in response to billboards that accused Black women of genocide for obtaining abortions. First appearing in a predominantly African American neighborhood in Atlanta, these billboards, with captions such as "Black Children

Are an Endangered Species" and "The Most Dangerous Place for a Black Child Is in the Womb," were soon also put up in Illinois, Texas, California, Florida, and Tennessee.[71] In its analysis of these conspiracies, SisterSong pointed to the campaign's exploitation of the historical "medical mistrust" within the Black community and the deployment of "social responsibility to claim that black women have a racial obligation to have more babies—especially black male babies."[72] The billboards were the opening salvo of a legislative campaign to introduce a bill in the Georgia legislature titled the Prenatal Non-discrimination Act (PreNDA). It provided a model for criminalizing race- or sex-selective abortions—in other words, for punishing Black and Brown women for seeking abortions—to allegedly prevent racial genocide.

In response to these attacks, SisterSong knew that it could not rely on the pro-choice movement, which "has not historically responded adequately to charges of racism and genocide" but instead has "visibly flinch[ed] when race and abortion are linked in narratives about ulterior motives."[73] Indeed, it is because of "our collective inability to thoroughly confront issues of race and power in the pro-choice movement" that SisterSong argued that these "accusations can sound credible."[74] Instead, SisterSong organized a "strong collaboration of organizations of diverse capabilities" that mobilized activism throughout the Atlanta area, with connections to national organizations, eventually defeating the bill.[75] Throughout this work, the #TrustBlackWomen campaign reiterated a powerful but radical point: "Black women deserve to be trusted."[76] In response to the accusations that abortion contributes to the violence against Black bodies, the campaign put forward a formal statement of solidarity with the Black Lives Matter campaign:

> We offer to the Movement for Black Lives the analysis that brought Trust Black Women into being: an analysis that centers Black women, low wage workers, LGBTQ people, and those living at the crossroads of these identities. We offer to the Movement for Black Lives our commitment to hold gender justice as dear as racial justice, with reproductive justice as the core of both these aspirations.[77]

It was incisive theory and imagination that allowed SisterSong to combat the creative racism of these billboard attacks. SisterSong reframed the situation, demonstrating the violence of these accusations, and boldly called for trust in the intellect, dignity, and moral authority of Black women.

It is in the real-life work and political responses of reproductive justice activists that the most nuanced and creative theorizing on abortion, colonialism, and patriarchy can be found. Consider, for example, the responses of various women in the Oglala Lakota Nation to a 2007 ban on abortion in South Dakota. As Carly Thomsen examines in her article "From Refusing Stigmatization Toward Celebration: New Directions for Reproductive Justice Activism," Cecilia Fire Thunder, the first woman president of the Oglala Lakota Nation, pushed back on this ban by promising to build a woman's health clinic on reservation land.[78] She was subsequently impeached by a nearly all-male council, which then proceeded to ban all abortions on tribal lands. The response by Lakota women and other Native American women around the country refuted rhetorics that tie abortion to genocide and require women to reproduce as saviors of their race. As Jacqueline Keeler argued, "We, as women, are more than our biology, we are more than just baby machines for a Lakota Nation, a Dakota Nation, or a Nakota Nation."[79] In the work of organizing, pushing legislative reform, or conducting public education, reproductive justice activists theorize abortion with imagination and without reservation. What is perhaps most inspiring about this work is how often it moves beyond, outside of, or to the side of pro-life and pro-choice debates. The work imagines differently what is at stake for health, survival, and dignity for bodies and lives who are pushed to the margins. The proclamations of Spark Reproductive Justice Now, also based in Atlanta, demonstrate this: "the South is ground zero for dangerous and restrictive legislative policies, but is also home to a number of fierce communities that intersect gender justice and LGBTQI sexual and health rights with racial justice and immigrants' rights."[80]

This is the place to listen to break out of the bind between religion and biopolitics.

Throughout this chapter, I have characterized the double bind between patriarchal religion and secular biopolitics that persistently distrusts and violates women; importantly, this is a bind that women experience differently depending on their biology, race, gender identity, class, region, religious background, etc. To believe in a theology that radically trusts women is not a declaration that they are all the same. In fact, it is quite the opposite. It is clear that women's reproductive choices are configured through dramatically different material possibilities and social meanings. In American nationalist rhetoric that prizes whiteness and wealth, the forty-year-old middle-class white woman who prioritized career and had an abortion in her twenties is labeled "selfish," whereas the poor single white woman or Black woman who is pregnant and on social welfare is deemed a "burden"; the immigrant woman who bears a child is accused of having an "anchor baby." In the face of such ethnonationalism, some womanist, mujerista, and queer voices have reclaimed mothering and caregiving as a revolutionary act.[81] Denied the social status of nationalist motherhood, poor, Black, Brown, and queer mothers claim the revolutionary possibilities of "the mothering of children in oppressed groups, and mothering to end war, to end capitalism, to end homophobia and to end patriarchy."[82] These identities and experiences emerge from a vast range of political and social configurations—making and unmaking such terms as *family, mother, parent, fetus,* and *baby*. We cannot fully know what women's choices will be or what meanings those choices will have. And so we need to trust women beyond our capacity to know them.

One way into radical trust is a practice of deep listening to the experiences of others. In this chapter, I have described listening to an abortion doctor and to reproductive justice activists. I continue to listen to those who seek abortions, to advocates of radical mothering, to queer approaches to reproductive justice, and to womanist and mujerista poets. I listen also to trans and intersex scholars and activists about how to use gender terms like "women" capaciously and gently. My call for a feminist theology of abortion is not an attempt to speak for or over those to whom I listen. My hope is only to bring these strands together—the work to destabilize the biological and social meanings of fetal life alongside the resolute confrontation of the racist history of American reproductive politics. To see all of this, in all of its complexity, violence, and promise—and to respond with a single declaration of a sacred act—we must trust women.

A FEMINIST THEOLOGY OF ABORTION

Why do I call for a theology of abortion rather than a feminist ethic or perspective on abortion? It is undoubtedly a strange choice, if for no other reason than the fact that theology vibrates with the respectability politics of patriarchal modes of being and knowledge. Certainly, there are theologies of the oppressed: those communities whose dispossession has given birth to new ways of understanding the life and message of Christ through Black, mujerista, liberation, or womanist lenses. These bold accounts of the faith of the marginalized are deeply inspiring. But my claim here is not that I have understood Christianity better or differently.

I have no new textual exegesis, no innovative hermeneutics. Rather, my hope for a theology is inspired by Donna Haraway's orientation to science in the opening of "A Cyborg Manifesto" when she suggests that "blasphemy is more faithful than direct worship and fealty."[83] She suggests that through blasphemy and deliberate irony, we can get free of the desire to construct purity and instead recognize our inherent complicity in violence. It is the fusion of biopolitics and Christian nationalism that has done so much harm to true liberation and flourishing with regard to abortion. How do we go beyond a critique of these entanglements to offer up a new ethic without reproducing a new taxonomy of certainty and righteousness? This is my attempt. My theology is a theology not because it is tied to a scripture or a specific metaphysic but because it is an effort to find a new sacred.

The 2016 Indiana state law imposes in the worst way on women, fulfilling the imperatives of Christian nationalism with an insistence on the grievability of a fetus while handling the eugenic manifestations of biopolitics through further governance, blame, and distrust of women. The academic and political liberal response unfortunately relies on confidence in the cleanly differentiated materiality of bodies and the possibility of an individual rights approach to abortion ethics. In this chapter, I have argued that abortion has both a history of being and the potential to be a tool of eugenics and that an abortion ethic cannot rest on a clear separation between woman and fetus or mother and child. A theology of abortion is therefore an argument for a radical trust in women that cannot guarantee either that abortion policies have no eugenic motivations or that the termination of a pregnancy is not the end of a life or of a person. It is an ethics of risk and a feminist embrace of that risk. A feminist theology of abortion is an openness to the ways in which our view of human life, suffering, and personhood has materially transformed through the history of technology. It suggests that our views of what is valuable, what is livable, and what is grievable have changed and may do so yet again in the future. It is an ethic of imperfection and humility and an acknowledgment that we may do something truly wrong. But is a theology that lives with these fears and believes that by setting aside a desire for epistemological and ontological certainty, we are free to radically trust that others will do the best that they can.

Trust is not jurisprudence. It is not a commandment. Trust happens beyond any proof that can be given, but it can also be violated beyond repair. The call to radically trust women is not an exhortation to view women as essentially good moral actors. It is a reorientation of the affective posture of lawmaking and public health policy, which is currently threaded through with the persistent belief that women are not to be trusted, that they are unaware of their own best interest and their decisions must be guarded against, lest they bring harm to themselves, their children, or the nation at large.

When I imagine a feminist theology of abortion, I picture a set of open hands: hands that might do the bloody work of obstetrics, hands that might tenderly hold a newborn, hands that might be held up in protest—but, most of all, hands that remain open, not grasping at control or certainty or authority. This feminist posture does not rest in the confidence of its own righteousness. Nor does it relax into an assurance of the purity of its intent. It sees work that must be done, lives that must be cared for, and communities that must be protected. And it is willing to be wrong in order to take the risk of trusting women entirely—without reservation.

NOTES

1. This chapter went to press shortly after the *Dobbs v. Jackson Women's Health Organization* decision on June 24, 2022, which overruled both *Roe v. Wade* (1973) and *Planned Parenthood v. Casey* (1992). The Supreme Court's decision makes the imperative of this chapter—a radical trust in women—even more compelling, even if some of its history is already out of date.
2. Indiana Omnibus Abortion Act, House Enrolled Act No. 1337 (2016).
3. Indiana Omnibus Abortion Act.
4. Box v. Planned Parenthood of Indiana and Kentucky, No. 18–483, 487 U.S. __ (2019) (Thomas, J, concurring, 2).
5. Alexandra Minna Stern, "Clarence Thomas' Linking Abortion to Eugenics Is as Inaccurate as It Is Dangerous," *Newsweek*, May 31, 2019, https://www.newsweek.com/clarence-thomas-abortion-eugenics-dangerous-opinion-1440717.
6. Adam Cohen, "Clarence Thomas Knows Nothing of My Work," *The Atlantic*, May 29, 2019, https://www.theatlantic.com/ideas/archive/2019/05/clarence-thomas-used-my-book-argue-against-abortion/590455/.
7. Eli Rosenberg, "Clarence Thomas Tried to Link Abortion to Eugenics: Seven Historians Told *The Post* He's Wrong," *Washington Post*, May 30, 2019, https://www.washingtonpost.com/history/2019/05/31/clarence-thomas-tried-link-abortion-eugenics-seven-historians-told-post-hes-wrong/?utm_term=.bfb186b6aeec.
8. Cohen, "Clarence Thomas Knows Nothing of My Work."
9. Rosenberg, "Clarence Thomas Tried to Link Abortion to Eugenics."
10. Karen Weingarten, "The Eugenicists on Abortion," *Nursing Clio*, July 2, 2019, https://nursingclio.org/2019/07/02/the-eugenicists-on-abortion/.
11. Importantly, abortions are a critical health service for nonbinary persons and transgender men. However, much of the literature on gender-inclusive and nondiscriminatory health

care practices for transgender and nonbinary persons tends to focus on general issues of access and medical services related to transition rather than specifically on abortion. The 2011 guidelines from the American College of Obstetrics and Gynecology Committee on Health Care for Underserved Women, for instance, make no reference to pregnancy or abortion in relation to serving transgender and nonbinary persons. Committee on Health Care for Underserved Women, American College of Obstetricians and Gynecologists, "Committee Opinion," December 2011. See also Alanna Vagianos, "Women Aren't the Only People Who Get Abortions," *HuffPost*, June 7, 2019, https://www.huffpost.com/entry/women-arent-the-only-people-who-get-abortions_n_5cf55540e4b0e346ce8286d3.

12. Gordon Lynch, *On the Sacred* (Durham, UK: Acumen, 2012), 1–2.
13. Kinsella Tringali and Izabela Martha, "Forced Sterilization Accusations at ICE Facility Fit with Trump's Poor Treatment of Immigrants," Brennan Center for Justice, September 18, 2020, https://www.brennancenter.org/our-work/analysis-opinion/forced-sterilization-accusations-ice-facility-fit-trumps-poor-treatment; Adam Liptak and Sabrina Tavernise, "After Silence from Supreme Court, Texas Clinics Confront Near-Total Abortion Ban," *New York Times*, September 1, 2021.
14. Daniel K. Williams, *God's Own Party: The Making of the Christian Right* (New York: Oxford University Press, 2010), 154–155.
15. In total, the Schaeffers made five films in the series. The film referenced here is the first entry, *The Abortion of the Human Race*, Francis A. Schaeffer, C. Everett Koop, and Vision Video, *Whatever Happened to the Human Race?* (Worcester, PA: Vision Video, 1979).
16. C. Everett Koop and Francis A. Schaeffer, *Whatever Happened to the Human Race?* (Westchester, IL: Crossway, 1984), 14.
17. Koop and Schaeffer, *Whatever Happened to the Human Race?*, 4.
18. Koop and Schaeffer, *Whatever Happened to the Human Race?*, 9.
19. Edward O. Wilson, *Sociobiology: The New Synthesis* (Cambridge, MA: Belknap Press, 1975), 562, quoted in Koop and Schaeffer, *Whatever Happened to the Human Race?*, 10.
20. Koop and Schaeffer, *Whatever Happened to the Human Race?*, 10.
21. Koop and Schaeffer, *Whatever Happened to the Human Race?*, 10.
22. Sharon Mara Leon, *An Image of God: The Catholic Struggle with Eugenics* (Chicago: University of Chicago Press, 2013), 3.
23. Christine Rosen, *Preaching Eugenics: Religious Leaders and the American Eugenics Movement* (Oxford: Oxford University Press, 2004), 11–14.
24. Committee on Cooperation with Clergymen of the American Eugenics Society, "Conditions of the Awards for the Best Sermons on Eugenics," 1926, American Eugenics Society Records, box 13, Sermon Contests, 1926 #1, American Philosophical Society.
25. "Eugenics," Eugenics Sermon Contest #7, 1926, 9, American Eugenics Society Records, box 12, Sermon Contests, 1926 #2, American Philosophical Society.
26. "Eugenics," Eugenics Sermon Contest #3, 1926, 4, American Eugenics Society Records, box 12, Sermon Contests, 1926 #2, American Philosophical Society.
27. Williams, *God's Own Party*.
28. Williams, *God's Own Party*, 154–155; Daniel K. Williams, *Defenders of the Unborn: The Pro-Life Movement Before Roe v. Wade* (New York: Oxford University Press, 2016), 237.
29. Jerry Falwell, *Listen, America!* (New York: Doubleday, 1980), 253–254.

30. "Pamphlets for Distribution to Local Churches 2008–2010," Institute for Creation Research, in the author's personal manuscript collection.
31. Concerned Women for America, "Our History," accessed April 1, 2019, https://concernedwomen.org/about/our-history/.
32. "Sunday School Curriculum 2007–2008," Skyline Wesleyan Church, in the author's personal manuscript collection.
33. Stephanie Strom, "The Political Pulpit," *New York Times*, September 30, 2011.
34. James Garlow, *Skyline Newsletter*, September 26, 2015.
35. James L. Garlow, *Well Versed: Biblical Answers to Today's Tough Issues* (New York: Simon & Schuster, 2016), 228, 209, 86.
36. Lynch, *On the Sacred*, 22.
37. Lynch, *On the Sacred*, 11.
38. Lynch, *On the Sacred*, 26.
39. Box v. Planned Parenthood of Indiana and Kentucky (Thomas, J., concurring, 2).
40. Michel Foucault, *Security, Territory, Population: Lectures at the Collège de France, 1977–78*, ed. Michel Senellart, trans. Graham Burchell (Basingstoke, UK: Palgrave Macmillan, 2007); Ann Laura Stoler, *Race and the Education of Desire: Foucault's History of Sexuality and the Colonial Order of Things* (Durham, NC: Duke University Press, 1995); Laura Briggs, *Reproducing Empire: Race, Sex, Science, and U.S. Imperialism in Puerto Rico* (Berkeley: University of California Press, 2003); Myrna Perez Sheldon, "Race and Sexuality: A Secular Theory of Race," in *A Cultural History of Race in the Age of Empire and Nation State*, ed. Marina B. Mogilner (London: Bloomsbury, 2021), 149–164.
41. Leslie J. Reagan, *Dangerous Pregnancies: Mothers, Disabilities, and Abortion in Modern America* (Berkeley: University of California Press, 2010), 2.
42. Reagan, *Dangerous Pregnancies*, 222.
43. Myrna Perez Sheldon, "Science, Medicine, and Technology: A Historiographical Survey," in *The Routledge History of the Twentieth-Century United States*, ed. Jerald E. Podair and Darren Dochuk (New York: Routledge, 2018).
44. Bill Ong Hing, *Defining America: Through Immigration Policy.* (Philadelphia: Temple University Press, 2012), 3–8.
45. Alan Frank Guttmacher, *Birth Control and Love: The Complete Guide to Contraception and Fertility* (New York: Macmillan, 1961), 216.
46. Guttmacher, *Birth Control and Love*, 217.
47. Guttmacher, *Birth Control and Love*, 273.
48. Alan Guttmacher, "The Place of Sterilization," in *The Population Crisis and the Use of World Resources*, ed. Stuart Mudd (New York: Springer, 1964), 279.
49. Jerome H. Kumer, "The Problems of Abortion and the Personal Population Explosion," in *The Population Crisis and the Use of World Resources*, ed. Stuart Mudd (New York: Springer, 1964), 274.
50. Kumer, "The Problems of Abortion," 192.
51. Johanna Schoen, *Choice and Coercion: Birth Control, Sterilization, and Abortion in Public Health and Welfare* (Chapel Hill: University of North Carolina Press, 2007), 143.
52. Alison Bashford, "Epilogue: Where Did Eugenics Go?," in *The Oxford Handbook of the History of Eugenics*, ed. Alison Bashford and Philippa Levine (Oxford: Oxford University Press, 2010), 546.

53. Bashford, "Epilogue," 544–546.
54. Roe v. Wade, 410 U.S. 113, 156 (1973) [emphasis added].
55. Planned Parenthood of Southeastern Pennsylvania v. Casey, 505 U.S. 833, 870 (1992).
56. Planned Parenthood of Southeastern Pennsylvania, 505 U.S. at 852.
57. Roe, 410 U.S. at 159 [emphasis added],
58. Christine Henneberg, "Opinion: When an Abortion Doctor Becomes a Mother," *New York Times*, June 27, 2019, https://www.nytimes.com/2019/06/27/opinion/abortion-doctor.html.
59. Krystal Redman, executive director of Spark Reproductive Justice Now, quoted in Anna North and Catherine Kim, "The 'Heartbeat' Bills That Could Ban All Abortions, Explained," *Vox*, June 28, 2019, https://www.vox.com/policy-and-politics/2019/4/19/18412384/abortion-heartbeat-bill-georgia-louisiana-ohio-2019.
60. Eleanor Klibanoff, "Supreme Court Again Declines to Intervene in Challenge to Texas Abortion Law," *Texas Tribune*, January 20, 2022, https://www.texastribune.org/2022/01/20/supreme-court-texas-abortion-law-challenge/.
61. "Stop Calling Them 'Heartbeat' Bills and Call Them 'Fetal Pole Cardiac Activity' Bills," *HuffPost*, December 12, 2016, https://www.huffpost.com/entry/dear-press-stop-calling-them-heartbeat-bills-and-call-them-fetal-pole-cardiac-activity-bills_b_584f19fde4b04c8e2bb14e15.
62. Christine Henneberg, "When an Abortion Doctor Becomes a Mother," opinion, *New York Times*, June 27, 2019.
63. Lynn M. Morgan and Meredith Wilson Michaels, eds., *Fetal Subjects, Feminist Positions* (Philadelphia: University of Pennsylvania Press, 2016), 7.
64. Lauren Berlant, "America, 'Fat,' the Fetus," *Boundary 2* 21, no. 3 (1994): 145–195, https://doi.org/10.2307/303603; Janelle S. Taylor, *The Public Life of the Fetal Sonogram: Technology, Consumption, and the Politics of Reproduction* (New Brunswick, NJ: Rutgers University Press, 2008); Sarah Franklin, *Biological Relatives: IVF, Stem Cells, and the Future of Kinship* (Durham, NC: Duke University Press, 2013); Carly Thomsen, "Crisis Pregnancy Centers and Sonograms," *The Immanent Frame: Secularism, Religion, and the Public Sphere*, August 7, 2020, https://tif.ssrc.org/2020/08/07/crisis-pregnancy-centers-and-sonograms/.
65. Karen Barad, "Posthumanist Performativity: Toward an Understanding of How Matter Comes to Matter," *Signs* 28, no. 3 (2003): 801–831.
66. Karen Barad, *Meeting the Universe Halfway* (Durham, NC: Duke University Press, 2007), 204.
67. Judith Butler, *Precarious Life: The Powers of Mourning and Violence* (London: Verso, 2004), and *Frames of War: When Is Life Grievable?* (London: Verso, 2009).
68. Butler, *Frames of War*, 13.
69. Butler, *Frames of War*, 20.
70. Henneberg, "When an Abortion Doctor Becomes a Mother."
71. Jael Silliman et al., *Undivided Rights: Women of Color Organizing for Reproductive Justice*, 2nd ed. (Chicago: Haymarket, 2016), x.
72. "First the Billboards," Trust Black Women, accessed April 25, 2019, www.trustblackwomen.org/our-roots.
73. "Analyzing the Political Movement," Trust Black Women, accessed April 25, 2019, www.trustblackwomen.org/our-roots.
74. "Analyzing the Political Movement," Trust Black Women, accessed April 25, 2019, www.trustblackwomen.org/our-roots.

75. "SisterSongs Response," Trust Black Women, accessed April 25, 2019, www.trustblackwomen.org/our-roots.
76. "Black Women Deserve to Be Trusted," Trust Black Women, accessed April 25, 2019, www.trustblackwomen.org/home.
77. "Solidarity Statement with BLM," Trust Black Women, accessed April 25, 2019, https://trustblackwomen.org/solidarity-statement.
78. Carly Thomsen, "From Refusing Stigmatization to Celebration: New Directions for Reproductive Justice Activism," *Feminist Studies* 39, no. 1 (2015): 152.
79. Jacqueline Keeler, quoted in Thomsen, "From Refusing Stigmatization to Celebration," 152.
80. "What Is Reproductive Justice?," Spark Reproductive Justice, accessed June 1, 2019, http://www.sparkrj.org/about/whatisreprojustice.
81. Alexis Pauline Gumbs, China Martens, and Mai'a Williams, *Revolutionary Mothering: Love on the Front Lines* (Oakland, CA: PM Press, 2016).
82. Gumbs, Martens, and Williams, *Revolutionary Mothering*, 20.
83. Donna Haraway, "A Cyborg Manifesto: Science, Technology, and Socialist-Feminism in the Late Twentieth Century," in *Simians, Cyborgs, and Women: The Reinvention of Nature* (New York: Routledge, 1991), 149.

CHAPTER 4

CAN ORIGINALISM SAVE BIOETHICS?

OSAGIE K. OBASOGIE

Arthur Caplan has been quoted as saying that the field of bioethics "was born out the ashes of the Holocaust."[1] This may seem odd to some, as ethical considerations concerning medicine date back to the times of Hippocrates. Over time, medical ethics has come to largely focus on the individual interactions between doctors and patients concerning issues of professional responsibility, appropriate clinical engagements and bedside manner, and approaches to decision-making. Yet bioethics has a much more recent history that breaks from this tradition. Many, such as Albert Jonsen, attribute the emergence of bioethics to the postwar technological transformations in science and medicine. Jonsen notes that the 1960s marked a turning point as "contributors to scientific progress ... worried that the old tradition of medical ethics was too frail to meet the ethical challenges posed by the new science and medicine."[2] It is certainly true that the postwar era was a time of profound technological advancement that raised new questions *at a societal level* that transcended the individualistic dynamics that dominated medical ethics. These postwar conversations included discussion on heart transplantation, brain death and its attendant considerations, assisted reproduction, and recombinant DNA and genetic engineering. Surely the ethical nuances of how to approach these issues could not be adequately captured by medical ethics' rather limited scope. Yet, to circle back to Caplan's often underappreciated observation, what is it about the end of World War II and public revelations of the horrific abuses in Nazi Germany, as historical and technological breaking points, that shifted ethical discourses in a manner that required a new field—one that we call bioethics? What roles did religion and secularity play during this moment to shape discussions of the appropriate use of science, medicine, and technology beyond the clinic—that is, how we think about our human future? How and why did bioethics move away from the values that originally gave rise to the field? What have been the consequences? And should we return to these founding values?

In this chapter, I will explore these questions and suggest that in order for bioethics to be a meaningful method for tackling serious and consequential issues regarding the dignity and integrity of humanity, it needs to revisit the context from which it emerged and develop modern approaches that keep these perspectives and values at the forefront of its evaluations. To do this, I will also explore originalism—a theory of constitutional interpretation that prioritizes fealty to the original intent and purpose of legal text over any mitigating circumstances that may arise when contemporary contexts are taken into consideration. Allowing the so-called dead hand of those who initially create rules and values to bind the actions of future generations has been deeply controversial in law and profoundly consequential for how we govern ourselves. And to be frank, originalism is an approach that, with regard to constitutional interpretation, often leads to inefficiencies, injustice, and the entrenchment of white supremacy, heteronormativity, and patriarchy. But bioethics is arguably a different matter, and in this chapter, I will advance the case that some form of originalism in the field ought to be considered.

BIOETHICS: ORIGINS AND EMERGENCE

Medical ethics, in its most traditional sense, largely understood itself as a field where researchers and physicians could be relied upon to regulate their own professional activities. The presumption that those trained in medicine would work only in the interest of patients animated the field and motivated many of the principles and frameworks it offered to resolve conflicts in clinical care and scientific research. Then, as Caplan suggests, the Holocaust changed the way that we understand physicians' and researchers' capacity to participate in evil. However, the birth of bioethics emerged not only out of a general awareness of the Holocaust and its brutality but specifically the public recitation of how the scientific method was central to torture and genocide, as revealed in the Nuremberg doctors' trial. To put an even finer point on this birthing moment, we can turn to Telford Taylor's "Opening Statement of the Prosecution" at this trial in 1946.

Taylor, the chief of counsel for war crimes and a brigadier general, begins his statement with an allegation that fundamentally upended the perception that medicine is an unquestionably isolated, autonomous, selfless, patient-centered, and self-regulatory field by noting that "the defendants in this case are charged with murders, tortures, and other atrocities committed in the name of medical science."[3] He then specifies why this trial is different from any regular trial:

> The defendants in the dock are charged with murder, but this is no mere murder trial. We cannot rest content when we have shown that crimes were committed and that certain persons committed them. To kill, to maim, and to torture is criminal under all modern systems of law. These defendants did not kill in hot blood, nor for personal enrichment. Some of them may be sadists who killed and tortured for sport, but they are not all perverts. They are not ignorant men. Most of them are trained physicians and some of them are distinguished scientists. Yet, these defendants, all of whom were fully able to comprehend the nature of their acts, and most of whom were exceptionally qualified to form a moral and professional judgment in this respect, are responsible for wholesale murder and unspeakably cruel tortures.[4]

For Taylor, the fact that individuals trained in medicine and the sciences used their specialized skills to harm and kill rather than heal is a particular evil that needed to be highlighted, exposed, and punished. And to be clear, the actions of the Nuremberg defendants were abominable: high-altitude experiments to determine at what point air pressure would kill, freezing experiments where prisoners were left outside in the cold for several hours to determine how best to rewarm people, malaria experiments where healthy people were deliberately infected with the disease, experiments with mustard gas to document the burns and potential remedies, seawater experiments to determine if certain filtration methods worked and how much could be consumed before death, sterilization experiments to find fast and cheap ways to wipe out large populations, and other human subjects research where people incarcerated in concentration camps were poisoned or harmed to analyze and measure their reactions.[5]

Taylor describes this gruesome and depraved approach to scientific research as *thanatology*, or "the science of producing death."[6] Exposing the morbid possibilities of science in this manner shattered the perception that it is intrinsically a humane and life-affirming practice that could hold itself accountable. This marked an important breaking point in the historical narrative of the medical sciences along at least three dimensions. First, Taylor identifies the

Nazi defendants' behavior as being outside of the realm of acceptable medicine and scientific research. Surely this was not the first time that powerful men in science exploited vulnerable populations in gruesome and unprincipled ways.[7] Yet the scale and visibility of Nazi medicine made indelible impressions on the public. Because the defendants sought death rather than life, Taylor framed their actions as illegitimate and deserving of the harshest scrutiny. Second, Taylor implicitly highlights the inability of medical ethics to respond to such actions. Confronted with a self-regulating field that is structured in a manner where ultimate accountability lies with no one, Taylor draws attention to the need for an ethical imagination for oversight that goes beyond what then existed—both as a substantive matter (beyond doctor/patient issues) and as a procedural concern (in terms of establishing external mechanisms and rules that can oversee research endeavors). Lastly, Taylor emphasizes the particular ethical problem that arises when professionals use their training and knowledge of the body to commit massive crimes against humanity. He notes:

> The thanatological knowledge, derived in part from these experiments, supplied techniques for genocide, a policy of the Third Reich, exemplified in the "euthanasia" program and in the widespread slaughter of Jews, Gypsies, Poles, and Russians. This policy of mass extermination could not have been so effectively carried out without the active participation of German medical scientists.[8]

This particular use of the methods, knowledge, and training of the medical sciences in a deliberate and systematic way to explore the most efficient ways to kill and eradicate populations raised new and profound questions about the role of science in society. By laying this bare, Taylor not only set out the terms of the defendants' prosecution but also created the conditions for a new field that looked beyond medical ethics' individualism to understand the roles of *social impact* and *social responsibility* in the research endeavor.

It is important to understand that the need for a different mode of assessment, which would ultimately become known as bioethics, was not a neutral endeavor but a deeply and unapologetically normative one. While the norms of bioethics would ultimately become known in the late part of the twentieth century through a principlist framework that focuses on respect for autonomy, beneficence, nonmaleficence, and justice, the norm at the birthing moment captured during the Nuremburg doctors trial can be succinctly described as *anti-eugenics*—i.e., protecting vulnerable populations from applications of the scientific method intended to destroy or undermine their humanity. The outrage giving rise to the trial itself is connected to the horrific ideology motivating the Third Reich's abuses: that people with disabilities, poor individuals, racial minorities, and other stigmatized groups are inherently inferior and should be eliminated in the name of racial hygiene and white supremacy. Eugenics is often thought of as a political ideology or a practice promoted by maniacal ideologues like Adolf Hitler. But Taylor's opening statement serves as permanent testimony that medical professionals and scientists were at the forefront of what many claim to be the most horrific moment in human history. Thus, bioethics must acknowledge that the normative value orienting the field from its inception was a commitment to understanding the centrality of science and medicine to this travesty and, going forward, to do everything possible to resist eugenics' reemergence.

ANTI-EUGENICS GIVES WAY TO PRINCIPLISM

In rendering judgment against the physicians and researchers in the Nuremberg doctors' trial, the court also provided ten rules for conducting ethical research that are known as the Nuremberg Code. By focusing on reorienting the clinical wrongs of Nazi research at the expense of a broader discussion of eugenics' *social and political harms*, the conclusion of the doctors trial created an immediate disconnect between the broader anti-eugenics sensibility in Taylor's opening statement for the prosecution and the end product, which focused on clinical aspects of research.

Nevertheless, the tenor of ethical conversations had been permanently changed. Jonsen notes in *The Birth of Bioethics* that in the early discussions of this new post–World War II ethical endeavor, "theologians were the first to appear on the scene"[9]—perhaps because of the relatively important role that religion and morality played in many conversations pertaining to traditional medical ethics before the war. Jonsen ties theologians' role in nascent conversations on bioethics to what he calls "the new biology" that was reshaping medicine and scientific research at this time. Then-emerging approaches in biology and genetics promised to give medical professionals and scientists remarkable power to not only heal people but also change the limits, possibilities, and future course of humanity in ways that had previously been thought to be the sole province of a higher being. Yet the moral quandaries concerning the new biology's ability to "play

God" had given way to a different domain and ethical inquiry: philosophy. Jonsen notes that "the bioethics that began to appear during the 1970s, while generated by the new medicine and science encountering human values and customs, was [philosophers'] creation."[10] With the rise of philosophy as the *lingua franca* of bioethics, the field began to break from its most immediate theological origins to become a secular and professional endeavor that resembled a consultancy for medicine more than a coequal branch of research oversight. This transition was facilitated by the publication of *The Belmont Report* in 1976 and Beauchamp and Childress's classic book *Principles of Biomedical Ethics* in 1979, now in its eighth edition.[11] While there is overlap between these two works, they are jointly seen as establishing principlism as the dominant framework of bioethical inquiry, which ultimately came to emphasize respect for autonomy, beneficence, nonmaleficence, and justice. Beauchamp and Childress argue that principlism emerged not from any one moral or religious tradition but from a "common morality," which "is not merely *a* morality, in contrast to other moralities. It is applicable to all persons in all places, and we rightly judge all human conduct by its standards."[12]

Principlism completed the secularization of bioethics by effectively claiming to distill the essence of all human cultures and societies into a singular moral code that did not need further contextualization and could therefore apply to all ethical situations. This secularization takes bioethics further away from its origins as a *normative* field concerned with countering eugenic temptations in science and medicine and turns it into more of a *descriptive* endeavor that focuses on isolated clinical and research interactions and distills the proper rules of engagement for physicians and researchers.

BEYOND PRINCIPLISM

John Evans describes principlism as part of a formally rational system that seeks commensuration, or the ability to measure "different properties normally represented by different units with a single, common standard of unit."[13] This produces "a method for discarding information in order to make decision making easier by ignoring aspects of the problem that cannot be translated to the common metric."[14] Evans views the appeal and dominance of principlism within bioethics as a function of its ability to "offer the lure of calculability and predictability," much like money offers more ease and flexibility when trying to determine the value of bartered goods.[15] Boiling down the ethical universe to four principles tied to a common morality turns the once normative behavior of enforcing values characterized by a commitment to countering and atoning for science's eugenic past into a cold, calculated description of the minimal obligations professionals owe patients and human subjects. The idea of having a singular (and predictable) way of reasoning through ethical problems has trumped the deeper anti-eugenic and theological contemplations that created the field.

Bioethics' emphasis on commensuration has resulted in a modern field beset with at least three endemic problems. First, it has led to a form of regulatory or institutional capture, where a field meant to be independent of the people and activities that it regulates has been unduly influenced by the politics, power, and prestige of the regulated group. The consultancy relationship that many bioethicists have developed with medicine and science has led to a familiarity and shared professional interests that have incentivized many bioethicists to not be as critical as one would think. Chapter 6 of Carl Elliott's *White Coat, Black Hat* provides a stunning look at how bioethics has, in many cases, been subsumed into health and science industries in ways that situate the field as more of a public relations mouthpiece than an independent watchdog acting in the public interest.[16] Other examples include the rise of for-profit institutional review boards, which enable researchers to pay outside reviewers to certify that their studies meet various human subject requirements—a dynamic that is inherently fraught with problems and conflicts of interest.[17] Principlist commensuration is another name for capitalist exploit, which makes bioethics ripe for industry manipulation where health care and research are multibillion-dollar endeavors.

A second problem tied to principlist commensuration is that bioethics has largely excluded the voices of marginalized groups during its development. Despite bioethics' emergence from a historical moment where minority groups were systematically persecuted, the field has grown to be an unabashedly white endeavor. Cat Myser argues in the *American Journal of Bioethics* that

> we have inadequately noticed and questioned the dominance and normativity of whiteness in the cultural construction of bioethics in the United States. By thus allowing the whiteness of bioethics to go unmarked, we risk repeatedly re-inscribing white privilege—white supremacy even—into the very theoretical structures and methods we create as tools to identify and manage ethical issues in biomedicine. In other words, by not seeing or locating the whiteness

in bioethics, by theorizing from this unself-reflective white standpoint and by extending its cultural capital into bioethics policies and practices, we risk functioning as cultural colonizers who do violence to social-justice concerns related to race and class.[18]

Similar critiques of bioethics can be made regarding its overwhelming male, class-privileged, and able-bodied sensibilities. But Myser's insight on how whiteness anchors bioethics is a stunning observation given the original intentions of the field to serve as a space of resistance against these types of racial normativities in how medicine and research are conducted. This also can be seen as a product of principlist commensuration; conflating the ethical universe to a common morality and four singular principles has predictably privileged the experiences and reasoning of those in power and obscured the experiences of the underprivileged and vulnerable that the field is meant to serve.

Lastly, principlism's pursuit of commensuration has saddled bioethics with a profound inability to understand how past concerns regarding eugenics are reemerging in new ways. When capital and market logics capture the moral reasoning of the field and whiteness becomes its normative subject position, the ability to see eugenic reasoning creeping in, as Troy Duster might say, through the backdoor[19] is compromised, exposing the most vulnerable subjects to violations that the field intended to never allow again. Whether it is the laissez-faire screening of embryos to eradicate certain disabled populations,[20] recent proposals to use persons who are incarcerated in clinical trials,[21] or the thinly veiled excitement surrounding proposals to use germline engineering on humans to irreversibly change their character and abilities,[22] the field of bioethics has been remarkably permissive in allowing such efforts to move forward even though this work contradicts many of the concerns expressed during the field's founding moments. The eugenic temptation is back. To be sure, it never quite left. Yet it is resurfacing now with the passive assistance of the professionals who were specifically charged with preventing this.

ORIGINALISM AS BIOETHICAL METHOD?

To be clear, there are many bioethicists who are concerned about the eugenic implications of technological advances in the reproductive sciences and genetic technologies. But the critique being offered in this chapter is that this concern is not nearly as prominent in the field as it ought to be. Anti-eugenics is a seemingly minority standpoint among bioethicists to the extent that eugenics is often defined so narrowly, as *state compulsion* of reproductive decision-making, that it appears inapplicable to modern (often market-based) practices.[23] Yet this constrained understanding of eugenics is dubious at best. If there is one thing that all bioethicists should agree on, it is this: efforts to use science and medicine to intervene in human reproduction and development—regardless of whether they are driven by the state, cultural preferences, or markets—and meant to give meaningful advantages to some while leaving others behind ought not be allowed. However, the field is far from consensus on this issue. The lack of agreement demonstrates how far modern bioethics has moved from the values that initially defined the purpose of the field.

Theorists of constitutional law have engaged in a debate that could be instructive for bioethics. When interpreting the U.S. Constitution to resolve disputes, federal judges can use different methods. On one end of the spectrum are living constitutionalists, or those who believe that the meaning of constitutional text can and should evolve with the society that it governs without requiring the formality of constitutional amendment. For example, the notion of *privacy* does not appear anywhere in the text of the Constitution but has been construed to be a core principle to protect certain rights, such as married couple's use of contraception.[24] The idea behind living constitutionalism is that as society evolves to acknowledge the importance of certain rights and concepts, judges may interpret constitutional text in a manner that extends important values to reach new ideals that may not have been fully intended by those who initially crafted the document.

On the other end of the spectrum are originalists, or those who advocate that the original intent or meaning of constitutional text—*at the time it was written*— should shape how judges apply the Constitution to contemporary legal matters. For example, in *District of Columbia v. Heller*, the U.S. Supreme Court took up the question of whether a District of Columbia statute banning the possession of unregistered firearms as well as all handguns (unless approved by the chief of police) comported with the Second Amendment.[25] In writing the majority opinion, Justice Antonin Scalia clearly states the court's preferred method for interpreting the Second Amendment:

> In interpreting this text, we are guided by the principle that "[t]he Constitution was written to be understood by the voters; its words and phrases were used in their normal and ordinary as distinguished from technical meaning." United States v. Sprague, 282 U. S. 716, 731 (1931). . . . Normal meaning may

of course include an idiomatic meaning, but it excludes secret or technical meanings that would not have been known to ordinary citizens in the founding generation.[26]

This is originalism, plain and simple: a hermeneutic time machine. Scalia then performs this originalist methodology by identifying key words and phrases in the Second Amendment (for example, "keep and bear arms"), scouring the historical record to divine their meanings at the time the amendment was created in the eighteenth century, and then putting those meanings in conversation with the contemporary D.C. statute to render a decision on its constitutionality.

Seeking consistency and predictability in legal decisions by encouraging fidelity to an original constitutional text that gave birth to a nation can be thought of as an admirable sentiment. Yet originalism presents many problems. Methodologically, it quickly buckles under its own weight. This is because the precise historical moments sought to serve as moral and interpretive reference points—a period of slavery, genocidal eradication of Native Americans, conquest and illegitimate land capture, patriarchy and deep gender bias, and other discriminatory attitudes—were so doggedly inequitable that it becomes hard to justify transporting eighteenth-century legal mores into the present as anything other than an attempt to maintain white male privilege.

Yet the problem with originalism as it is currently construed is not with the method itself. Rather, the issue is that it can lead to grave injustices when applied to American constitutionalism that still, to this day, has not yet fully grasped how deeply slavery and racial inequity shaped the founding of this country.[27] Originalists see the foundational sin of American slavery and its influence on the development of the U.S. Constitution as an isolated problem remedied by the Thirteenth, Fourteenth, and Fifteenth Amendments, along with other federal statutes. Yet this argument misses how the American legal system has continued to produce racial inequality—not only in spite of remedial efforts but also often through these mechanisms themselves.[28]

However, bioethics emerged from a radically different historical context than the ratification of the U.S. Constitution, which opens up different possibilities for an originalist methodology to serve the public interest. This chapter has put a finer point on Caplan's assertion that bioethics was born out of the Holocaust to identify Telford Taylor's opening statement at the Nuremberg doctors trial as the specific moment when medicine and human subjects research were exposed as having the ability to serve as weapons of terror—a realization that had not occurred to the public until the full scope of the eugenics program was laid bare during the trial. This created the corresponding need for greater oversight—hence, the field that later materialized as bioethics. It is this remarkably different historical context that gives originalism more legitimacy in the bioethics context than in law. Using an analytical method that seeks adherence to a historical moment of human liberation, professional accountability, and greater inclusion in the name of anti-eugenics is, by definition, more appropriate than using such approaches to prop up institutional values premised on conquest, exclusion, and rampant inequality.

CONCLUSION: WHY BIOETHICS NEEDS ORIGINALISM

Much like the technological moment following World War II that highlighted medical ethics' inability to resolve issues pertaining to what Jonsen called "the new biology," we are at a moment when emerging reproductive and genetic technologies raise the specter of a new eugenics that contemporary bioethical methods are wholly unequipped to manage. Germline engineering, embryo screening, and similar techniques give rise to the ability of science and medicine to enhance or eliminate populations in ways that would make nineteenth-century eugenicists utterly giddy. These new technologies are not inherently problematic. With appropriate oversight and regulation, we can move toward a future where responsible use might improve human health and reduce suffering without creating new disparities or inequities. But in order to achieve this future, bioethics must reorient itself away from principlism and back toward its original founding value: anti-eugenics. This path can be paved by embracing an originalist method that draws on the text of Taylor's opening statement as its interpretive guide for how to think through ethical challenges that might arise as these new technologies become a more common part of our lives.

NOTES

1. Arthur Caplan, quoted in George Annas, *American Bioethics: Crossing Human Rights and Health Law Boundaries* (Oxford: Oxford University Press, 2005), 161.
2. Albert Jonsen, *The Birth of Bioethics* (Oxford: Oxford University Press, 1998), 3.

3. Telford Taylor, "Opening Statement of the Prosecution, December 9, 1946," in *The Nazi Doctors and the Nuremberg Code*, ed. George J. Annas and Michael A. Grodin (Oxford: Oxford University Press, 1995), 67.
4. "Opening Statement," 68.
5. "Opening Statement," 71–86.
6. "Opening Statement," 70.
7. For examples of past malfeasance in medical research, see Allen Hornblum, *Acres of Skin: Human Experiments at Holmsberg Prison* (New York: Routledge, 1999), discussing the long history in the United States of using prisoners in medical research; and Deirdre Cooper Owens, *Medical Bondage: Race, Gender, and the Origins of American Gynecology* (Athens: University of Georgia Press, 2017), discussing the role that unethical research on Black slaves played in the rise of the field of gynecology.
8. Telford Taylor, Opening Statement of the Prosecution, December 9, 1946, in *The Nazi Doctors and the Nuremberg Code*, ed. George J. Annas and Michael A. Grodin (New York: Oxford University Press, 1995), 70–71.
9. Jonsen, *The Birth of Bioethics*, 34.
10. Jonsen, *The Birth of Bioethics*, 34.
11. National Commission for the Protection of Human Subjects of Biomedical and Behavioral Research, *The Belmont Report: Ethical Principles and Guidelines for the Protection of Human Subjects of Research* (Bethesda, MD: National Commission for the Protection of Human Subjects of Biomedical and Behavioral Research, 1978).
12. Tom L. Beauchamp and James F. Childress, *Principles of Biomedical Ethics*, 6th ed. (New York: Oxford University Press, 2009), 3.
13. W. Nelson Espeland, *The Struggle for Water: Politics, Rationality, and Identity in the American Southwest* (Chicago: University of Chicago Press, 1998), 24, quoted in John Evans, "A Sociological Account of the Growth of Principlism," *Hastings Center Report* 30, no. 5 (September–October 2000): 32.
14. Evans, "A Sociological Account of the Growth of Principlism," 32.
15. Evans, "A Sociological Account of the Growth of Principlism," 32.
16. Carl Elliott, *White Coat, Black Hat: Adventures on the Dark Side of Medicine* (Boston: Beacon, 2011).
17. Carl Elliott, "Poor Reviews," *Mother Jones*, September/October 2010.
18. Catherine Myser, "Differences from Somewhere: The Normativity of Whiteness in Bioethics in the United States," *American Journal of Bioethics* 3, no. 2 (2003): 2.
19. Troy Duster, *Backdoor to Eugenics* (New York: Routledge, 2003).
20. Adrienne Asch and Eric Parens, "Disability Rights Critique of Prenatal Genetic Testing: Reflections and Recommendations," *Mental Retardation and Developmental Disabilities Research Reviews* 9 (2003): 40.
21. Gina Kolata, "The Ideal Subjects for a Salt Study? Maybe Prisoners," New York Times, June 4, 2018.
22. Heidi Ledford, "CRISPR Babies: When Will the World Be Ready?," *Nature* 570, no. 7761 (2019): 293–296.
23. Osagie K. Obasogie and Marcy Darnovsky, introduction to *Beyond Bioethics: Toward a New Biopolitics* (Oakland: University of California Press, 2018).

24. Griswold v. Connecticut, 381 U.S. 479 (1965)
25. The Second Amendment states: "A well-regulated Militia, being necessary to the security of a free State, the right of the people to keep and bear Arms, shall not be infringed."
26. District of Columbia v. Heller, 554 U.S. 570, 577 (2008).
27. For a wide-ranging discussion of this, see "The 1619 Project," *New York Times*, August 14, 2019.
28. See Reva Siegal, "Why Equal Protection No Longer Protects: The Evolving Forms of Status Enforcing State Action," *Stanford Law Review* 49 (1996): 1111, discussing the concept of "preservation through transformation" as a way to understand how seemingly liberating legal concepts can be construed in ways that enforce the very status hierarchy they were thought to undermine.

PART II

BOUNDARIES

MYRNA PEREZ SHELDON, TERENCE KEEL, AND AHMED RAGAB

Propriety has long governed the analysis of the relationship between science and religion. The desire to know where one category begins and the other ends—and furthermore how they ought to relate to one another—reached its height at the end of the twentieth century. In his 1999 book *Rocks of Ages: Science and Religion in the Fullness of Life*, the evolutionary biologist and celebrity scientist Stephen Jay Gould proposed that science and religion constituted "Non-Overlapping Magisteria."[1] He coined this term to express his belief that science was the category for empirical knowledge, whereas religion was restricted to questions of "human purposes, meanings, and values."[2] In his view, as long as both science and religion kept to their appropriate spheres, unnecessary cultural conflict could be avoided. His proposition was roundly criticized by both religious leaders and science advocates as limited and impractical.[3] Nevertheless, it revealed an underlying intuition held by many Euro-American intellectuals that it might be possible and desirable to clearly define the proper relationship between these categories.

Closely related to the question of the boundary between science and religion is the so-called demarcation problem in the history and philosophy of science. This formal study of the distinction between science and other cultural activities was initiated in the early twentieth century by philosophers who sought to define science by a set of epistemological qualities. Karl Popper, for instance, argued that one of the hallmarks of science was its "falsifiability, or refutability, or testability."[4] Later in the century, sociologists and historians transformed the scholarly approach to the problem by recognizing that demarcation was as much a social activity as it was an issue of epistemic characteristics.[5] And, furthermore, the cultural distinctions made between science and its others—whether pseudoscience, superstition, or religion—had changed significantly over time.[6]

The contributors to this part approach boundaries through a different lens. From a critical vantage point, boundaries are understood as more closely akin to the concept of *border* or *borderland* in mestiza feminism. Gloria Anzaldua's 1987 book *La Frontera/Borderlands: The New Mestiza* used the image of the physical border between the United States and Mexico to reflect on the hybrid tensions in her identity as a Chicana lesbian.[7] Inspired also by the image of the cyborg in Donna Haraway's 1985 essay "A Cyborg Manifesto," queer and feminist literatures have expansively taken up the relationship between physical and material borders and the scandalous impurity of hybrid human and nonhuman bodies.[8] In this vein, the chapters in this part attend to the material and pragmatic entanglements through which science and religion emerge, intertwine, and redefine human existence. It is in hybrid objects—or in the entanglements of settler colonialism—that we witness the borders fashioned between science and religion, border that are created, reinforced, and weaponized but always fluid, transitioning, and transgressed. Ultimately, these chapters demonstrate the historical insights and critical vocabulary necessary to understand how science and religion are entangled in the power dynamics of patriarchy, colonialism, and slavery.

Tisa Wenger's chapter, "Spiriting the Johnstons: Producing Science and Religion Under Settler Colonial Rule," focuses on the history of the métis Johnston-Schoolcraft family of Sault Ste. Marie, Michigan. Wenger argues that new religious and scientific frameworks acted in an increasingly racialized settler society, and this simultaneously opened up and foreclosed possibilities for the members of the Johnston-Schoolcraft family. Her chapter powerfully demonstrates how both science and religion were entangled in colonial modes of governance and land theft. In this case, science and religion existed in a kind of harmony, not because of philosophical agreement but because of their partnership in settler colonialism. Nevertheless, she concludes this analysis with nuance, demonstrating how both categories could also function as sources of agency for the family, whether for job prospects or existential comfort.

Jason Ānanda Josephson Storm's chapter, "Dark Gods in the Age of Light: The Lightbulb, the Japanese Deification of Thomas Edison, and the Entangled Constructions of Religion and Science," focuses on the creation and global circulation of the incandescent lightbulb. In this history, Storm asks how it is possible that Thomas Edison, noted figurehead of American science and rationalism, became a Japanese deity, "the Bright God of the Electric Telegraph and Telephone." He uses the complex cultural history that lies behind the Edison bulb to demonstrate that this object was not merely the product of scientific invention, as is so often presumed. Instead, it came out of a history that includes everything from the American ghost machine to the Japanese god of warriors, as well as Buddhist monks and theosophists.

In "Questioning the Sacred Cow: Science, Religion, and Race in the United States and India," Cassie Adcock examines the material production of bovine animals, held to be sacred by Hinduism, at the intersection between secular colonial expertise and religious cultural knowledge. She argues that the British colonial infrastructure pushed for the scientific breeding of India's cattle as a solution to Indian poverty, even as it denigrated the Hindu view of cows as an obstacle to British modernization. She reveals the material contradictions within these colonial narratives, as the British shifted from encouraging Indian cattle breeding in the 1910s to deploring the overbreeding of cattle—and of the Indian people—in the eugenic discourse of the 1920s. Far from being a reflection of transcendent facts, Adcock argues that the colonial sacred cow was the result of specific embodied economic and social materialities.

And in his chapter, "'And God Knows Best': Knowledge, Expertise, and Trust in the Postcolonial Web-Sphere," Ahmed Ragab gives an account of the development and functioning of the "internet ecology" of fatwa-granting websites originating in various Muslim-majority countries, including Egypt, Qatar, and Saudi Arabia. He analyzes how the internet has reshaped the jurisprudential context of the postcolonial nation-state while also creating a new imaginary of the idealized global Muslim community. Importantly, he argues, many of the websites are not simply virtual representations of real-world entities but are themselves independent institutions. In this, they create a kind of hyperreality, in which they mirror an idealization of Muslim religious devotion while simultaneously creating a parallel reality to the nonvirtual world. By creating a transcendent Islamic community, Ragab contends that these technologies help to further integrate religious authority into the postcolonial state. But they also serve to essentialize Islam in relation to the West. Through this, Islamic individuals and communities are defined solely by their essential "Muslim-ness," such that they are rendered noncitizens in their own countries and potential terrorist threats in the securitized narratives of the West.

A critical approach to the study of science and religion requires more than attention to epistemic qualities. It also necessitates more than the recognition of the sociological dimensions of professional science and of sectarian religion.

It calls us to attend to the borders and boundaries that have been fashioned and then refashioned by histories of settler colonialism, patriarchy, and racism. Through this attention, we bear witness to the creation of these categorical borders as themselves ongoing technologies of Western identity. Nevertheless, we also find possibility in the transgressions of these borders as we learn how leaky, permeable, and limited the boundaries between these categories have been and will continue to be.

NOTES

1. Stephen Jay Gould, *Rocks of Ages: Science and Religion in the Fullness of Life* (New York: Ballantine, 1999), 5.
2. Gould, *Rocks of Ages*, 4.
3. Steve Paulson, *Atoms and Eden: Conversations on Religion and Science* (Oxford: Oxford University Press, 2010).
4. Karl Popper, *Conjectures and Refutations: The Growth of Scientific Knowledge* (1963; repr., New York: Routledge, 2002), 48.
5. Thomas F. Gieryn, "Boundary-Work and the Demarcation of Science from Non-science: Strains and Interests in Professional Ideologies of Scientists," *American Sociological Review* 48, no. 6 (December 1983): 781–795.
6. Michael D. Gordin, *The Pseudoscience Wars: Immanuel Velikovsky and the Birth of the Modern Fringe* (Chicago: University of Chicago Press, 2012); Peter Harrison, *The Territories of Science and Religion* (Chicago: University of Chicago Press, 2015).
7. Gloria Anzaldua, *Borderlands/La Frontera: The New Mestiza* (San Francisco: Aunt Lute , 1987).
8. Donna Haraway, "A Cyborg Manifesto: Science, Technology, and Socialist-Feminism in the Late Twentieth Century," in *Simians, Cyborgs, and Women: The Reinvention of Nature* (New York: Routledge, 1991), 149–182; Jasbir K. Puar, " 'I Would Rather Be a Cyborg Than a Goddess,' " *philoSOPHIA* 2, no. 1 (2012): 49–66; Margaret Rhee, "In Search of My Robot: Race, Technology, and the Asian American Body," *Scholar and Feminist Online* 13, no. 3 (2016), https://sfonline.barnard.edu/traversing-technologies/margaret-rhee-in-search-of-my-robot-race-technology-and-the-asian-american-body/.

CHAPTER 5

SPIRITING THE JOHNSTONS

Producing Science and Religion Under Settler Colonial Rule

TISA WENGER

Wabishkizzy may smile, Wabishkizzy may say,
I will teach you to read, I will teach you to pray,
But say, when our arts, manners, customs are lost,
What, then shall we cherish, what then shall we boast.
When the war flag is struck, & the war drum is still,
And the council fire glimmers no more on the hill . . . !

They tell me, that, blessings for me, are in store,
The sage's-saint's-poet's-philosopher's lore
The comforts that labor and science bestow,
The loom &, the compass, the sickle &, plough!
But ah! Can they tell me, where joy shall abide,
Without national customs, or national pride!

—"Algonac, A Chippewa Lament at Hearing the Reveille at the Post of St. Mary's," 1827

The "Chippewa Lament," though published anonymously, was most likely written by George Johnston, or Kahmentayha, son of an Ojibwe woman and a Scotch-Irish trader in the northern Michigan town of Sault Ste. Marie.[1] George's father had fought against the United States in the War of 1812 alongside the Ojibwe (Chippewa), part of the larger family of Anishinaabe nations, and the family was intimately involved in the changes that ensued. In 1823, George's sister Jane, Obabaamwewe-giizhigokwe, married the newly appointed U.S. Indian agent Henry Rowe Schoolcraft, who represented U.S. imperial authority on Anishinaabe lands. Henry's enlightened explorations in geology, ethnology, philology, and religion served the clear purpose of mapping resources—mineral, biological, and human—for colonial extraction and control. The lament thus captures an Anishinaabe perspective, however filtered, on science and religion as aspects of the larger imperial project of dismembering Indigenous worlds.

The lives of the Johnstons and the Schoolcrafts illuminate the intimate consequences and entwined configurations of science and religion at the edges of an expanding settler empire. Science and religion have typically been portrayed as naturally opposed. Recent scholarship has instead painted their relationship as complex and, in many ways, historically intertwined.[2] But the complexity thesis is insufficient. In distinct academic conversations, historians of science and scholars of religion have each identified their subjects as instruments for colonial control and, at times, for Indigenous survival and resistance.[3] Not unlike Cassie Adcock's history of the sacred cow in India (chapter 7), this chapter shows how science and religion were coconstituted in the interests of U.S. colonial rule. The distinctions between these fields—their collaborations as well as their apparent oppositions—developed historically within an imperial frame. European and U.S. colonizers initially defined both spheres as evidence of Western superiority, framing them in ways that excluded Indigenous people. Later they named the traditions of colonized peoples "religion" only to relegate them to an outmoded, primitive past while identifying themselves as modern because they had advanced either to the superior religion of Christianity or beyond it to master secular or scientific reason. Even as science and religion came to be seen as opposing forces, then, they were jointly configured for imperial rule.[4] Under the veneer of progress and enlightened reason, science facilitated the extraction of resources from colonized lands. Religion, meanwhile, fostered the illusion of settler benevolence even as it mopped up and papered over the violence of conquest. But this colonial frame cannot encompass everything. Within the constraints of a settler empire, Anishinaabe people struggled to (re)configure both science and religion for their own cultural and political survival. George, Jane, and their extended family offer glimpses of the potential, the limits, and the human stakes of these entangled configurations.

"A NEW WORLD APPEARED TO BE OPENING": UNSETTLING ARRIVALS IN AN IMPERIAL BORDERLAND

George and Jane Johnston were born into the leading family of Sault Ste. Marie. Their mother was Oshaguscodawaqua, the daughter of Waubojeeg, a prominent

Ojibwe chief at Chagoimegon (La Pointe) on the western side of Lake Superior, and an influential figure in her own right. When Oshaguscodawaqua was a girl, the entire Great Lakes region was Anishinaabe land, home to what the historian Michael Witgen calls "an infinity of nations." The French and the British could claim that God had granted them the right to rule. But the traders and agents who survived and thrived in what the French called the *pays d'en haut*, the backcountry of Upper Canada, had no option but to play by Indigenous rules. Many married Anishinaabe women *a la façon du pays*, without the formalities of church or state, producing Métis offspring who went on to become traders, interpreters, and cultural intermediaries in a rapidly transforming colonial world.[5] George and Jane's father, the Scotch-Irish fur trader John Johnston, followed in this tradition, moving to Canada in 1790 and eventually to the Lake Superior region. There he met and married Oshaguscodawaqua, who also became known as Susan Johnston. Together the Johnstons built a fur-trading business at the Ojibwe-Métis town of Baawitigong, Sault Ste. Marie, at the rapids between Lake Superior and Lake Huron. Their eight children—Lewis, George, Jane, Eliza, Charlotte, William, Anna Maria, and John McDouall—grew up thoroughly bicultural in an Anishinaabe world. Their mother spoke to them only in Ojibwe and taught them Ojibwe ways. Their father, who boasted of having "a library of the best English works," read with them in the evenings and took them on visits to Montreal, New York, and, in Jane's case, Ireland. John Johnston was also a devout Presbyterian who had each of his children baptized, held family devotions every evening, and hosted the occasional Protestant missionary in town.[6]

The Johnstons had no choice but to accommodate to an expanding U.S. empire. When Michigan's new territorial governor Lewis Cass visited Sault Ste. Marie in 1820, it was Oshaguscodawaqua who convinced the reluctant Anishinaabe headmen "of their own weakness and the strength of the United States." When she found out that some of them had hoisted a British flag and intended to prevent the governor from entering Lake Superior, as her son George later recalled, she convinced them to "suppress" all such anti-American action. The Anishinaabe must "receive the Americans with friendship," she warned, or else they would "bring ruin to the tribe." They would do better, she believed, by working with rather than against this new imperial power.[7]

Henry Rowe Schoolcraft first visited Sault Ste. Marie with the Cass expedition in 1820. Born near Albany, New York, the young Henry had learned glassmaking from his father and then had taught himself mineralogy and geology. He went west in 1817 to explore the Missouri territory and published several books on its geology and its prospects.[8] "A new world appeared to be opening for American enterprise there," he later recalled. "Its extent and resources seemed to point it out as the future residence of millions."[9] His writings—travelogues that doubled as natural histories and compendiums of Indian tales—were prolific but otherwise typical for the aspiring scientist of the time. Like the European explorers profiled by Johannes Fabian, he used the travelogue, "the genre that preceded, often contained, and sometimes paralleled the ethnographic monograph," to shape his image as a rational leader. Angling for a government appointment, he sent copies of his reports on the Missouri mines to prominent men in Washington. He saw "a new world" opening up in the west, and he intended to play a part in its conquest.[10]

Cass's goal for the 1820 expedition was to map the vast "northwestern territories" from Lake Superior to the headwaters of the Mississippi. With better information on this region's "animal, vegetable, and mineral kingdoms," he told the secretary of war, they could "ascertain its present and future probable value" to the United States.[11] Angling to "extinguish the Indian title to the country in the vicinity of Prairie du Chien and Green Bay," he also hoped to negotiate with the Winnebago (Ho-Chunk) nation and to dissuade them from further raids.[12] Schoolcraft readily signed on as expedition geologist, dedicating his *Narrative Journal of Travels* (1821) to Calhoun in gratitude for his "patronage . . . to the cause of science" and in admiration for his successes in "stretching our cordon of military posts, through the territories of the most remote and hostile tribes of savages." Military conquest and the progress of science went hand in hand.[13]

Schoolcraft's *Narrative Journal* also illuminates colonial ideas about religion. He dismissed the idea that Indians had any "religion" in the proper sense of the word. Always nominally Protestant, he associated true religion with interior piety and monotheistic belief. The Indians, he wrote, "believe[d] in the existence of a great invisible spirit" but did not "pay him the homage of religious adoration." Although they made offerings to countless minor spirits or "manitoes," he judged these "entirely optional." In place of religious devotion, he wrote, "they [had] jugglers and prophets, who predict events, who interpret dreams, and who perform incantations and mummeries." Enlightened Protestant critics had long used the word *mummery* to describe Catholic ceremonialism or any fraudulent performance that masqueraded as religion. Anti-Catholicism merged seamlessly into Schoolcraft's denigration of indigeneity. It was no wonder that the Catholics had been relatively successful with their missions in this region, he mused, since their "splendid ceremonies and external signs" would always appeal "to the illiterate and the vulgar." He thus mocked Indigenous ceremonies as fraudulent, "more or less superstitious," and so self-evidently ridiculous that "the spectator finds a difficulty in restraining his laughter." The Indians' susceptibility to the deceptions of masses, mummeries, and manitoes served as further evidence of their racial inferiority.[14]

Schoolcraft was very much invested in differentiating between the spheres of science and religion. In his eyes, the failure to distinguish between them served as a defining marker of the "primitive." He contrasted his own geological analysis of White Rock, a large limestone formation in Lake Huron, with the Anishinaabe "offerings" he observed at the rock. A "civilized" people, in his view, would have kept these spheres carefully distinct by praying to God—the interiorized realm of religion—while treating rocks as inert objects to be explored and excavated with the tools of science. Similar judgments appeared throughout his *Narrative Travels*. After a brief description of the "medicine dance" of the Ojibwe, for example, he wrote: "Such is the religion—the superstition, and the knowledge of medicine of the lake savages, blended as they appear. It is difficult to separate them, and to say how much may be considered religious, or mere mummery." Just as Indians mixed up religion and science at White Rock, the medicine dance showed how they failed to make the proper distinction between medicine and prayer. This failure made them "superstitious," as he judged them, in need of civilization, education, and moral reform.[15]

These judgments directly facilitated the theft of Native lands. In 1820, when Anishinaabe guides were reluctant to reveal the location of significant copper deposits—such as the massive Copper Rock on the Ontonagon River, south of Lake Superior—Schoolcraft attributed their attitude to their "superstition." The Ojibwe, he claimed, believed that copper held supernatural powers and that revealing its location would bring them misfortune. One of these guides had been shunned by his band for aiding the expedition because, they said, "this act was displeasing to the Great Spirit."[16] Schoolcraft considered their objections irrational and grounded in superstition. But these Ojibwe people understood the import of the expedition all too well. When the U.S. Senate inquired about the prospects for mining copper around Lake Superior and for "extinguishing [the Indians'] title," Schoolcraft replied that he knew exactly where the most valuable lands were and could easily convince these "chiefs and headmen" to part with them. Conveniently, he blamed any Indigenous objections on their immovable and irrational "superstitions." He saw such exchanges as inevitable, in the Indians' best interests, for science and the common good. Today, in an age of accelerating climate disasters, their relational approach to Copper Rock appears far wiser than the sort of scientific distancing that Schoolcraft exemplified.[17]

While Schoolcraft proclaimed religious-scientific distinctions as markers of civilization, he and his peers assumed the Christian foundations of U.S. settler society. In this colonial project, Christianity appeared in the foreground as a moral exemplar and an agent of civilization and, at the same time, receded into the unmarked privilege of the assumed norm. Schoolcraft's own prescriptions for "moral reform" identified Christianity as the foundation for progress. Far from keeping his own religion confined to an interior sphere, then, he joined a long tradition of European settlers who, under the doctrine of discovery, had used Christianity to facilitate the dispossession of Indigenous people. Later in the nineteenth century, settler science would more doggedly map religion on an evolutionary scale. A new generation of scholars would locate world religions on various hierarchical scales, positing a progression, for example, from the tribal to the national to the universal, part of a broader ranking of civilizations in which Europe and a rational, enlightened Christianity—or, alternatively, a post-Christian embrace of scientific reason—would inevitably come out on top. Settler colonialism elevated white settler Christianity even as it claimed an even playing field for all "religion," thus obscuring the privileges it conveyed and the dispossessions it enabled.[18]

"A WORLD OF SIN AND MISERY": ENGENDERING RELIGION IN SAULT STE. MARIE

In 1821, Governor Cass recommended Henry Rowe Schoolcraft to lead the U.S. Indian agency that President Monroe had just approved, along with a new military post, at Sault Ste. Marie.[19] The primary goals of an Indian agent were to pacify Native people while enabling their dispossession, or, in Schoolcraft's words, "to curb and control the large Indian population on this extreme frontier." The new agent arrived by steamboat on July 6, 1822, followed the same day by "all the troops under the command of Col. Brady."[20] The next day he reported to his closest colleague at Michilimackinac that "an unusual number of Indians" at the Sault had just visited the British outpost of Drummond's Island to receive "their annual presents." These people did not consider themselves subject to the United States, and Schoolcraft found their large numbers and obvious independence alarming.[21] In the years that followed, he deployed both science and religion as tools to aid in the project of their subordination.

The new agent's first order of business was scientific. Cass had recently sent a questionnaire to U.S. Indian agents and traders seeking detailed information on Indigenous histories, languages, and traditions. He advised Schoolcraft to seek lodging with "Mr. Johnston and his family," noting that they would also be able to provide "full and detailed answers to my queries" on "the language" and "polite literature of the Chippewas." Thus, it was with the Johnstons that Schoolcraft gained his orientation to Sault Ste. Marie.[22] Just as he had systematized his mineralogical findings, he wanted to pin down Indigenous languages and incorporate them into wider fields of knowledge. "By classifying . . . the terms relative to the different branches of natural history, topography, &c," he told Cass, "I am enabled to draw out many words that would otherwise escape me."[23] At stake in such philological research, as Sean Harvey has argued, was the future of U.S. Indian policy as politicians debated whether or not Native people had the capacity to become "civilized" and integrate into U.S. society. Cass believed that the structure of Indian languages demonstrated a permanent racial inferiority. His protégé Schoolcraft, who placed a greater emphasis on the deficiencies of what he called the Indians' "superstitions," entirely agreed.[24]

Soon after his arrival, Henry developed a new interest in the oldest Johnston daughter, the twenty-two-year-old Jane. Loved by all who met her, Jane was pious and kind, a talented writer who shared his interests in poetry and literature. Her modesty and literary knowledge more than fulfilled his feminine ideal. "I read your kind note last night at a time and place when I always think of you, if indeed you are ever absent from my mind," he wrote.[25] The next spring she informed him playfully "that her mother begs his acceptance of the accompanying little moccuks of maple sugar." Recalling a scene in *The Merchant of Venice* when Bessamio decried a world so easily "deceived by ornament," she noted that although "the unornamented moccuks" were "plain . . . the sweet they contain is not the less fine."[26] Perhaps she wanted to help Henry see, through the literary language they shared, just how much she also valued her Anishinaabe world. The young couple were married in October 1823 by a visiting Presbyterian minister, the Rev. R. M. Laird of Detroit, and settled into life as newlyweds in the comfortable embrace of her family's home. Nine months later their first son, William Henry Schoolcraft, was born.[27]

Henry and Jane shared characteristically Victorian conceptions of gender that assigned him the traits of leadership and rationality while expecting spiritual purity and fragility from her. Henry was not a particularly devout Christian at this stage in his life. Although he confided his "religious feelings" to the Rev. Laird, he did not want them publicly shared. Laird urged him to take a more "decided stand" for the gospel. "You will lose nothing of this world, that you really prize," he wrote. "Your reputation can in no respect be injured." Comfortable with a genteel and civilized Christianity, Henry may have feared that public displays of religious feeling would appear unseemly and perhaps even tarnish his image as a man of science. Hinting at his own temptations, he would later counsel his wayward younger brother James to avoid the sort of pride that considered it unworthy "of a man of liberal and high principles" to "follow the meek precepts of the gospel." He had to struggle to move beyond an image of masculinity that prioritized mastery and control, seeing only weakness in displays of piety. Christianity was a feminine virtue and a resource for governing inferiors; white men like Henry did not expect to be limited or humbled through its practice.[28]

When Henry did speak publicly about Christianity, he presented it as a measure of civilization and a way to secure the social order. In 1824, when William Ferry of the United Foreign Missionary Society became the first Protestant missionary in Michilimackinac, Henry and Jane jointly inspected his new school, and Jane reported positively on the "zeal and vigilance of the instructors" and the "moral and domestic conditions of the scholars."[29] Henry was less encouraging to a potential missionary whose ideas he judged to be speculative and impractical. Facts and experience were required, he explained, not just benevolent intent. A successful mission required a strategic pedagogy that tackled every aspect of Indigenous life. An Indian child, he wrote, "should be taught the value of labour and the mode of applying it; the superiority of agriculture and the productive arts of civilized life over the chase." Missions were a method of "subduing" or "reducing" Indians so they accepted the disciplinary practices of settler civilization. It is not surprising that Native people rarely responded to missions designed along such lines.[30]

Christianity was a haven for Jane and increasingly for Henry too, as together they weathered the storms of life. Jane worried about her husband's safety in the summer of 1825, when he took another trip to the western edge of Lake Superior for the Treaty of Prairie du Chien. She warned of growing Anishinaabe anger against the greed of the "Big Knives," the Americans, and informed Henry that an Ojibwe man named Keewanaquod had recently killed several Big Knives and threatened to kill more. Her own Anishinaabe people were losing land, and she knew that her beloved Henry was among those whom they might blame. She urged him to pray—and to keep his canoe "well armed." "Oh my Henry, what a world of sin and misery we live in!" she exclaimed.[31] That winter she was seriously ill and suffered a miscarriage. "I lost the little angel who would have been [Willy's] sister," she told her brother George, "but God's will be done."[32] A year later a sudden illness took the life of their beloved Willy, who was not yet three. The couple's overwhelming grief echoed through the poetry they circulated privately that year. Jane concluded one poem with hopes for heavenly reunion: "But soon my spirit will be free, / And I my lovely son shall see, / For God, I know, did this decree / My Willy."[33] They found new hope in the arrival of two more children: a daughter, Jane Susan, in October 1827 and a son, John Johnston, in October 1829.[34] Through these joys and sorrows, Henry treasured his wife's piety: "An even temper, mild, endearing kind / A sound discreet and regulated mind / Improved by reading, by reflection formed / By reason guided, by religion warmed / This have I often prayed 'heavens last best gift' to be / This have I oft, with joy, remarked in thee."[35] In his eyes, the (female) warmth of religion had to be guided by (male) reason and tempered by discretion, education, and the well-regulated mind. His reason balanced and directed her piety, and he expected her to cultivate the traits of resignation and submission as the Christian ideal.[36]

The religion of the white settler patriarch balanced reason, piety, and control. Henry was shaken by his brother James's imprisonment in November 1830 for stabbing a soldier from Fort Brady during a drunken brawl, following what locals called a "French ball." The victim barely survived, and James spent two months in the Michilimackinac jail. This was a raucous model of masculinity that Henry absolutely did not endorse, and he urged his wayward brother to abandon "laxity and irreligion." He searched his own heart—hence, the questioning letters to Jane—and early that February he formally joined the Presbyterian Church in Detroit. Briefly, he contemplated a new vocational commitment. "I design to fit and prepare myself for the ministry," he told James. "There is nothing in my whole life which I contemplate with greater satisfaction."[37] In the end, he decided he was better suited to a different leadership role. In March 1832, he founded the Algic Society to support "missionary effort to evangelize the northwestern tribes." His first address to this society extolled the triumphs of reason over "the prejudices and superstitions of former ages" and offered his ethnological and linguistic research as a resource for missions. As "inhabitants of the frontier," he asked, "can we not aid the evangelist" by "inform[ing] him in what language [the Indians] converse—with what people they war—with what superstitions they are debased?" In this work, once again, he jointly articulated science and religion as benevolent initiatives in the interests of U.S. rule.[38]

Henry and Jane eagerly joined in the evangelical revivals that came to northern Michigan, local manifestations of the Second Great Awakening. Jane read devotional tracts and attended worship services with the Baptist missionary Abel Bingham, described by a local wit as one who "thunders forth his orthodoxies, heterodoxies, and paradoxies, with as much energy and little success, as usual."[39] Jane and Henry were delighted when two Presbyterian clergymen showed up at the Sault in the fall of 1831, testing the waters for new missions. Henry reported fifty-two people, mostly soldiers at the fort, who professed their faith in Christ that year.[40] Among them were Oshaguscodawaqua; her daughters Jane, Eliza, Charlotte, and Anna Maria; and her son George. "All of us who love the Lord were much pleased at the indication of God's goodness and presence amongst you," wrote a friend from Michilimackinac, "and may the love of souls triumph over that of sectarianism."[41] Oshaguscodawaqua was so moved that she donated the funds to build a new Presbyterian church, further evidence of her prominence as a leading citizen of Sault Ste. Marie.[42]

As Oshaguscodawaqua's enthusiasm suggests, the revivals of the early 1830s also touched the local Ojibwe community. Indigenous evangelists were crucial to that work. When he arrived in Sault Ste. Marie, Bingham had employed Oshaguscodawaqua's third daughter, Charlotte, as his interpreter. Charlotte elaborated on his sermons, he reported, following "my discourse with an address of her own," and composed Ojibwe-language hymns that the congregation sang with great satisfaction.[43] She would continue in the role of interpreter-preacher alongside a young Anglican missionary named William McMurray, whom she married in 1833. Moving across the border to Canada, the McMurrays would continue their missionary labors together for many years. Charlotte was a rare female evangelist among a group of mid-nineteenth-century Ojibwe preachers who shaped a distinctively Indigenous Christianity. Abel Bingham recognized that he could not compete with the Ojibwe Methodist John Sunday, Shawundais, then preaching on the Canadian side of the river across from Sault Ste. Marie, who brought many Anishinaabe into the Methodist fold. This Native Christian movement so successfully revitalized Ojibwe identity and traditions that one historian has dubbed it an "Ojibwe Renaissance."⁴⁴ Some Anishinaabe communities were also able to leverage their Christianity (whether Protestant or Catholic) to protect their claims to at least a bit of land and protect themselves from the scourge of removal.⁴⁵ Native people thus adopted and adapted Christianity as a way to sustain their communities within the constraints of settler colonial rule.

Not all Ojibwes accepted Christian affiliations. Henry described how some people told him they did not "wish to be disturbed by the introduction of any religion, preferring, in their emphatic language, 'to follow the religion of their fathers.'"⁴⁶ Others tried to avoid such conversations. One headman welcomed a mission school so that his three children could learn English but politely deferred when Bingham offered "some instruction on the subject of religion." He was leaving for his winter hunting grounds, he said, but would reply upon his return.⁴⁷ There are countless examples of people like this headman who listened politely to missionaries and then simply changed the subject. Drawing their own lines in the sand, some Anishinaabe leaders entirely forbade missionaries from holding meetings in their villages. Many theorized that while Christianity was perfectly valid as the white man's religion, the Great Spirit had given them their own—the "religion of their fathers"—and did not want them to abandon it. Native American religions, defined and delineated as such, were forged in and through such oppositional encounters.

EXCAVATIONS AND SORROW SONGS: SETTLER SCIENCE IN MICHIGAN AND D.C.

Throughout his career, Schoolcraft conceptualized science and religion in ways that facilitated U.S. rule. In the early 1830s, he led several new expeditions designed to map the country for western expansion. In 1832, Cass, now serving as Andrew Jackson's secretary of war, asked him to travel across Lake Superior to the northern reaches of the Mississippi, where the Sauk chief Black Hawk was rallying Native nations for war. The primary goals of this expedition were to gather military intelligence and, as much as possible, to "pacify" the northern tribes. A small battalion of twelve soldiers accompanied them, and Schoolcraft ordered U.S. flags to fly on the boats.⁴⁸ Along with his duties as expedition commander, he assumed responsibility for historical, archaeological, and linguistic research. His accounts of the expedition were characteristically grandiose.

Ignoring the Indigenous guides on whom he relied, he boasted that he had finally "accomplished the discovery of the true source of the Mississippi River"—a task initiated but not completed by earlier heroic explorers De Soto, Marquette, Hennepin, Pike, and Cass. His name, he hoped, might one day be recited by schoolchildren as a great explorer who had opened up the continent for scientific progress and U.S. expansion.[49]

Christian missions served as a partner in these endeavors. Schoolcraft invited the Presbyterian missionary William Boutwell to join the 1832 expedition on behalf of the American Board of Commissioners for Foreign Missions. "All our attempts in the way of agriculture, schooling, and the mechanic arts," Schoolcraft explained, "are liable to miscarry and produce no permanent good, unless the Indian mind can be purified by gospel truth, and cleansed from the besetting sin of a belief in magic, and from idolatry and spirit-worship."[50] Christianity—true religion—would eliminate the "superstitions" that held Indigenous people in their thrall. And this spiritual transformation was essential to Indigenous progress in agriculture, science, and all the arts of civilization. Religion, science, and capital worked together as the handmaidens of colonial conquest. As Schoolcraft later put it, the "triple claims . . . of business, of science, and of religion created not the least distraction on my mind, but on the contrary, appeared to have propitious and harmonizing influences."[51] For him, these three spheres were distinct but complementary, working together in the interests of the settler colonial whole.

For Native people, the frameworks of science and religion had very real limits. Barely mentioned in Henry's accounts was Jane's brother George Johnston, Kahmentayha, who joined the 1832 expedition as baggage handler and interpreter. George had worked over the years as a trader and as a subagent, under Henry's supervision, for the U.S. Indian Service. From 1826 to 1829, he had directed a subagency at La Pointe, his mother's birthplace, where he distributed supplies, monitored the fur trade, and developed his own interests in geology. Obliging Henry's requests for copper samples, he aimed to identify and procure samples of "every rock and mineral the products of this lake." George also supplied various artifacts—minerals, stuffed birds, examples of Native craftsmanship, etc.—for the curiosity cabinets of Cass and for an earlier secretary of war, Thomas McKenny, who had grown very fond of the entire Johnston family when he visited the region in 1826. Geology and botany were compelling intellectual pursuits for George in the relatively isolated environs of La Pointe. They were also ways to establish credibility as a man of science in the eyes of his brother-in-law and of more distant authorities like the secretary of war.[52]

Occasionally, George expressed his frustrations at the pretentions of white settler society. He had grown up in the leading family of Sault Ste. Marie but increasingly found himself dependent on the goodwill of his imperious brother-in-law for a series of low-level jobs. And so he lauded the purity he saw in Indian life and in the natural world, contrasting these to the corruptions of so-called civilization. He told Henry that at LaPointe, he felt "more independent and at ease and in my element, breathing a pure native air." In a journal entry signed "An Aborigine," he bemoaned his family's misfortunes and the "calumny, malice, hatred and all uncharitableness" of "the civilized world." "Better it is to remain among the sons of nature," he wrote, "and contemplate on the folly of the world and mankind." George was personable, an effective administrator, and well-educated in the English language, law, and literature. He acted in most of his official communications like any other U.S. Indian agent, even offering up patronizing descriptions of the Indians' "miserable" and "filthy" conditions. As racial distinctions grew sharper, perhaps he hoped that such language would aid him professionally. Yet increasingly he was judged a "half-breed," limited to subordinate positions and not even considered for new posts in the Indian Service, which went to arriving white settlers instead.[53] His position at La Pointe was eliminated in 1829 during an Indian Service reorganization. After the expedition in 1832, he became subagent at Michilimackinac but was demoted to agency interpreter after Henry moved his own headquarters there. The next year he resigned, informing Cass that he could no longer submit to Schoolcraft's "haughty mien" and "tyrannical" behavior. As the father of a growing family, he always needed more work: he and his first wife, Louisa, had three children, and after her death, he remarried and fathered four more. He struggled to make a living in mining, land speculation, and the declining fur trade. Eventually, he secured minimal annuity payments as Kahmentayha, son of Oshaguscodawaqua, and filed for a share of the proceeds from the cession in 1836 of his late wife Louisa's ancestral (Ojibwe) lands. His life reveals, with a kind of stop-motion clarity, how his Métis identity gradually became untenable in Jacksonian America.[54]

Through these years, George also helped shape Indigenous Christianity in the region. Traveling between Lake Superior and Lake Huron, he wrote, "I gave the Methodist-Indian [John Sunday] an opportunity of revealing God's holy word to Cacoqish's band."[55] In 1840, Henry assigned him and his wife Mary positions at a new subagency at Grand Traverse, on the eastern shores of Lake Michigan. There George gathered new "historical facts and [Anishinaabe] traditions" for Henry's "collections."[56] While at Grand Traverse, he joined the Episcopal Church—a departure from his family's Presbyterian affiliation—and translated an Episcopal prayer book and hymns into Ojibwe for the Diocese of Michigan. He also penned an Odawa petition accusing the local Presbyterian missionary of leaving the people "ignorant of the true religion."[57] Perhaps, on these Canadian borderlands, the Church of England appealed as a counterweight to the U.S. nationalism of the Presbyterians. The constraints of settler colonialism thus shaped Indigenous choices and the limits and possibilities of Anishinaabe religion. Much like the scientific research in which George engaged, religion could not restore the status his family had once enjoyed or the lands and livelihoods of the Anishinaabe people.

Henry had no qualms about publishing the tales he gleaned from the Johnston family to build his own renown. His scorn for "Indians" only grew stronger with his reputation and the changing times. Theories of religion were central to the civilizational hierarchies he helped build. In order to understand "the philosophy of the Indian mind," he explained in *Algic Researches* (1839), he needed to "examine the mythology of the tribes." Their stories showed that they were "polytheists," he wrote, "not believers in God or Great Spirit, but of thousands of spirits; a people who live in fear, who wander in want, and die in misery." Once again, he concluded that Indians had no true religion but only "superstition" and "necromancy." This lack proved their racial inferiority, and for Henry, it was no surprise that such a people were in decline. "The great Anglo-Saxon stock," he wrote, living by its "notions of liberty and justice," had brought the Indians Christianity and treated them with sympathy. The unfortunate outcome was not the fault of white settlers or the United States, he concluded, but was "to a very great extent, owing to [the Indians'] own idle habits, vices, and short-sightedness." Like so many other colonial authorities, Henry associated the values and virtues of liberty, reason, and religion with the "Anglo-Saxon," while the "Indian" appeared to be enslaved to superstitions. Here, the emerging science of comparative religions blamed Indigenous people for their own dispossession.[58]

Henry's position as a U.S. Indian agent made his theories immediately relevant to federal Indian policy. These theories, shared by many of his colleagues and superiors, helped justify the genocidal policy of Indian removal. Henry had long since concluded that all Indians would be better off west of the Mississippi, and his 1840 *Annual Report* as Michigan's acting superintendent of Indian affairs offered a detailed rationale. He warned of the potential for Indigenous attacks on white settlers and noted that all the lands south of Michilimackinac had a "highly favorable" climate and soil for white settlers. The Indians' racial characteristics, he claimed, made coexistence impossible. "It is apparent that the power of ratiocination [reasoning] in this race is feebly developed," he explained, "while all past observation proves that the desire of present good, or gratification of a merely physical character, have uniformly predominated over considerations of the past and future." Thanks to his analysis of what he called Indigenous "mythology" and "delusions," he had concluded that Indians simply did not have the racial capacity to reason, to plan ahead, or to coexist with white settlers. In his view as U.S. Indian agent, the only real solution was removal.[59]

Meanwhile Jane's health continued to decline. She was desperately lonely after they moved to Michilimackinac in 1833 and even more so in Detroit after Henry became state superintendent of Indian affairs.[60] When she went with Henry in 1839 to the Pictured Rocks of Lake Superior, part of a journey to inventory Indian "improvements" in recently ceded territories, she wrote a poem that came closer than ever to questioning her husband's judgment and authority. On a visit to Granite Island, at the western border of the ceded lands, she wrote an Ojibwe-language poem that she translated this way:

> Ah nature! Here forever sway
> Far from the haunts of men away
> From here there are no sordid fears,
> No crimes, no misery, no tears
> No pride of wealth; the heart to fill,
> No laws to treat my people ill.

This verse offers a biting critique of the forces that were wreaking such havoc in Anishinaabe lives. In another poem, she mourned her "forefathers" who once lived "in liberty free . . . Ere Europe had cast o'er this country a gloom." Henry may not have written the "laws" that oppressed her people, but he implemented them with a self-righteous fervor that must have become harder and harder for her to take.[61]

Her gloom only temporarily lifted when her beloved children, Jane Susan and John Johnston, came home from boarding school for the winter of 1840–1841. Henry had faced nepotism charges earlier that year and only barely kept his position in the face of a hostile investigation. George, his wife Mary, and the youngest Johnston brother, John McDouall, all at Grand Traverse, lost their jobs instead.

When William Henry Harrison of the Whig Party claimed the presidency in 1842, Henry, too, was dismissed from the Indian Service. He decided to move the family to New York City, where he could advance his literary and scientific career. That summer he took a long-desired voyage to Europe. Jane was too weak for the journey and decided to spend the season with her sister Charlotte in Ontario. A few weeks later, Charlotte found her unresponsive in her bed. At the age of forty-two, after many years of ill health, Jane's life had come to an end.[62]

After her death, Henry gradually disentangled himself from the Johnston family, even as he established his professional reputation through their unacknowledged expertise. He was already gaining renown as a founding father of the new settler science of ethnology. In 1842, he helped create the American Ethnological Society in New York; the next year he helped set the agenda for the newly established Smithsonian Institution in Washington, D.C. In 1847, he married Mary Howard, a member of a wealthy slave-owning family from South Carolina, who is now best remembered as the author of a sentimental proslavery novel, *The Black Gauntlet*, published in 1860.[63] Meanwhile he continued to write books about Indians, many of them recycling old materials he had gleaned from the Johnstons. And he continued to solicit new materials and translations from George. In March 1845, he wrote to thank George "for the Indian names of the rivers of the east and south shores of Lake Michigan" and offered advice on the land disputes in which the family was embroiled. He had worked to ensure that the Johnston family, including his own children, would receive their share of tribal lands in the treaties that had gradually dispossessed the Anishinaabe people. At the same time, as an Indian agent and settler scientist, he viewed the Anishinaabe as racially inferior and the identity of an "Indian" as simply untenable in the modern world.[64]

Schoolcraft had once rejected the idea that Indians had anything meriting the term *religion*. Now, though, he identified the "religion of the Red Man" as the "most powerful source" of their "character. "By it he preserves his identity, as a barbarian," he wrote, "and when this is taken away, and the true system substituted, he is still a Red Man, but no longer, in the popular sense, an Indian—a barbarian, a pagan." He had not changed his view of Indians but rather expanded his definition of religion to include what he called "superstition." Indians in his view remained improperly religious. They did not know how to keep spiritual phenomena properly distinct from a natural world that, in his view, should be viewed through the lens of science. "The native tribes who occupy the borders of the great lakes, are very ingenious in converting to the uses of superstition, such masses of loose rock ... as have been fretted by the action of water into shapes resembling the trunks of human bodies, or other organic forms," he wrote. The "natives, who are not prone to reason from cause to effect," could only attribute "all that is past comprehension ... to the supernatural agency of spirits." Nor did they keep religion properly distinct from the sphere of government. "The two principal causes" for their "barbarism," he wrote, "are a false religion, and false views of government." One sprung from the other: their false religion made good governance impossible. Indians remained uncivilized, outside of modernity, because they did not keep the spirits in their proper places. The novelty here, in keeping with the emerging European imperial science of *Religionswissenschaft*, was simply that he had expanded his use of the category religion to include such false pretenders.[65]

Henry's work demonstrates how U.S. authorities organized knowledge about religion for the purposes of settler colonial rule. In 1847, he moved to Washington to work as an administrator for the U.S. Indian Service and initiated an early effort to systematically collect information on all Indians. Soon after this appointment, he circulated a fifty-five-page questionnaire, *Inquiries Respecting the History, Present Condition, and Future Prospects of the Indian Tribes of the United States*, to which he hoped government agents, missionaries, and other officials would respond. He provided detailed, leading questions under eight broad headings. His first set of questions on the topic "Religion" followed an implicitly Christian model. "Do they believe there is a deity pervading the universe, who is maker of all things? What ideas do they possess of the Great Spirit? Is he believed to be self-existent, eternal, omnipresent, omniscient, omnipotent, or invisible?" From there, he moved on to questions that assumed the existence of a globally homogenous paganism. "Is there reason to believe the Indians to be idolators? Are images of wood or stone ever worshipped?" he asked. "Have you observed any traces of the Ghebir worship, or the idea of an eternal fire?" The goal of this exercise, he explained, was simply to obtain accurate information so that the government could formulate more effective policy. As David Chidester has also shown, the scientific study of religion reinterpreted knowledge about religion from circulating Indigenous and colonial sources, repackaging it as a tool for colonial management and control.[66]

The Johnston and Schoolcraft families reveal a great deal about how science and religion were jointly enacted upon a settler colonial stage. It is not enough to identify the relationship between science and religion as complex or multifaceted. Henry's settler science served alongside settler Christianity as twinned mechanisms for U.S. imperial expansion. As a white settler patriarch in Michigan and later in Washington, D.C., he had the power to advance and articulate sciences—geology, ethnology, and comparative religion—that supported U.S. conquest and facilitated U.S. rule. These settler sciences not only partnered with Christianity but also worked to delineate religion in ways that relegated Indigenous people and their traditions to the overlapping, racialized categories of primitive, pagan, and heathen. This regime mobilized science and religion together to enable the seizure of Indigenous lands, the expulsion of Indigenous people, and the destruction of Indigenous worlds.

At the same time, the lives of Jane, George, and their Anishinaabe kin hint at how Indigenous and Métis people could reconfigure science and religion as tools for survival under the constraints of U.S. rule. Christianity offered a source of hope and meaning. When white settler laws and institutions stripped Indigenous people of tribal lands and sovereignty, Christian affiliations could help them hold onto tribal identities and sometimes land as well. As a respected and legally protected category in the United States, religion—as claimed both by Anishinaabe Christians and by those who continued to practice "the religion of our fathers"—became a way to define and maintain Anishinaabe traditions and collective identities. Science served different but tactically related purposes in the Johnstons' lives. George engaged in geological and ethnological research and reported his findings to Henry not only to satisfy their mutual intellectual curiosities but also to prove his own scientific expertise in a losing battle to claim his own full humanity in a society that dismissed him as a "half-breed." His occasional discussions of Anishinaabe expertise on the flora, fauna, and geology of the Great Lakes also hinted at Indigenous worlds of knowledge production that—while they might sometimes be categorized as science or religion—have always managed to exceed these settler colonial configurations.

NOTES

1. The poem appeared in a hand-written literary journal edited by Henry Rowe Schoolcraft in 1825–1826, published as Henry Rowe Schoolcraft, *The Literary Voyager, or, Muzzeniegun*, ed. Philip P. Mason (East Lansing: Michigan State University Press, 1962), 87–89, 178. Robert Dale Parker attributes this poem instead to Henry Rowe Schoolcraft. See Jane

Johnston Schoolcraft, *The Sound the Stars Make Rushing Through the Sky: The Writings of Jane Johnston Schoolcraft*, ed. Robert Dale Parker (Philadelphia: University of Pennsylvania Press, 2007), Appendix 4, p. 258.
2. Peter Harrison, ed., *The Cambridge Companion to Science and Religion* (New York: Cambridge University Press, 2010); Peter Harrison, *The Territories of Science and Religion* (Chicago: University of Chicago Press, 2015); Bernard Lightman, ed., *Rethinking History, Science, and Religion: An Exploration of Conflict and the Complexity Principle* (Pittsburgh, PA: University of Pittsburgh Press, 2019).
3. Tisa Wenger, *We Have a Religion: The 1920s Pueblo Indian Dance Controversy and American Religious Freedom* (Chapel Hill: University of North Carolina Press, 2009); Peter Gottschalk, *Religion, Science, and Empire: Classifying Hinduism and Islam in British India* (Oxford: Oxford University Press, 2013); Cameron B. Strang, *Frontiers of Science: Imperialism and Natural Knowledge in the Gulf South Borderlands, 1500–1850* (Chapel Hill: University of North Carolina Press, 2018); Britt Rusert, *Fugitive Science: Empiricism and Freedom in Early African American Culture* (New York: New York University Press, 2017).
4. Talal Asad, *Genealogies of Religion: Discipline and Reasons of Power in Christianity and Islam* (Baltimore: Johns Hopkins University Press, 1993); Talal Asad, *Formations of the Secular: Christianity, Islam, Modernity* (Stanford, CA: Stanford University Press, 2003); David Chidester, *Empire of Religion: Imperialism and Comparative Religion* (Chicago: University of Chicago Press, 2014).
5. Michael J. Witgen, *An Infinity of Nations: How the Native New World Shaped Early North America* (Philadelphia: University of Pennsylvania Press, 2012); Sylvia Van Kirk, *Many Tender Ties: Women in Fur-Trade Society in Western Canada, 1670–1870* (Winnipeg: Watson & Dwywer, 1999).
6. Augusta Rohrbach, *Thinking Outside the Book* (Amherst: University of Massachusetts Press, 2014), 13; Henry R. Schoolcraft, *Personal Memoirs of a Residence of Thirty Years with the Indian Tribes on the American Frontiers* (Philadelphia: Lippincott, Grambo, 1851), 93.
7. Thomas Loraine McKenney, *Sketches of a Tour to the Lakes* (Baltimore: Fielding Lucas, 1827); Henry Rowe Schoolcraft, *Narrative Journal of Travels Through the Northwestern Regions of the United States* (Albany, New York: E. & E. Hosford, 1821), 505–508.
8. Henry Rowe Schoolcraft, *A View of the Lead Mines of Missouri: Including Some Observations on the Mineralogy, Geology, Geography, Antiquities, Soil, Climate, Population, and Productions of Missouri and Arkansaw* (New York: Charles Wiley, 1819); Henry Rowe Schoolcraft, *Transallegania, or, The Groans of Missouri* (New York: J. Seymour, 1820).
9. Schoolcraft, *Personal Memoirs*, xxxvi.
10. Johannes Fabian, *Out of Our Minds: Reason and Madness in the Exploration of Central Africa* (Berkeley: University of California Press, 2000), 7.
11. Lewis Cass to John C. Calhoun, November 18, 1819, Lewis Cass Papers, 1780–1909, Burton Historical Collection, Detroit Public Library (hereafter Cass Papers, DPL).
12. Lewis Cass to William Woodbridge, March 10, 1820, Cass Papers, DPL.
13. Schoolcraft, *Narrative Journal*, i.
14. Schoolcraft, *Narrative Journal*, 88–92.
15. Schoolcraft, *Narrative Journal*, 87–88, 91.
16. Schoolcraft, *Personal Memoirs*, 260; see also Bernard C. Peters, "Wa-Bish-Kee-Pe-Nas and the Chippewa Reverence for Copper," *Michigan Historical Review* 15, no. 2 (1989): 47–60;

Erik M. Redix, " 'Our Hope and Our Protection': Misko-Biiwaabik (Copper) and Tribal Sovereignty in Michigan," *American Indian Quarterly* 41, no. 3 (August 22, 2017): 224–249.

17. Henry Rowe Schoolcraft (HRS) to John C. Calhoun, "Facts and Remarks on the Copper Mines of the Region of Lake Superior," October 1, 1822, Henry Rowe Schoolcraft Papers, Microfilm Edition, reel 2, Library of Congress Manuscript Division, Washington, DC (hereafter Schoolcraft Papers); Dina Gilio-Whitaker, *As Long as Grass Grows: The Indigenous Fight for Environmental Justice, from Colonization to Standing Rock* (Boston: Beacon, 2019).
18. Chidester, *Empire of Religion*; Sarah Dees, "The Scientific Study of Native American Religions, 1879–1903" (Ann Arbor, MI: ProQuest, 2015); Tisa Wenger, "Sovereignty," in *Religion, Law, USA*, ed. Joshua Dubler and Isaac Weiner (New York: New York University Press, 2019).
19. Lewis Cass to HRS, November 4, 1821, Schoolcraft Papers, reel 1.
20. Henry R. Schoolcraft, "Private Journal of Indian Affairs," 1849, 1, Schoolcraft Papers, reel 2.
21. HRS to George Boyd, July 7, 1822, George Boyd Papers, Part 3, 1821–1826, Detroit Public Library; Schoolcraft, *Personal Memoirs*, 89–90.
22. Lewis Cass to HRS, September 26, 1822, Schoolcraft Papers, reel 2; HRS to John Johnston, July 6, 1822, Schoolcraft Papers, reel 2.
23. Schoolcraft, "Private Journal of Indian Affairs," Schoolcraft Papers, reel 2; HRS to Lewis Cass, October 25, 1822, Schoolcraft Papers, reel 2.
24. Sean P. Harvey, " 'Must Not Their Languages Be Savage and Barbarous Like Them?' Philology, Indian Removal, and Race Science," *Journal of the Early Republic* 30, no. 4 (2010): 505–532.
25. HRS to Jane Johnston, October 24, 1822, Schoolcraft Papers, reel 2.
26. Jane Johnston to HRS, May 19, 1823, Schoolcraft Papers, reel 18.
27. Marjorie Cahn Brazer, *Harps Upon the Willows: The Johnston Family of the Old Northwest* (Lansing: Historical Society of Michigan, 1993), 156–167, 177–179.
28. R. M. Laird to HRS, May 1824, Schoolcraft Papers, reel 2; HRS to James Schoolcraft, February 11, 1831, Schoolcraft Papers, reel 5.
29. Jane Johnston Schoolcraft (JJS), "Having Visited the Mission School at This Place," October 4, 1824, Schoolcraft Papers, reel 2.
30. HRS to Albert Hodge, October 6, 1824, Schoolcraft Papers, reel 2.
31. JJS to HRS, July 4, 1825 Schoolcraft Papers, reel 3.
32. JJS to George Johnston, January 19, 1826, George Johnston Papers, box 1, Burton Historical Collection, Detroit Public Library (hereafter Johnston Papers, DPL).
33. Schoolcraft, *The Literary Voyager*, 152, 157–158.
34. Brazer, *Harps Upon the Willows*, 198–202, 208.
35. Schoolcraft, *The Literary Voyager*, 49.
36. HRS to JJS, December 25, 1830, Schoolcraft Papers, reel 5.
37. Brazer, *Harps Upon the Willows*, 212–216; HRS to James Schoolcraft, February 11 1831, Schoolcraft Papers, reel 5.
38. Algic Society and Henry R. Schoolcraft, *Constitution of the Algic Society: Instituted March 28, 1832, for Encouraging Missionary Effort to Evangelizing the North Western Tribes* (Detroit: Cleland & Sawyer, 1833), 17.
39. Abel Bingham to Lucius Bolles, March 16, 1829, Church History Documents Collection, box 1, folder 2, Special Collections Research Center, University of Chicago Library

(hereafter CHDC); Unsigned letter to James Schoolcraft, January 5, 1832, Schoolcraft Papers, reel 5.
40. Schoolcraft, "Revival of 1832 in the Pres. Ch. Commenced in January, Continued Four or Five Months," 1832, Schoolcraft Papers, reel 5.
41. Robert Stuart to HRS, March 29, 1832, Schoolcraft Papers, reel 5.
42. Brazer, *Harps Upon the Willows*, 230–231; Schoolcraft, *Personal Memoirs*, 431.
43. Abel Bingham to Lucius Bolles, October 20, 1828, CHDC, box 1, folder 2. On Ojibwe hymnody, see Michael David McNally, *Ojibwe Singers: Hymns, Grief, and a Native Culture in Motion* (New York: Oxford University Press, 2000).
44. Robert Penner, "The Ojibwe Renaissance: Transnational Evangelicalism and the Making of an Algonquian Intelligentsia, 1812–1867," *American Review of Canadian Studies* 45, no. 1 (March 2015): 71–92.
45. Christopher Wetzel, *Gathering the Potawatomi Nation: Revitalization and Identity* (Norman: University of Oklahoma Press, 2015), 29–31; Matthew L. M. Fletcher, *The Eagle Returns: The Legal History of the Grand Traverse Band of Ottawa and Chippewa Indians* (East Lansing: Michigan State University Press, 2012), 28–31.
46. Schoolcraft, *Narrative Journal of Travels*, 91–92.
47. Abel Bingham to Lucius Bolles, October 20, 1828, CHDC, box 1, folder 2.
48. Elbert Herring to HRS, May 3, 1832, Schoolcraft Papers, reel 5; C. C. Trowbridge to HRS, April 28, 1832, Schoolcraft Papers, reel 5.
49. Schoolcraft, *Personal Memoirs*, 405–406, 419–421.
50. HRS to David Greene, February 25, 1832, Schoolcraft Papers, reel 5.
51. Schoolcraft, *Personal Memoirs*, 408.
52. George Johnston to HRS, February 12, 1827, Schoolcraft Papers, reel 4; George Johnston to HRS, August 20, 1827, Schoolcraft Papers, reel 4; Brazer, *Harps Upon the Willows*, 184192.
53. George Johnston to HRS, September 26, 1826, Schoolcraft Papers, reel 3; Brazer, *Harps Upon the Willows*, 188–190.
54. George Johnston to Lewis Cass, December 12, 1834, George Johnston Letterbook, Bentley Historical Library, University of Michigan, Ann Arbor (hereafter Johnston Letterbook, BHL); Brazer, *Harps Upon the Willows*, 298; Robert E. Bieder, "The Unmaking of a Gentleman: George Johnston and a Mixed-Blood Dilemma," *Hungarian Journal of English and American Studies* 16, no. 1/2 (2010): 125–135.
55. George Johnston to HRS, January 22, 1833, Schoolcraft Papers, reel 6.
56. HRS to George Johnston, July 6, 1839; George Johnston to HRS, December 25, 1840, Johnston Letterbook, BHL.
57. Eshquagowaboy and Nahwahquagwzhia to Bishop McKroskey, June 29, 1842, Johnston Letterbook, BHL. The prayer book, *The Morning and Evening Prayer of the Protestant Episcopal Church*, is listed in Henry R. Schoolcraft, *A Bibliographical Catalogue of Books, Translations of the Scriptures, and Other Publications in the Indian Tongues of the United States* (Washington. DC: C. Alexander, 1849), 12.
58. Henry R. Schoolcraft, *Algic Researches: Comprising Inquiries Respecting the Mental Characteristics of the North American Indians* (New York: Harper, 1839), 9–10, 41, 174, 35.
59. Henry R. Schoolcraft, *Annual Report of the Acting Superintendent of Indian Affairs for Michigan* (Detroit: Asahel S. Bagg, 1840), 4, 6.

60. JJS to HRS, June 24, 1833, Schoolcraft Papers, reel 6.
61. Schoolcraft, *Personal Memoirs*, 632–633; Maureen Konkle, "Recovering Jane Schoolcraft's Cultural Activism in the Nineteenth Century," *The Oxford Handbook of Indigenous American Literature*, ed. James H. Cox and Daniel Heath Justice (New York: Oxford University Press, 2014), https://doi.org/10.1093/oxfordhb/9780199914036.013.024.
62. Brazer, *Harps Upon the Willows*, 310–316.
63. Mary Howard Schoolcraft, *The Black Gauntlet* (Philadelphia: J. B. Lippincott, 1860).
64. HRS to George Johnston, March 19, 1845, Johnston Papers, DPL, box 4, folder 6.
65. Henry R. Schoolcraft, *Onéota, or, The Red Race of America* (New York: Burgess, Stringer, 1844), 2:126–127, 132, 81.
66. Henry R. Schoolcraft, *Inquiries, Respecting the History, Present Condition and Future Prospects of the Indian Tribes of the United States* (Washington, DC: s.n., 1847), 18–24, 53–55; Chidester, *Empire of Religion*.

CHAPTER 6

DARK GODS IN THE AGE OF LIGHT

The Lightbulb, the Japanese Deification of Thomas Edison, and the Entangled Constructions of Religion and Science

JASON ĀNANDA JOSEPHSON STORM

THE BRIGHT GOD OF THE ELECTRIC TELEGRAPH AND TELEPHONE

The foothills of the Arashiyama Mountains on the western outskirts of Kyoto, Japan, are well-known to Japanese sightseers, who make their way to the region for its famous bamboo groves, medieval temples, hot springs, and seasonal cherry blossoms. But nestled amongst these quaint tourist attractions, not so far from the hermitage of a renowned haiku master, sits a very unusual Shinto shrine, the Denden-gu Shrine, dedicated to Denden-Myōjin 電電明神, the Bright God of the Electric Telegraph and Telephone—a deity that from its title alone seems to occupy a vexed position among religion, science, and technology.[1]

In reverence to this god, people in formal attire descend on the shrine every year on May 23 for a festival. Typically, a group gathers under the covered awning in front of the "radio tower" stupa (電電塔), which is dedicated to incarnations of the Bright God. This group is mostly made up of Japanese salarymen in dark suits, but usually there are women pilgrims as well who are attired in similarly stark formal dresses. What these disparate individuals share is that most work in consumer electronics, the electrical power industry, or other broadcast media. When asked in 2016 about what makes the shrine unique, the administering priest said that "it was founded on the idea that electricity is not a human creation, but a blessing (*megumi*, 恵み) granted by nature."[2]

The yearly ritual includes Buddhist sutras chanted by Shingon-sect priests from the adjoining Hōrinji Temple (法輪寺). The head priest also leads the congregation in lighting incense and leaving flower offerings to various aspects of the Bright God. In a short set of prepared remarks, a representative of the Kyoto Broadcasting System Company memorializes the techno-scientific deity, thanks it for its inspirations, and asks for its protection for the coming year. Various protective talismans blessed by the Bright God are available for purchase at the shrine-temple's gift shop, which, at least in 2015, included laptop stickers, cell phone decorations, and microSD cards, all of which are touted for their ability to prevent computer viruses or power surges, maintain the safety of electrical cars, and otherwise facilitate the functioning of electrical technologies.

According to the shrine-temple's promotional pamphlets and official website, the shrine dates back to the ninth century. The founding legend is that one day a Buddhist monk named Dōshō (道昌, 798–875) was engaging in a meditative practice (Gumonji-hō, 求聞持法) intended to give its practitioner perfect memory. The rite is typically performed while viewing the planet Venus (understood as the Morning Star) and repeating mantras dedicated to the Bodhisattva of Boundless Space (Kokūzō Bosatsu, 虚空藏菩薩, Sankrit Ākāśagarbha Bodhisattva).[3] But on this occasion, Dōshō was said to have seen a shooting star and then to have received a personal vision of the bodhisattva, who instructed the monk to build a shrine to him in avatar form as the god of lightning. The shrine is said to have lasted for most of a thousand years and to have played a small but important part in Kyoto area pilgrimages before it was burned down during a rebellion against the Tokugawa Shogunate in 1864.[4]

But there is a contemporary twist, as in 1956, the director of the Kinki Bureau of Telecommunications, Hirabayashi Kinnosuke 平林金之助, had a new vision that the Bright God had possessed a different additional incarnation—the American inventor Thomas Alva Edison.[5] The shrine was refocused on electricity more generally instead of merely lightning. Even today part of the appeal of the shrine is the dedicatory Buddhist stupa with an informational plaque that explains that the shrine guards the soul (tama, 霊) of Thomas Edison as one of the deity's incarnations—and reveals that his true identity is the Divine Patriarch of Electricity (Denki soshin, 電気祖神). This might sound like an idiosyncratic rebranding of an old shrine but for the fact that it is not the only shrine in the area dedicated to Thomas Edison.

About twenty kilometers to the south of Arashiyama, at the Iwashimizu Hachimangū Shrine 石清水八幡宮 on Mount Otokoyama, a dedicatory plaque to Thomas Edison has stood in place since 1934. Among other things, it states: "Who can deny the grace of the Hachiman gods in [Edison's] undertaking?

There are no cities in the world where electric lamps are not lighted . . . [thus electric lights are] proclaiming to the world the divine virtues of the [Great God] Hachiman Daijin."[6] The theology of the Iwashimizu Hachimangū Shrine is slightly different insofar as it is basically jointly attributing the electrification of the globe to Edison and the deities of the shrine, but the shrine also distributes special prayer plaques with Edison's visage on them.

The main task of this chapter will be to answer two questions: First, how did Thomas Edison become a Japanese god? And, second, what does this hybrid figure (Shinto-Buddhist, Japanese-American, archaic-contemporary, religious-scientific) tell us about the entangled constructions of religion and science as global categories?

Before we get down to business, I want to be clear at the outset as to why I've chosen this particular topic. In keeping with the theme of this edited volume, Edison's apotheosis will let me revisit one of the classic debates in postcolonial theory from a new vantage point. For those of you with training in other forms of critical theory, it is worth explaining that one of the early significant controversies in this burgeoning subfield concerned a very different, but also supposedly "deified," Westerner.

For about a decade, two anthropologists—Marshall Sahlins and Gananath Obeyesekere—engaged in an increasingly polemical public dispute about whether the British explorer Captain James Cook had actually been worshipped as a god by the Indigenous Hawaiians. I return to this debate in the chapter's conclusion, but the core of the quarrel hinged on a disagreement about the universality of rationality. According to Sahlins, who launched the controversy, Cook had been deified by native Hawaiians, and this was a key piece of evidence that Indigenous peoples possessed nonrational ways of knowing. Rationality, in this account, has unique cultural contours. Obeyesekere, however, argued that the whole idea that Hawaiians ever worshipped a European explorer as a god was a myth formulated by Europeans rather than Indigenous peoples, and he concluded that the Indigenous Hawaiians had the same kind of "practical rationality" as contemporary Europeans today.[7]

The Japanese case presents us with a similar set of issues in part because until recently, both Euro-American ethnographers and Japanese nationalists promoted the view that the Japanese had a particular and unique way of thought that was supposed to be distinctively different from European rationality.[8] Although it may seem like an old debate with little contemporary purchase, versions of the question of the putative Eurocentrism of rationality have returned alongside recent conversations in decolonizing the discipline.[9] While I support decolonizing efforts, the debate has left certain questions unanswered. At its depth, the question is, Is practical rationality universal? And if so, what does it entail? Who actually deifies figures like Cook and Edison? Are these the myths of "white people" or of Indigenous peoples? And, finally, what does this all tell us about the relationship between religion and science as discursive systems? To answer these questions, I'll sketch out the transcultural cocreation of a divine Edison.

This explanation of Edison's apotheosis will require an account of the history of an everyday object, the lightbulb. This chapter will show that a complex cultural history lies behind the simple Edison bulb, a history that includes everything from an American ghost machine to the Japanese god of warriors, from inventors and capitalists to revolutionary samurai turned politicians, from Buddhist monks with imperial aspirations to theosophist seekers of ancient wisdom.

THE EDISON BULB

The incandescent lightbulb is an artifact whose simplicity obscures a convoluted history. There is a standard narrative about the history of the lightbulb. It is told in two different national variants, but the American one is as follows.

On September 8, 1878, the young inventor Thomas Edison was invited to a Connecticut factory by an engineer named William Wallace who had installed a set of eight newly designed electrical arc lights. Arc lights were then a fairly new technology; they were capable of providing a dazzling radiance that could illuminate large areas, but they flickered like a strobe light, and Wallace's model required each set to be powered by massive dynamos, which made them difficult to scale. Edison was at that time looking for a new project, having recently completed his work on the phonograph and telegraph to much public acclaim. According to an eyewitness, "Edison was enraptured" by the lights, and he rushed around making a set of energy calculations before he told Wallace triumphantly, "I believe I can beat you making the electric light."[10] A month later Edison was reporting to the *New York Sun* newspaper that after only a couple nights of experimentation in his home lab, he had "discovered the necessary secret" and would soon be bringing an affordable lightbulb to the public market.[11] This statement was sufficient to inspire a run on gas stocks in both the United States and Europe and helped to secure Edison funding from the likes of J. P. Morgan and the Vanderbilts.[12]

But in many respects, Edison was bluffing. He had not, in fact, discovered "the secret" of electric lighting insofar as a crucial problem remained unsolved—namely, finding a good material for the wick. He had been trying platinum, which worked well enough but was prohibitively expensive for the consumer artifact he had in mind. Moreover, Edison was not alone in the search for a functioning incandescent bulb to replace the arc light. He had competitors in the United States like Moses Farmer, Hiram Maxim, and William Sawyer. But his real rival was the English physicist Sir Joseph Swan.[13] According to what used to be the British nationalist account (although it has likely become the standard one), Swan was a step ahead of the Americans, having demonstrated a working lightbulb in February 1879 that used a filament made from carbonized paper.[14]

In July 1880, two years after Edison had announced his breakthrough, he had little to show for his work. His creditors were getting anxious, and the press was getting increasingly critical. So Edison made up his mind that Swan's focus on carbon filament was a good idea but that the specific filaments Swan was using had serious limitations due to their type of wick. The Englishman's bulbs simply did not last long enough to be commercially viable. Edison decided to throw some of his dwindling resources at the issue and sent out agents to India, Japan, China, Brazil, and Florida in the hopes that they would find the perfect natural material to be carbonized. As a minor digression, it is worth noting that this search was a fraught and sometimes dangerous enterprise. For instance, Edison's scout in Florida wrote back: "What makes this job extremely interesting is the strong probability of getting bitten by a snake," but having failed to find a good source, that agent was then sent to Cuba, where he then caught yellow fever and died.[15]

Edison's breakthrough is often attributed to his assistant William H. Moore, who was sent to China and then to Japan. In Japan, Moore is supposed to have discovered a particular variety of bamboo (*Phyllostachys bambusoides*), the fibers of which provided the ideal filament for a lightbulb. This breakthrough meant Edison was able to begin successfully manufacturing bulbs that vastly outperformed those of his rivals by the end of 1880. The following year Edison was able to put these bulbs into a larger electrical lighting system he had designed, and the so-called War of the Currents to electrify America had been launched.

If this is the popular story, the main intervention by historians of science has been to note that Edison was late to the game and there had been experiments aimed at producing incandescent light for much longer than this narrative suggests. Historians have also observed that Edison's financial backing and his lawyers' skill with deploying quick patents were much more crucial than the legend lets on.[16] Perhaps Edison's most important insight was not about the specific wick material but about the automation of the discovery process; a group of assistants could undertake a mass of parallel research and testing efforts, matching the efficiency of one of Henry Ford's assembly lines. In effect, Edison's Menlo Park laboratory was a key step on the way to industrialized or Big Science.[17] But on a technological note, historians have regularly granted that the bamboo filaments that Edison's team discovered *did* work better than the alternatives because they were thinner, required less electricity to produce light, and lasted longer. So Edison's subsequent ushering in of a new age of mass illumination (and of further scientific research into electricity) was thus made possible by insights that relied on Japanese bamboo wicks. This story has been told several times and is interesting on its own account. But the transnational backstory of the birth of the lightbulb turns out to be even weirder than is usually countenanced.

To foreshadow, here are two key things the conventional (American) historiography omits: (1) Moore cannot claim sole credit for the discovery, as the species of bamboo was suggested to him by members of a particular faction in the newly forming Japanese government, and (2) it wasn't just any old bamboo that the Americans brought back from Japan but a particular type of bamboo from the precincts of the shrine of a Japanese deity—the so-called tutelary god of warriors, Hachiman.

THE BAMBOO ARROWS OF HACHIMAN

This section will first provide some background on Hachiman, one of the two supposedly "Shinto" deities that became partially entangled with Thomas Edison, and then it will recover the forgotten Japanese figure who recommended the bamboo from that particular shrine to Edison's assistant.

Our first clue to the identity of the god Hachiman and his importance comes from the fact that in 1868, twelve years before Moore arrived on the scene, the Japanese government officially changed the deity's title. A single governmental edict promulgated in that year read: "Henceforth, the title Hachiman Great Bodhisattva is not to be used. Instead, the title Hachiman Great God (*Hachiman Ōkami*, 八幡大神) is to be adopted instead."[18] This brief line—about the substitutions of one noun for another—is itself a microcosm of one of the crucial Japanese ideological struggles of the last half of the nineteenth century. To explain, I want to provide a brief sketch or perhaps thumbnail "biography" of Hachiman.

The earliest reference to Hachiman in history occurs in the *Shoku Nihongi* 続日本紀 (Continuing chronical of Japan), a dynastic history compiled under the direction of the burgeoning Japanese state and completed in 797. Before we get to the entry, you should know that the name of the deity has somewhat baffled Japanese historians. *Hachiman* (八幡) literally consists of the characters for "eight flags." Since this seems to be basically meaningless on its own, it has lent weight to the speculation that the name comes from a phonetic transcription of an older Japanese word, probably originally pronounced *Yahata*, *Yawata*, or *Yabata*.[19] The problem is that most Indigenous Japanese deities have names that are either clearly meaningful (e.g., the Great Sky Illuminating Deity) or correspond to a place name, and Yahata seems to do neither.[20]

The first entry about Hachiman in the *Shoku Nihongi* is dated 737 and appears in a set of rituals performed at several shrines at different locations around the island to protect Japan (or at least the Yamato court) from an invasion by the Korean kingdom of Sila. Although Hachiman is now typically described as a Shinto deity, all subsequent entries in the *Shoku Nihongi* refer to Buddhist deities or institutions.[21] For instance, in the next entry, dated 741, Hachiman's original shrine receives donations of Buddhist sutras and a three-story stupa (a Buddhist structure). Subsequent entries describe a set of oracles that Hachiman issued in subsequent years (by way of a Buddhist nun who served as the main attendant) and the increasing role that these pronouncements played at the Nara court. Basically, Hachiman issued repeated proclamations for the next hundred years that raised his profile.

Eventually, Hachiman came to be particularly associated with the Buddhist temple Tōdaiji (東大寺, Eastern Great Temple) opened in 752 in Nara. According to the temple's early official chronicle (*Tōdaiji Yōroku*, 東大寺要録, 821), in 783 "Hachiman spoke," revealing that he had been born in the distant past; that before he had taken shape as a Japanese god. he had been Daijizai-ō-bosatsu 大自在大菩薩 (the Great Self-Existing Bodhisattva); and that from that time forward, he wanted to be referred to as a bodhisattva—in other words, as a "pre-Buddha"—and furthermore to have the epithet Daijizai-bosatsu added to his official title.[22] The title *Daijizai* is interesting because it seems to be directly taken from Daijizaiten 大自在天, the Japanese name for the Indian deity Maheśvara, or Shiva. In essence, this oracle seems to have been suggesting both an Indian and a Buddhist pedigree for the god.[23]

In the same chronical, in an entry dated 821, Hachiman revealed another earlier incarnation—namely, as Emperor Ōjin (応神天皇, supposedly 200–310 CE).[24] Although somewhat controversial today in the popular arena, the figure of Ōjin has almost no historical basis. The name first appeared in imperial ancestry lists as the alleged fifteenth emperor of Japan, but nothing more is recorded about this obscure and largely legendary figure. Shortly after Hachiman was identified as his imperial ancestor, however, a Buddhist priest named Gyōkyō had a dream in which Hachiman told him to erect a shrine closer to the capital to protect the imperial household (*chingo kokka*, 鎮護国家).[25] The resulting shrine-temple complex established in 859 CE was known as Iwashimizu Hachimangū. It was located just outside the city of Kyoto, which was then the imperial capital.

Iwashimizu was the key site dedicated to Hachiman, both as Emperor Ōjin and as a bodhisattva who took the form of a Buddhist monk. Rhetorically, Hachiman not only was drawn to the center of the Buddhist establishment but also came to be perceived as one of the origins of the imperial family. These processes reinforced each other. As Hachiman became centered at court and in the Buddhist institution, the court (and imperial family) became enmeshed in another set of Buddhist rituals. Essentially, what happened was that the imperial family would visit Iwashimizu to perform Buddhist rituals to their ancestors. This particular shine therefore became increasingly important to Hachiman as such, and, interestingly enough, as this shrine came to eclipse the one that was the original location of Hachiman rites, Hachiman stopped issuing oracles. With one major exception, Hachiman's last "speech" occurred in 878, after which the shrine priests interpreted Hachiman's will in omens like the movement of doves or noises like "the bleating of a calf" and the thunder clap.[26]

All that is to say, until roughly 1867, Hachiman was centrally associated with the origins of the imperial family, the Indigenous gods of Japan, and the Buddhist institution. Crucially, it was from the grounds of the richly symbolic Iwashimizu Hachimangū Shrine that the bamboo was harvested to serve as Edison's lightbulb filaments.

The renaming of Hachiman was but one indicator of governmental attempts to separate the gods and the buddhas—fracturing Shinto from its Buddhist context. To make a long story short, while Shinto is today portrayed as primitive nature-worship and Japan's "oldest indigenous religion," for most of its history Shinto was understood to be an extension of Buddhism.[27] At stake in the 1868 edict was an attempt to retain particular symbolic associations with Hachiman while rejecting others. This amounted to the symbolic stripping of Hachiman's Buddhist robes to reveal his (supposed) Shinto figure, thus constructing an Imperial Shinto separate from its Buddhist foundations and associations.

But in 1880, when Moore arrived in Japan, the Japanese state was involved in a long process of purging Hachiman's Buddhist associations. To explain something of why he was sent there, it would be helpful to know who pointed Moore to this particular grove. Fortunately, we do.

According to the historical records of the Iwashimizu Hachimangū Shrine, the man who suggested Edison's assistant utilize that particular bamboo was Makimura Masanao 槇村正直.[28] To provide a short biographical sketch, Makimura was born in 1834 in a minor samurai family in the Domain of Chōshū in southwestern Japan. Perhaps the most fateful thing that happened to the young Makimura was that he became acquainted with a local boy one year older than him named Katsura Kogorō 桂小五郎. Katsura was from a once wealthy samurai fallen on hard times. As young men, Makimura and Katsura both became increasingly hostile toward the ruling shogunate and critical of its handling of trade with the Euro-American powers. They were both ultimately drawn into a group of samurai who plotted radical reform and then later revolution. Katsura must have had significant leadership potential, as under the new name Kido Takayoshi 木戸孝允, he went onto become one of the three main samurai leaders of the Japanese revolt that ultimately deposed the ruling shogun in 1868. Although Kido died fairly young, he is still famous in Japan (he has been the subject of numerous anime—e.g., Kido is the protagonist's youthful mentor in the anime series *Rurouni Kenshin*) and has been the subject of significant volumes of scholarship.[29] But while Makimura is much less well known, as a young man he was also a key player in the Meiji restoration, or Japanese revolution, of 1868 and saw action alongside Kido in Kyoto against forces of the shogunate.

For those unfamiliar with Japanese history, after about 250 years of the "pax Tokugawa," a combination of internal and external pressures led to a revolution, framed as a restoration, that put the Japanese "Emperor" Meiji on the throne (the title of emperor is a modern translation of what had been a largely ceremonial role in the Tokugawa period). The early ideology of the new revolutionary Meiji government suggested that it would be a return to the divine age, when the workings of the state harmonized with rituals devoted to the imperial ancestors and the great gods of Japan. Simultaneously, the Meiji state advocated policies—from funding new educational establishments and scientific research to encouraging industrialized manufacturing and modernizing agriculture—that were radically proscience and protechnology.

While there are plenty of historical examples of either progress-oriented revolutions or reactionary restorations, one of the most striking features of the Meiji restoration is that, like the Puritan revolution, it seems to have been a synthesis of both impulses. I have discussed these in detail elsewhere—suffice it to say many Japanese thinkers of the mid-nineteenth century believed that Shinto and science were uniquely compatible.[30] What we have preserved of Kido and Makimura's correspondence suggests that they were committed partisans of this particular project.[31]

After the revolution, Makimura built a career as a local politician in Kyoto, where he ultimately became prefectural governor. There he attempted to follow Kido's lead and actualize the reforms central to the early Meiji ideological project. Although he ran afoul of other forces in the new government and was briefly imprisoned, Makimura was able to launch a national political career that took him from serving in a basically senatorial position to heading the administrative court (*Gyōsei saibansho chōkan*), and in 1890, shortly before he died, he was ennobled as a baron.

At the time, Moore had been sent to Kyoto—and to Makimura, in particular—by one of Edison's contacts in Japan, Itō Hirobumi 伊藤博文, a Japanese politician who had studied in London and was part of the same regional faction as Kido and Makimura.[32] When Moore arrived in Kyoto in 1880, Makimura was the governor.

Makimura's policies as a Kyoto regional official were pretty consistently pro-Shinto and promodernization. In 1872, his government led an early anti-Buddhist campaign: demolishing smaller temples; urging the physical destruction of Buddhist statues, transforming them into building materials for new schools; and encouraging the appropriation and sale of Buddhist "art" on the private market.[33] In 1879, he moderated his attitude somewhat and wrote a letter to the local newspaper in which he encouraged his constituents to use their private funds to contribute to the upkeep of Kyoto's ancient shrines and temples and to patronize local festivals.[34] We do not know Makimura's personal attitudes toward Hachiman, but we have his friend Kido's diaries, and Kido was a devotee who made multiple pilgrimages to the Hachiman shrine.[35]

So all that is to say Makimura and his friends were basically endorsing Hachiman as part of promoting both emperor and Japan as "the nation of gods" (*shinkoku*, 神国). Part of the explanation for why he directed Moore to Hachiman in particular is likely that patrons of the Iwashimizu Hachimangū Shrine had long associated Hachiman with bamboo grown in the area, which was used to make special bamboo arrows that were used in rituals intended to ward off demons. I discuss other reasons in a later section, but I want to observe first that the burgeoning Shinto institution seems to have endorsed this conflation of technology and divine power.

For instance, Iida Takesato 飯田武郷 (1828–1900), a Shinto priest and professor of Japanese classics at the Tokyo Imperial Institution, wrote an essay shortly thereafter titled *Denkitō* 電氣燈 (Electric lamp, published posthumously in 1903), in which he provided a Shinto rationale for electrical power. In it, he remarked: "The people of the West, being industrious in everything, developed a way to replace the oil-lamp [by] assembling light and manufacturing this into a marvelous apparatus that may also be called a 'small sun'. . . . Nevertheless, as might be suspected, their people have not the faintest idea that [in the lightbulb filament] there exists the fire of the deity Kagutsuchi's blood. . . . Do they know that the electric lamp hinges on the fire deity's august deeds?"[36] There isn't space to get into the intricacies of Iida's text as a whole, but basically he argues that Westerners' use of Japanese bamboo in the lightbulbs is effectively a miracle, demonstrating the supremacy of the Japanese gods. This was not an argument unique to Iida; a number of Japanese thinkers from the period argued that Euro-American science and technology were really rooted in the power of the Japanese gods.[37] But in brief, for Iida, Japan's rapid entrance into the age of light—kicked off with the first public electric lighting in Tokyo in 1882—was proof that the Golden Age of the Gods had returned.

Iida's peers often thought that they had Edison partially to thank for that. Moreover, as it happened, Edison's birthday was February 11, which falls on the same date in the Gregorian calendar associated with the founding of Japan by the legendary Emperor Jimmu (神武天皇, Jinmu-tennō). Accordingly, Iwashimizu Hachimangū Shrine celebrates February 11 each year with a ceremony commemorating the birth of Japan and then a separate ritual honoring the birth of Thomas Edison. All this contributed to a double apotheosis of Edison because he continues to be revered at two sites in Kyoto: the Iwashimizu Hachimangū

Shrine (where Edison is celebrated as an important figure, if not quite a god) and the Denden-gu Shrine (where we began).

So one version of the story about Edison's place in Japan is the result of an excess of reverence. This might seem to be a twist on Marshall Sahlins's account of the mythologization of Captain Cook, but instead of reflecting mere European propaganda, Japanese leaders impressed by specific Western figures went on to mythologize them. But that isn't the whole story.

To return to the postcolonial debate that animates this case study, both Sahlins and Obeyesekere associate the modern West with a form of practical rationality. What they differed about was whether Indigenous peoples shared that rationality. Moreover, even as they differed, both thinkers presumed an account of rationality that precluded belief in spirits or gods. This should not be a surprise. As I have explored in detail elsewhere, there is a long history of classical anthropological and sociological theories that often promoted notions of an opposition between primitive animist irrationalism associated with a belief in spirits and modern materialistic rationalism associated with a dispirited cosmos.[38] Moreover, this binary opposition was often spatialized to suggest a contrast between a primitive, spiritual East and a modern, rational West.[39] In the following, I show how a paradigmatic figure in the United States fails to fit these (artificial) notions of rationality.

EDISON'S GHOST MACHINE

On a dark winter's night in 1920, Edison assembled a small group of scientists and engineers in his laboratory to test a most unusual invention. According to *Modern Mechanix* magazine, the lights were dimmed, and Edison fired up a machine that he had been working on for more than a decade. It flickered to life, bathing the room in a feeble electrical glow. Its activity coincided with that of an amplifier intended to capture the subtlest of sounds. The purpose of the machinery was to discover haunting spirits. The press would later dub this device "Edison's Ghost Machine."[40] I will return to this machine and how it fit into Edison's project in a moment, but first I want to talk about why it might be a surprise.

Today in the United States (as well as Japan), Edison's reputation as an inventor persists in the popular imagination. He appears in elementary school textbooks and popular accounts of American innovation and would seem to be a paragon of secular and/or common-sense rationality. Indeed, even today, the clichéd lightbulb appearing over a character's head is often used to symbolize a stroke of genius or flash of insight.

As I have argued elsewhere, the putative opposition between séance and science is largely an anachronism. But one significant reason that Edison's spiritualist researches never became fully common currency is that they were suppressed, not in his own lifetime but later. Strikingly, Edison himself was open about this research and his quasi-spiritualistic beliefs, and he discussed them at length in the very first printing of his posthumously published diary (*The Diary and Sundry Observations of Thomas Alva Edison*, 1948). But his heirs intervened after the very first printing (not the first edition, the first printing) and had these expunged and all unsold copies of the first printing destroyed. Fortunately, a handful of libraries had already purchased first copies. Hence, I was able to get ahold of the rare "lost chapter" in which Edison confesses what amounts to his spiritualist beliefs, paranormal research, and even occult interests.

The first thing to note is that Edison was pretty explicit about his attempt to develop a scientific instrument capable of verifying the existence of the survival of the personality after death. As he argued, "If we can evolve an instrument so delicate as to be affected, or moved, or manipulated . . . by our personality as it survives in the next life, such an instrument, when made available, ought to record something."[41] In this respect, he was aiming to provide paranormal researchers with the technology necessary to definitively prove the existence of ghosts. Edison thought that if anyone could crack this problem, it would be him and his laboratory. But while he claimed that he had produced some tantalizingly results, he never asserted that he had categorically verified the issue.

The ontology behind Edison's ghost machine is a particularly fascinating case in point for purposes of this chapter. To get us there, I need to make a confession. Readers coming to an account of the lightbulb may have been anticipating a gesture toward some version of New Materialism, as expressed in Jane Bennett's vibrant matter, Karen Barad's agential realism, or Bruno Latour's actor-network-theory, all of which share a commitment to tracing the "agency of things." But while this literature has some notable insights, for reasons I have elaborated elsewhere, I think its shared theory of agency is largely vacuous.[42]

But that said, Edison would probably have been a New Materialist, albeit a particularly quirky one. This is significant in part because it pushes against much of the environmental politics of much of the contemporary movement. Many influential New Materialists (e.g., Bennett) have promoted what amounts to the claim that vitalism it itself an answer to climate change denialism and the exploitation of the natural world. This is as though an appreciation for living nature or the vital materiality of things should be enough to preclude cruelty to animals and attempts to dominate nature. But Edison was on the one hand, an exploiter of natural resources who famously electrocuted a circus elephant for a marketing ploy, and on the other hand, a committed believer in a form of a vibrant matter.

In the lost chapter of his diary, Edison argued for a kind of vitalist monism. He suggested that "our bodies are made up of myriad of units of life," and, indeed, all living things are composed of what amounted to mind-matter monads (think Leibnitz), which Edison suggested come together in swarms to produce multicellular beings. Moreover, these life units are neither created nor destroyed, and they possess memory and what amounts to limited agency. As he put it elsewhere, "I do not believe that matter is inert, acted upon by an outside force. To me it seems that every atom is possessed by a certain amount of primitive intelligence: look at the thousand ways in which atoms of hydrogen combine with those of other elements.... Do you mean to say they do this without intelligence?"[43] Edison is here endorsing a very literal notion of vibrant matter. It was this monism that he suggested had convinced him of the possibility of ghostly survivals. It was also this monism that drew him to occult circles.[44]

In particular, Edison was a member of the Theosophical Society. For those who've never heard of it, in the briefest of brushstrokes the Theosophical Society was founded in 1875 in New York under the leadership of Helena Blavatsky and Henry Steel Olcott. It was a very influential New Religious Movement whose members engaged in comparative religion with the intension of recovering lost wisdom and magical powers.[45] The society was crucial to the Euro-American adoption of yoga and Buddhism and was the dominant source of many features of globalized metaphysical religion, from auras to spiritual evolution to the idea of ascended masters. It is often described as the source of New Age religion. Indeed, the popularization of the very expression *New Age* as a near synonym for metaphysical religion can be traced to a particular theosophist, Alice Bailey.

I still have more research to do on Edison's engagement with theosophy. I've reached out to the New Jersey lodge where he was a member, and while they were supportive of me doing research, the pandemic has precluded access to their archives. The historical record also contains tantalizing references to Edison's parents as spiritualists who taught Edison automatic writing as a child, but I haven't been able to verify them.[46]

What I have found out so far is that Edison joined the Theosophical Society in 1878, making him the 162nd official member of an organization that would grow to something like 45,000 in the ensuing decades.[47] He does not seem to have been a very active member, but he met with Olcott, one of the cofounders of the society, on several occasions. According to Olcott's diaries, they discussed "occult forces," telekinesis, and other issues.[48] Edison also had one of his assistants record the voice of Blavatsky on an early phonographic disc (unfortunately, it became damaged). In summary, the very years during which Edison was "inventing" the Edison bulb seem to have also been those when he was most engaged with theosophy.

Edison also joined Olcott's mission to promote Buddhism in the West. In 1902, Edison produced the first American film about India titled *Hindoo Fakir* (a prime example of "mystic" orientalism). He also may have been one of the financial sponsors of the Sinhalese Buddhist activist Anagārika Dharmapāla.[49] There is more research to do here, but Edison seems to have played at least a minor role in the early promotion of Buddhism as a religion in the United States.

Before we move off of Edison and rationality, it is worth noting that in his own day, Edison was given the epithet "the Wizard of Menlo Park" (first appearing in the press in 1878). I don't want to read too much into the title, but it keys into something important—namely, his capacity for autohagiography. As several of Edison's biographers have noted, he was a relentless self-promoter and mythologizer. He claimed to have personally discovered things that were actually the products of his lab of technicians. He gave misleading interviews in which he told invented anecdotes for purposes of self-aggrandizement. He bragged about capacities his technology did not yet have and generally cultivated his reputation as a genius to whom things came in a flash of insight.[50] To be sure, Edison's public relations campaign produced what could be described as a *secular* apotheosis. But it is important to note that his self-mythologizing was constructed to baffle not some colonized population but the American people.

In the last section, I show how this Edison, as apostle of techno-science, literally appeared in Japan.

EDISON AND JAPANESE INDUSTRY

Edison arrived in Japan in 1922, just in time to celebrate his seventy-fifth birthday. His sponsor was not a Shinto shrine, although he did go to Iwashimizu, but rather the Japan Industry Club. This is a clue about the way in which Edison's reputation in Japan was as much economic or techno-scientific as Shinto. I don't want to present a false binary between economic and religious motivations, but let me provide instead some more evidence of the former.

To return to Makimura Masanao, he clearly had economic motives. In addition to promoting Shinto, he was consistent in his emphasis on encouraging the regional economy, and he played a crucial role in the 1871 establishment of the Office for the Promotion of Industry, which provided grants to facilitate education and research in fields like physics and medicine, as well as encouraging the growth of a manufacturing sector.[51] He also engaged in legislative maneuverings to keep existing employers from moving to Tokyo, and he promoted Kyoto as an early tourist site.[52] So promoting Kyoto to Edison was a straightforward way to help the regional economy.

Makimura and his peers clearly saw it as a way to build a further connection between Japan and Edison's company. Indeed, after Makimura had facilitated the production of a pipeline between Japanese bamboo harvesting and Edison's lightbulb manufacturing, budding Japanese inventors took advantage of those connection to travel to the United States and work with Edison.[53] Perhaps the most famous was Iwadare Kunihiko 岩垂邦彦 (1857–1941), who left Japan in 1886 to work in Edison's laboratory in Manhattan. After a decade in the United States, Iwadare came back to Japan and initially worked as a sales agent for Edison's company. But soon he went into business for himself, founding what would become the NEC Corporation, which today is the largest personal computer manufacturer in Japan and something like the fourth-largest manufacturer of computer components in the world. If you have an Android smartphone, you are probably using a technology NEC patented. Another was Fujioka Ichisuke 藤岡市助 (1857–1918), a Japanese electrical engineer who studied with Edison briefly in 1884 but then came back to Japan intending to beat Edison at his own game by designing a better bulb. Ultimately, Fujioka adapted and modified Edison's designs to produce a dual-coil electric bulb (which also used bamboo filaments) in 1890. The company he cofounded to promote it ultimately became the consumer electronics giant Toshiba.

NEC and Toshiba were vitally important to the post–World War II Japanese consumer electronics industry. While the linkage between Edison's company and Japanese companies wasn't the only (or even the main) reason for Japan's successes in these arenas, Edison came to symbolize techno-scientific success, and deification became one way to celebrate and, in that respect, to indigenize these industries. Even today NEC and Toshiba are important donors to the Denden-gu Shrine. Again, it is worth remembering that Hirabayashi Kinnosuke, who in 1956 "discovered" that Edison was the avatar of the god of lightning, was the director of the Kyoto regional Telecommunications Bureau (電波監理局) and the shrine was rebuilt with donations from NHK, Asahi Broadcasting, and Kyoto B, so the deification of Edison also functioned as a way to boost the local telecommunications industry.

Again, I don't want to reinforce false notions of essential oppositions between Shinto and economics, religion and science, or faith and rationality. Perhaps one of the main reasons that Makimura picked the Iwashimizu Hachimangū Shrine was that it was already in business. Indeed, before the revised Japanese law codes of the late nineteenth century that distinguished businesses from religious practices, the priests (*jinin*, 神人) of Iwashimizu had long been associated with trade. In the medieval period, they even formed a guild focused on selling fish and oil. Before the restoration, the shrine priests traded in "silk, cotton, cloth, cloth-dyeing, fresh produce, indigo, cakes, medicine, sesame, and salt," and they leased out farmlands for cultivation, as well as serving as moneychangers.[54] Bamboo was a simple diversification for an institution with a long history in commerce.

CONCLUSION

I return, for the last time, to the debate between Marshall Sahlins and Gananath Obeyesekere. To recap their argument in nutshell, Sahlins argued that Captain Cook had been deified and that this was evidence of a fundamental divergence between Indigenous rationality and modern Euro-American thought. Obeyesekere differed, saying: "I doubt that the native created their European god, the Europeans created him for them. This 'European god' is a myth."[55] Moreover, Obeyesekere argued that the Indigenous Hawaiians and Europeans possessed the same *practical rationality*, which he defined as "the process whereby human beings reflectively assess the implications of a problem in terms of practical criteria" and which he suggested was in some sense incompatible with deification[56]—an oddly secularist notion of rationality he qualifies somewhat elsewhere.

I think the history of Edison's deification suggests a few things: (1) deification can be the product of practical rationality; (2) Edison mythologized himself, and both American and Japanese people bought into that mythologization in distinctive but comparable ways (if I wanted to hit the point home ,I could provide more detailed accounts of the various American spiritualists who invoked Edison as a spirit guide); (3) unlike Obeyesekere, we now have a plethora of historical accounts of the deification of people, often by Indigenous actors enmeshed in complex systems of colonial domination (e.g., the deification of Britain's Prince Philip in the South Pacific);[57] (4) of the various people involved in his deification, it was Edison who explicitly described a belief in an animated, vibrant cosmos; and (5) all of this provides evidence that there is not a vast gulf between Japanese and Euro-American rationalities. This latter point fits the broader psychological research into the issue, which suggests that while different cultures have different conceptual categories and disagree about the kinds of entities that populate our universe, there is not a great difference in human rationality or cognitive modes.[58] Broadly speaking, we are all perhaps equally rational or irrational.

This tells us something about the entanglement of religion and science. Many nonspecialists portray religion and science as unproblematic entities locked in a state of continual struggle. This idea of a religion-versus-science conflict is typically charted via a set of (largely fictitious) flashpoints, such as the voyages of Christopher Columbus that purportedly disproved a biblical flat earth with Jerusalem in its center, Galileo's supposed imprisonment and torture for promoting heliocentrism, and the imagined destruction of Charles Darwin's faith by his discovery of evolution. All of these are myths, and, indeed, the whole conflict model was refuted many decades ago.[59] But these are myths widely regarded as true by nonspecialists, which means that those of us who work on the relationship between religion and science are often forced to act like zombie killers, defeating a long-refuted model again and again.

Part of the reason the conflict model lives on is that the main contemporary successor model about the relationship between religion and science is the complexity or entanglement model, which basically amounts to the statement that religion and science have historically interacted in complex ways about which no grand generalizations can be made. For this reason, the complexity model adds little to our general understanding, but it does allow historians to do what they do best, which is to chart out the specific relationships between specific putatively religious and scientific agents in their particular contexts.

The problem is that the complexity model is a nonmodel. It doesn't really tell us anything about either religion or science. It functions primarily as a rejection of generalizations that does little to advance broader conversations. Indeed, it gets most of its mileage from refuting the zombie conflict model mentioned above. Religion and science are often found entangled. That should not be a surprise. Cases like the lightbulb above illustrate the point. So what is the alternative?

To confess to my own aspirations, I have been working toward a novel alternative to the classical theories (and anti-theories) about religion and science. It rests not on this case study of Edison's deification, but on a grand theoretical reorientation toward the basic subject matter in the humanities and social sciences. While I elaborate the theoretical edifice in much greater detail in *Metamodernism: The Future of Theory*, in these concluding pages I would like to provide you with what I hope will be a tantalizing glimmer of part of this theory and its implication as a novel critical approach to science and religion—namely, a process social ontology approach to social kinds.

The first step in this theorizing is to grant all the evidence that drove the formation of the complexity model in the first place. Depending on how a given scholar defines *religion* and *science*, they can almost necessarily find the two in any number of possible constellations: conflicting, harmonizing, entangled, mutually implicated, overlapping, and so on. More importantly, religious studies and science studies have now spent decades relativizing their respective objects. Scholars of religion now know that religion is not a universal part of human nature but a culturally specific category that initially took shape in Western Christendom at the end of the seventeenth century and then was radically transformed through a globalization process over the course of the long nineteenth century, producing both world religions and the assertion that "religion" is an autonomous domain of human experience. Although the issue is more controversial, scholars of science and technology studies have argued that there is no single universal scientific method, that demarcating science off from pseudoscience is nigh impossible, and that *science* functions largely as a prestige term to indicate value to the capitalist military-industrial complex. To this, some have added the postcolonialist observation that "modern science" emerged in the long nineteenth century with a radical reformulation of European natural philosophy and expanded through globalization and the selective absorption and disintegration of local knowledge systems. (The dissolution of "religion" as a category is not likely to be troubling to most readers of this chapter, but criticisms of science might seem more reckless in an age of vaccine and climate skepticism, and I'll return to "science" later.)

These aren't the full extent of the critiques of science and religion, and I could rehearse them ad nauseum, but it might seem that in different sectors of the academy, we have stopped believing in their definitional independence. To this, I would add that one can find similar critiques of a whole host of master categories from art to culture to society itself. All told, there is a somewhat extensive literature suggesting that all such concepts are fundamentally problematic.[60]

Looked at from a significant remove, much of the evidence for these arguments is that categories such as religion and science (1) appear to be impossible to define clearly in ways that permit clear demarcation and an absence of gray areas, (2) lack stable content insofar as they have changed significantly over the course of history, (3) also lack clear cross-cultural instantiation insofar as their application outside the historical horizons of Euro-American history is heavily fraught, (4) are normative concepts that encode values as well as epistemic claims, and (5) as a consequence of the previous two, have often been entangled with patriarchy, slavery, race, and colonialism (although to be fair, what hasn't?). Finally, because of how these categories have been critically historicized and provincialized, scholars have often concluded that they lack essences and have instead been reified and/or socially constructed. To be clear, I find most of these critiques persuasive. Indeed, I think that rather than undercutting the critical approaches, we need to grant them, but instead of merely terminating our arguments in the negative gesture, we need to move forward.

But before we do so, we need to radicalize the critical movement even further. I mean two concrete things by this: First, these critiques are typically presented in isolation as though a given term (e.g., *religion*) was uniquely flawed. But we can radicalize the movement by starting from the recognition that all such master categories are equally problematic. It is often said that "a language is a dialect with an army and navy." There is a grain of truth in this inasmuch as no language is truly exempt from the legacies of power. But this means that we will never be able to fully decolonize our conceptual vocabulary. Likewise, there is no Archimedean point for knowledge outside the horizon of history and geography. Moreover, fact and value are often mutually imbricated. (That said, neither of these is the problem it is often taken to be.[61])

Second, the critique of the master categories can be pushed further because the very subject matter they address is best understood not in terms of fixed categories but rather as constantly chaotic and interlocking systems, which not only have historically changed but will also continue to do so. Restated, we need to reject the idea that the social world consists of clearly delineated categories, describable by means of necessary and sufficient condition definitions and consisting in roughly fixed cross-cultural and cross-temporal constellations. For instance, no definition of "religion" can capture all the future things on which people may base religions and so on.

It might seem that by granting all of these critiques, we have gone too far. Indeed, claims like the previous one have often been taken to suggest that the systematic study of culture and society is impossible, but this analysis of their findings is precisely wrong—we can actually come to understand a social world in motion.

The second significant move of this theorizing is that instead of presuming fixity, we presume change—hence, process. Many of the deconstructive criticisms of the disciplinary master categories amount to identifying errors stemming from reification, atemporality, and misplaced concreteness—in other words, faults rooted in misidentifying the processual nature of their subject matter. These criticisms have purchase because, while few would dispute the processual aspect of society, the prevalent forms of analyzing human affairs are often geared toward substance thinking rather than process thinking.

To drive the point home, we frequently refer to religion or science as if they were bounded objects with distinct borders and even as if they were subjects rather than dynamic processes. But processual kinds are change by definition. Thinking about our work in terms of an analysis of heterogenous and changing but nonetheless roughly describable processes permits us to both grant the critique of the categories and produce new forms of analysis. We may draw generalizations while simultaneously acknowledging that we are discussing processes in flux and that whatever we say today can and will certainly alter in the future. Nonetheless, our observations may be very useful—they can help us obtain a more accurate image of the stage of unfolding we are currently in, as well as those that have gone before.

The third—and crucially significant—step is to recognize that while the social world is constantly changing, it isn't all changing the same amount or at the same speed. Systems of power can produce transitory stabilities and imperfect homogeneities. Hence, as part of the process social ontology, I posit a notion of temporary zones of stability called social kinds. Succinctly put, *social kinds* are socially constructed (or mind-dependent), dynamic clusters of properties that are demarcated by the causal processes that anchor the relevant clusters. Social kinds tend to be highly entropic or varied both temporally and spatially (hence, they tend to be historically contingent), their properties emerge via their relationships to other social kinds, and they cross-cut each other, so that the same entity can be the intersection of different kinds.

Different categories are stabilized by different anchoring processes. These anchoring processes are also important because, since I am presuming change and difference, relative stability and similarity become the things to be explained. Put differently, we need an account of the multiple distinctive causal, anchoring, or stabilization *processes* that give social kinds their shared properties. These processes are the reason there is any stability in the social world. The reduplication of process language in this section is completely intentional. Social kinds are processes that require other processes as catalysts. Elsewhere I discuss a set of different, if often interwoven, types of anchoring processes and show that these processes are the things that actually produce what is often called social construction (but without the implication that just because something is socially constructed, it is unreal).[62]

The anchoring process that is likely to be most familiar to scholars reading this volume is the *dynamic-nominalist* process. In the case of many of the kinds in the human sciences we are most interested in, they have come to share the properties/capacities they do because of specific classificatory processes that have become entangled with mechanisms of enforcement (cultural, social, institutional, legal, and so on). Examples are bountiful: legal processes that confer particular capacities on certain groups (such as the ability to vote) that have built-in mechanisms to attempt to ensure compliance, boundary policing that goes on in academic disciplines, social norms of shaming or reciprocity that encourage or prevent certain kinds of behaviors, tax codes that provide religious exemptions to organizations that take on properties, and so on. But dynamic-nominalist processes are not all there is.

I also discuss *ergonic convergence* by which I mean that sometimes kinds share properties through a process of selection or design intended to fulfill a certain function or purpose. Think of an electric lightbulb. All such lightbulbs—from the original incandescent Edison bulbs with their carbonized bamboo filaments to fluorescent lights to the LED lamp I have over my desk—share a capacity to transform electrical power into visible light. But there are many different ways of putting physical components together to make light from electricity: e.g., heating up a carbon filament in a vacuum until it glows, producing an electrical arc through a low-pressure mercury-vapor gaseous medium, and passing an electrical current across an energy gap in a semiconductor composed of a layered compound containing gallium arsenide. Each of these utilizes a different physical mechanism to produce the result. So why do these very different physical components all produce the same effect? Because they were designed to fulfill a common function. All lightbulbs were designed to transform electrical current into luminescence because they were all selected for that function (even if they eventually burn out). The main point is that if you know that something is a lightbulb, you can produce more robust generalizations about it than you could if you know only the material out of which it has been composed. For this reason also, shared technological preconditions with an attendant economic market for such would incline independent researchers to come up with different solutions to the same functional problem. All that is to say artifacts (and some other social kinds) tend to embed teleology and, as such, tend to be subject to what amounts to Weberian rationalization toward specific ends.

To return to religion and science, I would argue that both of these are primarily dynamic-nominalist categories. There is a lot more to be said here (especially as terms like "religion" can refer to more than one social kind). Nonetheless, most entities labeled as *religions* only share properties with other *religions* as a result of dynamic-nominalist anchoring processes. Various entities are designated as *religions* by being named or categorized as such by international diplomacy, domestic legislation, scholarly treatments, and so on. This is not a teleological or transhistorical process but one that arose from a certain logic at a specific point in history, and its globalization was inherently normative, selective, and, to some degree, arbitrary. Furthermore, it was produced via negotiation rather than unilateral imposition or hegemony. There is always a remainder inasmuch as contestation never ends. And as I mentioned earlier, there must be a variety of incentives or enforcement mechanisms in place to encourage the categorization to take root.

One could tell a variation of the same narrative about science. In the very barest of brushstrokes, the various individuals who have been retroactively connected with the Scientific Revolution did not typically consider themselves to be pursuing a single endeavor; nor was there a common idea of science, much less an epochal discovery of a unified scientific method. The term *science* entered English and French in the twelfth century as a synonym for *knowledge* and took on a new meaning, combined with autonomy and institutional authority, only over the course of first half of the nineteenth century.[63] No sooner had this category taken hold than various works such as Andrew Dickson White's "The Battle-Fields of Science" (1869) and John William Draper's *History of the Conflict Between Religion and Science* (1874) began to promote the idea that religion and science were in conflict. The idea of a conflict between religion and science was then projected backward into human history.

Against standard reading of the contextualist model, this is a clue that the histories of religion and science are not two independent narratives but rather the tale of conjoined twins born late and in the moment of their mutual contrastive self-definition. Religion and science were formulated as respective discursive domains alongside the notion that they were in opposition, even as various fugitive hybrids were produced that claimed to overcome this antagonism.[64] The issue of science's relationship to religion became one of the "big questions" of European and American thought precisely because these two categories were being formulated in mutual entanglement. It also became a question asked of different cultures, as if the twin categories were ahistorical abstractions rather than historically conditioned, unfolding processes.

Strikingly, this means we can historicize the background presuppositions that define not only this volume but also the entire field. This whole volume of critical approaches to religion and science was made possible only by various institutional and material preconditions. Publishers know that discussions of religion and science sell. Academics know that they can base their careers on overcoming (or sometimes exacerbating) the zombie conflict between religion and science. The long history of this goes at least as far back as the nineteenth century, when the genesis of the idea of a conflict between religion and science became one of the motivating forces behind the secularization of the university system itself.[65] This contributed to the divergence of the academic disciplines, the exile of theology from natural history, and then, in an effort to counter these developments, the numerous attempts to restage the reconciliation of science and religion. We might think of the Templeton Foundation and their funding for the very workshop that gave birth to this volume as an attempt to promote exactly the latter. It also has meant a thin trickle of academic positions and recognized journals focused on religion and science, even as the field has been historically dominated by Anglo-American Protestant and Catholic thinkers and problematics drawn from Christian history. But to be clear, all of this has been dwarfed by the vast funding available to science and the vast institutional apparatuses that have gone into maintaining its boundaries.

To translate the insight into the theory I've been putting forth in these last pages, the very attempts to produce a scientific method and a scientific community are best thought of as aspirations toward anchoring science as an ergonic kind directed toward the production of knowledge. But, historically, Euro-American thinkers have tended to be overconfident about the status of present "scientific" knowledge, and they have tended to exclude too much, especially to the degree to which they have precluded the knowledges of the non-European other and consigned them to categories like religion or, worse, superstition. Moreover, as I have been arguing in this chapter, they have often conflated rationality with the ontology of a particular reductionistic version of "scientific" materialism, such that alternate ontologies (including those populated by deified American inventors or ghosts) are seen as irrational even if many scientists (including Edison) were often open to a less reductionistic and more "enchanted" cosmology.[66] But, again, this should not surprise us as there was never a fully closed or unified scientific world picture, only a myth of one.

It might seem especially rash to criticize the unity of science in this particular historical moment, when antiscientific movements run wild. But if science is not a unity, then it is actually freeing. Instead of discussing the scientific cosmology as such, we should really be thinking of the longevity of different localized theories. Science often stands in for knowledge in general. For this reason, arguments about the social construction of quarks and the like are often taken to spell trouble for science as such. But one straightforward implication of dissolving the unity of science is that some truth claims have better justifications than others. Science does not stand or fall together. Some of the better-justified claims are probably associated with sciences, but many others are not.

For instance, the assertion that "Amelia Earhart crossed the Atlantic Ocean in 1928" is a well-justified knowledge statement, as we have many eyewitness testimonies and it is unlikely (although not impossible) that any possible future evidence will change historians' minds about that fact. But it not a scientific claim (as we use the word *science* in English). Some "scientific" claims (e.g., anthropogenic climate change) have lots of robust evidence for them, whereas others (e.g., particle supersymmetry) do not. Some claims in the humanities and social sciences have more or less evidence for them as well. All that is to say we do not actually need an inflated and universalized category of science to support knowledge.[67] We need instead a theory of knowledge (which I aim to provide elsewhere).[68] Demarcation fails, but this does not mean that we throw out all scientific claims. We still have statements that are more or less epistemically well-justified. Well-grounded statements seem to cluster in scientific disciplines. This is no small matter and is worth celebrating as a significant achievement, even as we have to be careful not to fall victim to scientism.

To return to the subject at hand, if science and religion are themselves second-order categories used to aggregate various putative sciences and religions, we need a third-order category to understand how this aggregation occurs and what it means. For that reason, I have been working toward a broader account of *social kinds*, which, in turn, will permit a fresh grand reappraisal of the higher-order categories of religion and science and how they fit together as a dynamical formation of socially constructed kinds. Rather than merely gesturing at complexity and cutting off further theorizing, this suggests that we should grant a nonessentialist account of the dynamically formulated categories of religion and science as historically conditioned process kinds, as this will enable us to make progress in understanding their properties and the causal processes that anchor them.

But these processes are what we should study. Processes are often presented as unknowable flux, but this is a mistake. In the first instance, we may track how things have changed in the past without having to worry about how their current dynamics undermine our interpretation. For example, there has been a lot of solid work done that investigates the historical and cross-cultural unfolding of "religion" as a category, which is no less helpful given the category's variability and inclination to change. In the second case, the minimal precondition required to begin to understand categories like religion and science is that the pace of knowledge gain has to be greater than the rate of change. Difference and variation make prediction difficult (and prediction has never been one of the human sciences' strong suits), and generalizations need to focus on what anchors them in a given context rather than (ahistorical) essences, but none of these prohibit examination of the past or the present. Moreover, understanding the processual nature of these kinds has the potential to make our investigation of them more accurate.

In summary, religion and science are equally complex categories but so are other social kinds (e.g., art). Their complexity and regional variation might seem to imply that nothing can be said about them. But instead, I have been arguing that the reverse is true—once we understand their processual nature, we can turn Foucauldian genealogy on its head. We presume the interplay of powers, heterogeneity, and change. This lets us focus on the anchoring processes that produce even temporary zones of stability. We can also trace backward the history and spread of social kinds, their interrelation, their variation, and globalization. Doing so allows us to see that the history of science is the history of religion and vice versa. Recast, ever since the period in which "religion" and "science" first became reflexive dynamic nominalist kinds, they have been fundamentally entangled.

As I have been arguing here, the lightbulb itself, like many other artifacts, can be seen as the intersection of various open systems or unfolding processes. As noted, lightbulbs as functional artifacts are ergonic kinds that tend to converge on their specific purpose. But this is not the limit of what the theory discussed here would tend to anticipate. It would also encourage us to look at the way that patent law works to stabilize subkinds, such as Edison bulbs.

All that is to say, according to the theory I've been sketching here, we should expect to see artifacts entangled in different global categories because the categories cross-cut. So the question is not "Is this lightbulb religious or scientific?" but "When is this lightbulb religious?" and social kind theory gives us the capacity to answer that. Indeed, the same type of Edison bulb is exhibited at the Iwashimizu Hachimangū Shrine in Japan and is included in the permanent collection of the Smithsonian National Museum of American History in Washington, D.C. The context of the shrine means that it is then going to take on a different set of attributes (people will tend to see it in terms of its symbolism, of, say, Shinto doctrines, and so on). In contrast, the bulb exhibited in an American museum alongside other artifacts will tend to be read (as the exhibit card encouraged) as a symbol of scientific research and technological progress (people will also tend to see it in relation to other lightbulbs, the history of electricity, and so on). Jasper Johns could have taken the same bulb and placed it in an art gallery, where it would have been read as an art object with some of the attendant attributes of art objects (people will attribute it to a specific creator and attempt to read it aesthetically or formally or place it in its broader art historical context). All that is to say a specific artifact's membership in a broader set of kinds is situationally and contextually dependent. Moreover, it is never complete, as other meanings/contexts continue to be available.

Had I space to do so, the narrative could be continued in the other direction, tracing not just the historical processes that produced the lightbulb and its attendant symbolisms but also the effects that the electrification of light had on Japanese culture, aesthetics, and even vision. As the Japanese novelist Tanizaki Jun'ichiro 谷崎 潤一郎 reminds us, for all its advantages, the adoption of mass illumination also contributed to banishing the mysterious wonders of the shadowy world, once populated by gods, ghosts, and monsters. The bright gods rose at the expense of dark deities—so for every figure deified by the age of electric light, others were lost, and sometimes it is necessary to speak "in praise of shadows."[69]

NOTES

1. For additional discussion of a semiotics inspired by this ethnographic research, see Jason Ānanda Josephson Storm, *Metamodernism: The Future of Theory* (Chicago: University of Chicago Press, 2021), 149–152.
2. See Munakata Aiko 宗像藍子, "Kyōto Arashiyama ni Denki Denpa Sangyō no Jinja 京都・嵐山に電機・電波産業の神社," *Nihon Keizai Shinbun Newspaper* 日本経済新聞, September 3, 2016.
3. See Yamasaki Taikō 山崎泰廣, "A Morning Star Meditation," in *The Life of Buddhism*, ed. Frank Reynolds and Jason Carbine (Berkeley: University of California Press, 2000), 99–108.
4. https://www.kokuzohourinji.com/dendengu.html, accessed July 1, 2019. The shrine was partially rebuilt in 1884.
5. According to a shrine worker interviewed by the author on March 15, 2015.
6. See "Edison Monument Erected at Kyoto," *Denki-Gakkai* 54, no. 6 (June 1934): 82. Edison's postmortem status is less explicitly articulated at Iwashimizu.
7. Marshall Sahlins, *How "Natives" Think: about Captain Cook, for example* (Chicago: University of Chicago Press, 1995), and Gananath Obeyesekere, *The Apotheosis of Captain Cook* (Chichester, UK: Princeton University Press, 1992). See also Robert Borofsky, "Cook, Lono, Obeyesekere, and Sahlins: Forum on Theory in Anthropology." *Current Anthropology* 38, no. 2 (1997): 255–282.
8. See Morimoto Tetsurō 森本哲朗, *Nihongo Omote to Ura* 日本語表と裏 (Tokyo: Shinchōsha, 1985).
9. For example, Wael Hallaq, *Restating Orientalism: A Critique of Modern Knowledge* (New York: Columbia University Press, 2018).
10. Matthew Josephson. *Edison: A Biography* (New York: Wiley, 1992), 178.
11. Josephson, *Edison*.
12. William Broad, "Rival Centennial Casts New Light on Edison," *Science* 204, no. 4388 (1979): 32–36.
13. Broad, "Rival Centennial," 32.
14. Broad, "Rival Centennial."
15. Broad, "Rival Centennial," 36.
16. This account can be seen in both professional and popular histories. See Theresa Collins, Lisa Gitelman, and Gregory Jankunis, *Thomas Edison and Modern America* (Boston: Palgrave Macmillan, 2002); Ernest Freeberg, *The Age of Edison: Electric Light and the Invention of Modern America* (New York: Penguin, 2013); Robert Friedel, Paul Israel, and Bernard

Finn, *Edison's Electric Light: The Art of Invention* (Baltimore: Johns Hopkins University Press, 2010); Andrew Hargadon and Yellowlees Douglas, "When Innovations Meet Institutions: Edison and the Design of the Electric Light," *Administrative Science Quarterly* 46, no. 3 (2001): 476–501; and Jill Jonnes, *Empires of Light: Edison, Tesla, Westinghouse, and the Race to Electrify the World* (New York: Random House, 2004).

17. Although some scholars like to date Big Science from the Manhattan Project, Edison had pioneered a form of automated or industrialized discovery backed by big capital investment much earlier. See Peter Galison, *Big Science: The Growth of Large-Scale Research* (Stanford, CA: Stanford University Press, 1999).

18. Dajōkan 260, 1.4.24 (May 19, 1868) and 366, 1.5.3 (June 22, 1868), in *Hōrei Zensho* 法令全書, 139 vols. (Tokyo: Naikaku Kanpōkyoku, 1887–1912).

19. Nakano Hatayoshi 中野幡能, *Hachiman Shinkō* 八幡信仰 (Tokyo: Hanawa Shobō, 1985), 10–11.

20. Scholars such as Motoori Kiyozō dismiss this problem by suggesting that the original place name has been lost. Motoori Toyokai 本居豐穎 and Motoori Kiyozō 本居清造, *Motoori Norinaga Zenshū* 本居宣長全集 (Tokyo: Yoshikawa Kōbunkan, 1926), 5:122. (As cited in Bender page 175)

21. This first reference to a Hachiman shrine is suggestive insofar as it relates to the protection of the kingdom. But in treating this as evidence for the early state-protecting role of Hachiman, most scholars seem to ignore the fact that the Hachiman shrine is not singled out but is instead one of several shrines where offerings were made. Thus, it seems here to be nothing more than one of many equivalent apotropaic institutions that received offerings in a crisis. In the historical record, the first reference to Hachiman was made in connection to the Usa Shrine in Buzen Province in Kyushu. According to archeological and documentary evidence, this region of Buzen Province seems to have been associated with Buddhism from sometime early in the sixth century. Nishida Nagao 西田長男, *Shintōshi No Kenkyū* 神道史の研究 (Tokyo: Risōsha, 1957), 2:162; Christine Kanda, *Shinzō: Hachiman Imagery and Its Development* (Cambridge, MA: Harvard University Press, 1985), 37. For the importance of jewels in the symbolism of Hachiman, see Bernard Faure, *The Fluid Pantheon: Gods of Medieval Japan* (Honolulu: University of Hawai'i Press, 2015), 1:281–285.

22. Nakano, *Hachiman Shinkō*, 134–135.

23. This history is not particularly unique, however, insofar as many Shinto deities turn out to have associations with non-Buddhist Indian gods referred to in imported sutras. Thus, Shinto functions in some cases like an indigenized "Hindu" remainder. See Iyanaga Nobumi 彌永信美, "Honji Suijaku and the Logic of Combinatory Deities: Two Case Studies," in *Buddhas and Kami in Japan: Honji Suijaku as a Combinatory Paradigm*, ed. Mark Teeuwen and Fabio Rambelli (New York: Routledge Curzon, 2003), 145–176.

24. See also Miyaji Naokazu 宮地直一, *Hachimangū No Kenkyū* 八幡宮の研究 (Tokyo: Risōsha, 1956), 136.

25. Miyaji, *Hachimangū No Kenkyū*, 140.

26. Ross Bender, "Metamorphosis of a Deity: The Image of Hachiman in Yumi Yawata," *Monumenta Nipponica* 33, no. 2 (1978): 165–178.

27. For a more about this, see Jason Ānanda Josephson [Storm], *The Invention of Religion in Japan* (Chicago: University of Chicago Press, 2012), esp. 98–102.

28. See Ohsawa Yukio 大澤幸生 and Nishihara Yoko 西原陽子, *Innovator's Marketplace* (New York: Springer, 2012), 33–34. I fixed the romanization of Masanao's surname.
29. See Sidney Brown, "Kido Takayoshi and the Meiji Restoration: A Political Biography, 1833–1877" (PhD thesis, University of Wisconsin, 1962); Sidney Brown, "Kido Takayoshi (1833–1877): Meiji Japan's Cautious Revolutionary," *Pacific Historical Review* 25, no. 2 (1956): 151–162; Albert Craig, *Choshu in the Meiji Restoration* (Cambridge, MA: Harvard University Press, 1978); and Momiji Saitō 齊藤紅葉, *Kido Takayoshi to Bakumatsu Ishin* 木戸孝允と幕末・維新 (Kyoto: Kyōtodaigaku Gakujutsu Shuppankai, 2018).
30. See Josephson [Storm], *The Invention of Religion in Japan*.
31. See Chūta Tsumaki 妻木忠太, ed., *Kido Takayoshi monjo* 木戸孝允文書 (Tokyo: Nihon Shiseki Kyōkai, 1929).
32. When William Moore arrived in Japan, he had a list of contacts provided for him by Edison. The one with the highest profile was Itō Hirobumi. Itō suggested to Moore that the best bamboo in Japan was probably located in Kyoto, and he sent Moore there with the recommendation that he speak to the governor of Kyoto.
33. Hayashiya Tatsusaburō 林屋辰三郎, ed., *Kyoto no rekishi* 京都の歴史 (Tokyo: Gakugei shorin, 1974), 7:524–545; Martin Collcutt, "Buddhism: The Threat of Eradication," in *Japan in Transition: From Tokugawa to Meiji*, ed. Marius Jansen and Gilbert Rozman (Princeton, NJ: Princeton University Press, 1986), 143–167; Patricia Jane Graham, *Faith and Power in Japanese Buddhist Art, 1600–2005* (Honolulu: University of Hawai'i Press, 2007), 199–200.
34. See Steven Christopher Bullard, "Celebrating Kyoto, 1895," (PhD thesis, Australian National University, 2003), 43–44, 69–70.
35. Kido Takayoshi 木戸孝允, *Kido Takayoshi Nikki* 木戸孝允日記 (Tokyo: Matsuno Shoten, 1996), 1:282.
36. Iida Takesato 飯田武郷, *Hōshitsushū* 蓬室集 (Tokyo: Iida Suehara, 1903), 272–273. I also translate the same passage in Josephson [Storm], *The Invention of Religion in Japan*, 160.
37. See Josephson [Storm], *The Invention of Religion in Japan*.
38. See Jason Ānanda Josephson Storm, *The Myth of Disenchantment: Magic, Modernity and the Birth of the Human Sciences* (Chicago: University of Chicago Press, 2017).
39. For how a mystical reading sometimes went both ways, see Jason Ānanda Josephson Storm, "The Mystical 'Occident' or the Vibrations of Modernity in the Mirror of Japanese Thought," in *Spirits and Animism in Contemporary Japan: The Invisible Empire*, ed. Fabio Rambelli (London: Bloomsbury, 2019), 29–44.
40. B. C. Forbes, "Edison Working on How to Communicate with the Next World," *American Magazine*, October 1920, 10–11, 84–85.
41. Thomas Edison, *The Diary and Sundry Observations of Thomas Alva Edison*, 1st printing (New York: Greenwood, 1948), 235.
42. See Josephson Storm, *Metamodernism*, 154–162.
43. Quoted in Neil Baldwin, *Edison: Inventing the Century* (Chicago: University of Chicago Press, 2001), 95.
44. For occult and other monist movements, see Jason Ānanda Josephson Storm, "Monism and the Religion of Science: How a German New Religious Movement Birthed American Academic Philosophy," *Nova Religio* 25, no. 2 (2021): 12–39.

45. For the role theosophy played as a vanished mediator in the history of the disciplinary formation of religious studies, see Jason Ānanda Josephson Storm, "A Theosophical Discipline: Revisiting the History of Religious Studies," *Journal of the American Academy of Religion* 89, no. 4 (2021): 1153–1163.
46. J. M. Peebles, *What Is Spiritualism? Who Are These Spiritualists? and What Can Spiritualism Do for the World?* (Battle Creek, MI: Peebles, 1910), 112.
47. See Josephson Storm, *The Myth of Disenchantment*.
48. Henry Steel Olcott, *Old Diary Leaves: First Series* (Adyar, India: Theosophical Publishing House, 1974), 466–468.
49. Tara N. Doyle, "Liberate the Mahabodhi Temple! Socially Engaged Buddhism Dalit-Style," in *Buddhism in the Modern World*, ed. Steven Heine and Charles Prebish (Oxford: University of Oxford Press, 2003), 259.
50. David Nye, *The Invented Self: An Anti-Biography from Documents of Thomas A. Edison* (Odense, Denmark: Odense University Press, 1983).
51. Bullard, "Celebrating Kyoto," 43–44; Akamatsu Toshihide 赤松俊秀 and Yamamoto Shirō 山本四郎, eds., *Kyōto-fu no rekishi* 京都府の歴史 (Tokyo: Yamakawa Shuppansha, 1982), 246.
52. Ohsawa and Nishihara, *Innovator's Marketplace*, 33–34.
53. One such example is Yoshiro Okabe 岡部芳郎, the Japanese engineer who was crucial to building the Edison monument in 1934.
54. Helen Hardacre, *Shinto: A History* (New York: Oxford University Press, 2016), 174.
55. Gananath Obeyesekere, *The Apotheosis of Captain Cook* (Chichester, UK: Princeton University Press, 1992), 3.
56. Obeyesekere, *The Apotheosis of Captain Cook*, 19.
57. For this example and many more, see Anna Della Subin, *Accidental Gods: On Men Unwittingly Turned Divine* (New York: Metropolitan, 2022).
58. N. Y. Louis Lee and P. N. Johnson-Laird, "Are There Cross-Cultural Differences in Reasoning?," in *Proceedings of the 28th Annual Conference of the Cognitive Science Society*, ed. Ron Sun and Naomi Miyake (Wheat Ridge, CO: Cognitive Science Society, 2006), 459–464; Philip Johnson-Laird, *How We Reason* (Oxford: Oxford University Press, 2011); Jean-Baptiste Van der Henst, Yingrui Yang, and P. N. Johnson-Laird. "Strategies in Sentential Reasoning," *Cognitive Science* 26, no. 4 (2002): 425–468.
59. See Ronald Numbers, ed., *Galileo Goes to Jail: And Other Myths About Science and Religion* (Cambridge, MA: Harvard University Press, 2009).
60. See Storm, *Metamodernism*, 49–84.
61. See Storm, *Metamodernism*, 247–253.
62. Storm, *Metamodernism*, 29–47.
63. Storm, *The Myth of Disenchantment*, 59–60. I also discuss similar developments in the German-speaking world.
64. Peter Harrison, *The Territories of Science and Religion* (Chicago: University of Chicago Press, 2015).
65. See Julie Reuben, *The Making of the Modern University: Intellectual Transformation and the Marginalization of Morality* (Chicago: University of Chicago Press, 1996).
66. For other examples, see Storm, *The Myth of Disenchantment*.

67. Knowledge is itself a social kind, and epistemology continues to shift.
68. Storm, *Metamodernism*.
69. Tanizaki Jun'ichiro 谷崎 潤一郎, *In'ei raisan* 陰翳礼讚 (Tokyo: Paintanashonaru, 2018). Tanizaki would also remind us that even electric light can be racialized or at least impregnated with notions of Western chauvinism, although Tanizaki could himself be criticized for restaging an auto-orientalist binary.

CHAPTER 7

QUESTIONING THE SACRED COW

Science, Religion, and Race in the United States and India

CASSIE ADCOCK

Sacred cow:

1. The cow as an object of veneration amongst Hindus.
2. figurative (originally U.S.).
   a. Journalism. (a) someone who must not be criticized; (b) copy that must not be altered or cut.
   b. An idea, institution, etc., unreasonably held to be immune from questioning or criticism.[1]

That the sacred cow is integral to Hindu religion is an idea virtually immune from questioning. In the United States, many people still know—or think they know—little about India beyond Hinduism and little about Hinduism beyond "caste, cows and karma."[2] In its narrower sense, the phrase *sacred cow* references a Hindu taboo against killing cattle or eating beef. The phrase also captures the fanatical spirit with which proponents of the Hindu Right in India today profess to defend national cultural and religious institutions, not least the cow, from perceived attack.[3] Since 2015, self-proclaimed "cow protectors" in India have perpetrated egregious acts of violence against their fellow citizens. With a Hindu nationalist government in power at the center and antislaughter laws tightened in many states, cow protection has provided a pretext for Hindu vigilantes to intimidate, assault, and murder Muslims, Dalits, and others. Commentary on this politics by members of the international press—including U.S. journalists—habitually leads with the well-known fact of Hindu veneration of the sacred cow. Hindu reverence for the cow provides a ready explanation for cow protection; political developments in India reinforce the "caste-cows-and-karma" understanding of Hinduism. And so the sacred cow stands unquestioned.

The phrase *sacred cow* derives from a colonial history. As the *Oxford English Dictionary* teaches us, perhaps the earliest use of the phrase in print can be attributed to the father of the famous British imperialist writer Rudyard Kipling. Kipling turned the phrase in India, in 1892, amidst a rising swell of cow protectionist agitation.

Addressing the Indian politics of cow protection, officials in British colonial India, too, posited the sacred cow as a certainty—an unquestionable fact of Indian religion and culture. Colonial officials portrayed cow protection as self-evidently rooted in a Hindu religious interdiction against killing the sacred cow. As a religious interdiction, they understood cow protection to be intractable and absolute, dead set on a total stop to cattle killing in India.

In the eyes of British officials and experts in colonial India, another incontrovertible fact was that Hindus' commitment to preserving the sacred cow was contrary to the laws of nature and economy. This fact was enshrined in Indian politics after independence from British rule. In 1948, the National Planning Committee reported that the Republic of India suffered from a large number of useless cattle that consumed food better reserved for more productive animals. The committee pinpointed religious sentiment against cow killing as the cause of the problem—and the chief obstacle to its resolution. It noted that, on account of this religious sentiment, "on the whole the yield and utility of the enormous cattle population of the country goes on, instead of improving, progressively deteriorating."[4] The committee concluded, in short, that the sacred cow posed an obstacle to the development of the country. When Hindu nationalists endorsed the identification of cow protection with a total and inflexible interdiction on cow slaughter, these arguments became associated with secularist opposition to cow protection. And so colonial common sense lives on, in India as it does in the United States.

This chapter challenges common sense to pose the sacred cow as a question. It approaches the sacred cow not as certainty or established fact but as a figure of British colonial discourse. The *colonial sacred cow* purported to name a Hindu religious interdiction against killing the cow that defied reason.[5] Crucially, this chapter treats the imputed opposition to the laws of nature or economy as intrinsic to the sacred cow. Building on the insight that religious and secular are mutually constitutive categories, it takes the constant reference to immutable religion as a provocation and asks how changing configurations of nature's laws re-formed the sacred cow of colonial discourse.[6] What had appeared to be an intrinsic feature of the Indian subcontinent when regarded as an essentially religious icon reveals its transnational scope and heterogeneous provenance when encountered through a critical history of science and religion.

The sacred cow appears to be an essentially religious, Hindu icon when viewed through the lens of cow protection. To loosen the hold of cow protection on our historical understanding, this essay looks backward to the nineteenth century, and forward to postcolonial India, from a starting point in the 1920s. In much of the subcontinent, and certainly in north India, draft cattle were vital to all phases of agricultural production: clearing land, preparing soil, irrigating crops, harvesting, threshing, and milling. In the 1920s, colonial discourse of the sacred cow was unyoked from the politics of cow protection and put to work in debates over agricultural development. From the vantage point of the 1920s, it becomes clear that the colonial sacred cow constellated themes of (excess) population, (bad) breeding, and (limited) territory that had global circulation. The sacred cow was put to work in debates over poverty and imperialism, culture and eugenics, feeding and breeding; not tethered to the subcontinent, it ranged from the politics of white supremacy in the United States to the politics of Hindu extinction in India. Sharpening our picture of Hindu nationalism, a critical history of the sacred cow uncouples the contemporary politics of cow protection from essentialist understandings of Hinduism.

SACRED COWS OF COLONIAL DEVELOPMENT

Already in the nineteenth century, the colonial sacred cow was a figure of Hindu religion in contrast with rational government. By the 1920s, the sacred cow signified the intractable cultural resistance of (Hindu) India to the scientific husbandry of India's material resources by the British colonial state.

The question of Indians' poverty had now been the focal point of public debate for decades, with the British government posing as its solution, while Indian nationalists decried colonial rule as its cause. Moreover, Indian nationalism was in full swell, having grown into a mass movement of protest and organized action. The magnitude of the effort required to justify British colonial rule in India as an ameliorative force is evident in the 1928 *Report* of the Royal Commission on

Agriculture in India (RCAI). A "textual monument to empire," the multivolume *Report* was designed to make "agrarian India appear as a field for progress and as a production system under scientific management" by the British colonial state.[7] The RCAI weighed in on the condition of Indian cattle in the context of its claims about "agriculture and rural economy" in India and "the promotion of the welfare and prosperity of the rural population."[8]

The 1928 *Report* portrayed the sacred cow as a direct impediment to the improvement initiatives of the colonial state. Deploring "the miserable condition of so many" Indian cattle, the RCAI invoked the determining influence of the sacred cow: "The religious veneration accorded to the cow by the Hindu is widely known. To at least half of the population of India, the slaughter of the cow is prohibited, and this outstanding fact governs the whole problem of the improvement of cattle in this country."[9] The "problem" of cattle improvement was said to be, in brief, an oversized population of cattle of inferior quality—a result, the *Report* implied, of Indians' aversion to killing or culling. Indian cultivators' indifferent approach to animal husbandry was said to have caused "progressive deterioration" in the quality of stock—a bovine population of visibly reduced weaklings—as more and more cattle competed for limited feed. The RCAI determined that improved "feeding and breeding" was the solution.[10]

In vowing to undertake the improvement of India's cattle, the RCAI expressed the colonial government's grim determination to shoulder an almost Sisyphean task: "The process having gone so far, India having acquired so large a cattle population and the size of the animals in many tracts having fallen so low, *the task of reversing the process of deterioration and of improving the livestock of this country is now a gigantic one*; but on improvement in its cattle depends to a degree that is little understood the prosperity of its agriculture and *the task must be faced*."[11] The agricultural advisor to the government of India used the idiom of accumulated time to stress the scale of this task: "the systematic improvement of Indian cattle, by the gradual repair of *the results of centuries of neglect*, is a most formidable undertaking."[12] Expressing confidence in the scientific expertise of the British government in India, the RCAI *Report* performed the determination of the colonial state to overcome the nearly insurmountable obstacles of Hindu religious tradition and Indian lethargy.

The RCAI *Report* posited the sacred cow as a perennial feature of (Hindu) Indian culture, and it attributed the impoverished state of agriculture in India to the cumulative effect of centuries of Indians' refusal to cull their cattle. In this

1920s iteration, the sacred cow was an obstacle to scientific agriculture. Hindu religious sentimentality was described as having reduced Indian cattle to a particularly backward condition, and British government, founded in scientific know-how and expertise, was described as necessary to redress the situation.

HINDU RELIGION AS ANTISCIENCE

The RCAI *Report*'s conflation of all Indians with Hindus was not a reflection on India's demographics. Instead, it reflected a characteristic bias in British colonial accounts, which, beyond presuming that religion was the primary social identifier on the subcontinent, were often vague on details. When imposing colonial rule in the mid-eighteenth century, the British had portrayed the ruling powers as Muslim tyrants and the populace as Hindu, the better to pose as liberators. Even after documenting the enormous diversity of the subcontinent, the British often found this shorthand formula useful. But it was not long before they portrayed colonial rule as liberating Hindus from themselves—or, more precisely, from the material consequences of Hinduism, represented as an impossibly spiritual, impractical religious tradition that rendered Indians powerless against nature's onslaught.[13]

British imperialist discourse between 1880 and 1930 commonly portrayed colonized populations as terminally out of step with science and nature and therefore destined to be either governed or supplanted by the British.[14] In India, British "good government" was depicted as inevitable, or even providential, for helplessly otherworldly (Hindu) Indians. Signifying a Hindu religious interdiction on killing cattle that was the opposite of good, rational practice, the colonial sacred cow drew together these powerful themes in British imperialist discourse. If the sacred cow's signification has seemed to remain constant—since the nineteenth century, it has continued to stand for a Hindu religious interdiction on cow slaughter—its contours have significantly altered to accommodate the different oppositional work it was called to do.

The British first summoned the sacred cow, around 1893, to counter the new Indian politics of cow protection. Nineteenth-century cow protectionists protested the unrestricted slaughter of cows under British rule among a variety of colonial conditions that, in their view, were harming cattle—including shrinking grazing grounds, high land revenue, and unregulated export trades. In so doing, they protested the effects of colonial rule on the Indian people, for cow protectionists made a direct connection between impoverished cattle and Indians' poverty, hunger, and mortality.[15] By figuring the sacred cow as a Hindu religious taboo that did not answer to reason, the British dismissed cow protectionists' material charges against colonial government. They also portrayed Hindu attitudes toward the sacred cow as opposed to nature's laws—and by the end of the nineteenth century, nature's laws paired Thomas Malthus and Charles Darwin, population and natural selection.[16]

The RCAI *Report*'s account of the material consequences of attachment to the sacred cow was premised on a simple point: India's cattle population outstripped food supply. Repeated down the decades, this Malthusian critique was as characteristic of the sacred cow as was its imputed religious nature. Malthus had asserted it to a be a law of nature that animals reproduced at a more rapid (geometrical) pace than the fixed and steady (arithmetical) rate of vegetal growth, so that there was a natural "tendency for population to outstrip subsistence."[17] He portrayed natural calamities such as famine and epidemic disease as serving the positive function of righting the balance between human or animal populations and food supply.

In the nineteenth century, the British marshaled Malthusian reasoning to justify the colonial policy on cattle slaughter as a positive check that trimmed the excess population of Indian cattle. An early example was *The Cow Question in India*, which was published in India in 1894. Evidently composed in answer to Indian cow protection, *The Cow Question* presented an extended refutation of cow protection's putative aim to put a total stop to all cattle slaughter in India.[18] Front and center among its "reasons why the slaughter of oxen and cows should be permitted as in . . . modern Europe" was the Malthusian argument that "all living cows require food," whereas the "land can maintain only a certain number of cattle."[19] Indians' putative refusal to slaughter their cattle logically implied a superfluity of "useless" cattle whose maintenance must come at "enormous expense" to the country.[20]

By positing the existence of an excess population of Indian cattle, nineteenth-century British colonial officials and experts justified cattle slaughter in the name of scientific government, even as they portrayed Hindu religious sentiment for the sacred cow as fundamentally opposed to nature's laws. This embedded Malthusian perspective made the colonial sacred cow available to British reflections on economics.

Malthus had published his *Essay on the Principle of Population* in 1803[21] as an argument in political economy—specifically, as a case against providing material assistance to the English poor. He claimed that poor relief would "[stimulate] population growth without promoting any corresponding expansion in agricultural production," ultimately increasing suffering and hunger.[22] Married to the Darwinian law of natural selection, the Malthusian law of population facilitated arguments in India about the scientific breeding and feeding of livestock. The Malthusian critique constitutive of the colonial sacred cow was reworked and repurposed down the decades until, by the 1920s, it supported the portrait of Indians' poverty as a direct consequence of centuries-old habits so far from in tune with nature's laws that they led directly to degeneration and decline.

After around 1910, the sacred cow was regularly deployed in the field of agricultural economics to argue that India's excess cattle caused active harm to the cattle population at large. Just as Malthus had contended that poor relief would mean taking food from industrious members of society, colonial officials now asserted that Indian cultivators' vast number of useless animals consumed the food of productive livestock.[23] The point was still repeated a decade later: "A remarkable feature of the supply of cattle is the vast number of old and worn out animals. As the cow is a sacred animal . . . , it is not possible to use these animals for food. . . . A severe drain is in this way imposed on the slender fodder resources of the country."[24] Hindus' religious feeling for the sacred cow was said to yield an excess population so great that any reduction in their number must constitute an improvement to Indian livestock and agriculture.

As indicated by the RCAI's 1928 *Report*, by the 1920s the problem of breeding—or population quality—was given prominence alongside that of feeding—or population size. "Two great obstacles in raising the standard of the work cattle in India must be faced at the outset. The cow is a sacred animal. In the improvement of the breed by modern selection methods, there is therefore no method available, as in Europe, for the disposal of individuals which fall below a certain standard."[25] Experts had sounded this Darwinian note in 1911, citing "the impossibility of eliminating the unfit owing to the religious objections of Hindus to kill cattle" as one of the main obstacles to progress in cattle improvement.[26] The RCAI *Report* described the poor condition of Indian agriculture as the result of a centuries-long, ongoing process of breed deterioration, which British experts could only gradually reverse and repair. If the amelioration of Indians' impoverished condition under colonial rule was imperceptibly slow, the *Report* implied, the fault lay not with British government but with the heavy drag exerted by Indian culture itself, as accumulated over centuries in the inherited qualities of Indian livestock.

QUESTIONING COLONIAL IMPROVEMENT

The colonial sacred cow posited a radical contrast between colonial science and Hindu religion and—by the 1920s—between colonial progress and Indian decline. Attributing the stalled development of Indian agriculture to Hindu religious tradition, the colonial sacred cow effectively erased more proximate historical change, including the disruptive and often unintended effects of colonial "improvements." After more than a century of colonial rule, the RCAI *Report* portrayed British government as poised to close the book on India's benighted tradition of animal husbandry and to open a new era of science and prosperity. Yet colonial science did not stand outside the sacred cow. Colonial science had a hand in producing the cattle decried by the RCAI *Report*: by the late 1920s, Indian cattle certainly embodied colonial science. And whereas the sacred cow figured an unbridgeable divide between British science and Indian religion, a closer look at cattle improvement in India shows colonial science joined with Hindu tradition.

The colonial sacred cow of the 1920s provided a Malthusian diagnosis of flawed Hindu tradition that exempted the British in India from displaying their vaunted skills in the breeding of improved livestock.[27] The RCAI *Report* described the relationship between land and cattle in India as critically unbalanced—a "vicious circle" very difficult to break. That vicious circle conjoined breeding and feeding: the *Report* portrayed Indian cultivators as fanatical in their opposition to culling but apathetic about caring for their cattle.[28] Yet even as colonial experts faulted Indian culture, they acknowledged another source of the imbalance between land and cattle: grazing land had shrunk.[29]

Colonial improvement privileged settled agriculture. In north India throughout the nineteenth century, the colonial government pushed to expand cultivation through the construction of monumental irrigation canals, the clearing of forest wilds, and the settling of itinerant people. Agricultural improvement transformed the landscape—but with many uncounted costs.[30] By 1910—just as agricultural economists were arguing that excess cattle were damaging India's cattle population—British experts in Punjab and the United Provinces were reporting with some alarm that Indian cattle breeding was in crisis, and cattle supply was threatened.[31]

Arid lands that colonial improvement turned into garden had been breeding grounds for cattle; forests had provided important sources of seasonal fodder. Many professional, itinerant cattle breeders, who had formerly reared large herds of animals, were abandoning the trade. Surveys brought news of the extinction of highly valued local varieties of cattle and the threatened extinction of others. Whatever the recognized importance of cattle for Indian agriculture, for decades the colonial government had given them little consideration. Far from a target of British science, initiative, and expertise, cattle had been omitted from government plans.[32] Agricultural improvement had disrupted the conditions of rearing and maintaining agricultural cattle in many parts of north India.

For colonial officials, these conditions manifested as economic problems: high prices and high costs. The colonial sacred cow of the 1920s foregrounded the deterioration or negative breeding of Indian cattle, which was portrayed as the result of Indian cultivators' irrational religious sentiment with regard to culling their animals and their apathy about caring for them. But in north India in the years around 1910, the problem of breeding was a different one: that of inducing Indian cultivators to breed or rear their own cattle at all. Far from maintaining excess animals, many cultivators had abandoned cattle rearing as too costly. Now, breeding grounds worked by professional cattle breeders were overdrawn, and livestock prices were too high. The solution, colonial experts concluded, was to induce Indian cultivators to rear their own replacement animals. But how could cattle rearing be made economical? The cost of feeding cattle was high. If farmers were to be induced to divert their land and labor, cattle rearing had to be made more profitable.

Colonial experts envisioned a solution through scientific breeding—the engineering of a new, "improved" breed with enhanced productive qualities. Among most Indian varieties, cows were "mothers of bullocks," and they produced scant milk—enough to feed their offspring. Colonial experts determined that the female cow had to pay her way. Whereas mature working bullocks were more or less indispensable, cows now had to be engineered to produce milk for sale to cover the cost of rearing the next generation of work animals. From around 1910, many government experts advocated a breeding strategy that would generate a new, "dual-purpose breed" of Indian cattle from indigenous stock.[33] Indian cow protectionists embraced the prospect of this new cow, whose economic value might obviate her sale for slaughter, and partnered with underfunded government scientists to disseminate this improvement through the countryside. The improved animals were unpopular with farmers, although this did little to temper the professed enthusiasm of either government policy makers or cow protectionists.[34] And because colonial science was scarcely invested in the welfare of the Indian population,[35] the scale of this effort was small. But the point remains: even as scientific breeding was trumpeted as the colonial antidote to the degenerative effects of Indian tradition, government scientists were allied with upper-caste Hindu cow protectionists in the effort to upgrade the cattle population.

In 1928, the RCAI erased this history when it posited that British scientific expertise would be needed to liberate Indians from the material effects of Hindu religious sentimentality. Attributing the causes of Indians' suffering to static Hindu tradition, the colonial sacred cow named scientific breeding as the solution to Indian poverty. But a critical history of science and religion tells a different story: The sacred cow was an amalgam of transnational discourses centered on population, class, and race.

SACRED COWS OF RACE AND CULTURE

On the whole, the colonial state did little to implement the RCAI *Report*'s recommendations for improving Indians' working cattle. Of far more immediate value was the *Report*'s detailed accounting of the Indian impediments to colonial science. In colonial discourse of the 1920s, the sacred cow was the icon of a vicious circle of starvation and degeneration that impeded British progress in India; it merged Malthusian configurations of population, class, and feeding with new ways of talking about race and breeding. The sacred cow encapsulated British attitudes toward Indians' cattle; it also reflected colonial attitudes toward Indians.[36] Official accounts of the poor condition of Indian cattle replicated government responses to the poverty and suffering of its human subjects.

The 1928 RCAI *Report* drew on new understandings of population growth—human population growth—as a government problem. In colonial India, excessive population growth was cited as a reason to delay the introduction of government improvements, whether in public health or in agriculture.[37] One of the most influential public health officials articulated the new government reasoning: "until [Indians] can be induced to restrict their rate of reproduction there is no hope of doing much good by medical relief or sanitation, as the population is very nearly up to the possible limit."[38] Officials contended that government improvements would only cause population growth to outpace food production. In similar tones, the RCAI *Report* stated that the impossibility of feeding an oversized cattle population was the primary obstacle to the scientific improvements of livestock that could remedy India's agricultural economy. By the close of the 1920s, Indians' unrestrained fertility was described as an obstacle staying the improving hand of government. Colonial discourse of the sacred cow refracted official attitudes toward the human colonized in arguments about Indian livestock.

The colonial sacred cow kept pace with changing official discourses about human poverty and mortality. In the late nineteenth and early twentieth centuries, recurring famines in India, accompanied by extraordinary losses of human life, were met with "official complacency" and written off as merely "Nature's ordinary programme for the restriction of excess population."[39] In these years, officials were little exercised about famines and epidemics that held Indian population in check. When fodder famine struck agriculturalists' cattle—as it did in the United Provinces in 1905–1906—officials justified inaction in similar terms, contending that famine served the Malthusian function of culling worthless animals. Official discourse began to shift in the 1920s. Just as the RCAI spoke of progressive deterioration in the quality of Indian cattle, the 1931 *Report on the Census of India* expressed concern that "the wrong sorts [of people] were reproducing"—namely, "poor agricultural labourers" who "were unable to contribute their share to India's productivity."[40] Among humans as among cattle, after 1920 colonial officials in India worried that "useless" members of the population were consuming too great a share of limited food.

Far from being an unchanging feature of the cultural landscape of the subcontinent, the sacred cow participated in a "shared language" with "phenomenal" transnational circulation and currency: the language of eugenics.[41] Eugenics was characterized, above all, by an interlocking set of concerns with the quantity and quality of human populations, territory, and national welfare. Francis Galton, who gave eugenics its name, drew together the Malthusian notion of excess population and the Darwinian theory of natural selection to advocate active control over the natural cycles of reproduction and evolution in human beings. Galton's eugenics consisted in "bringing no more individuals into the world than can be properly cared for, and those only of the best stock."[42] It therefore aimed to regulate human reproductive habits: "to check the birth-rate of the Unfit, instead of allowing them to come into being."[43] But eugenics was as likely to be concerned with puericulture—the manner of rearing offspring—as with biological inheritance. Eugenic arguments connected poverty to the inherited habits of the poor, to their genetics, or to both.[44]

Eugenic perspectives on class and race were often mixed, transposable, and mutually reinforcing. In twentieth-century Britain, eugenics addressed itself to the population increase of "paupers" relative to the middle and upper classes. Poor Laws or sanitation were seen as worsening the national poverty problem by "subsidiz[ing] reproduction only of the lowest social types."[45] Malthusian understandings of "the poor as irresponsible breeders, fecund beyond their limited resources," traveled colonial circuits, feeding eugenic assumptions that "inferior races bred at a higher rate."[46] In India, eugenics arguments focused on differential fertility among the poor; in the United States, they warned of white race suicide, stoking fears that white America would be overrun by Asian population explosions.[47]

In the 1920s, colonial discourse of the sacred cow replicated and reinforced wide-ranging eugenic arguments about how to ameliorate social problems through the designation and reproductive management of "fit" and "unfit" human beings. To drive home this point, we cross the Atlantic, to where the sacred cow—untethered—was put to work in debates on race and immigration in the United States.

SACRED COWS OF EUGENICS

Nearly simultaneously with the RCAI *Report* in India, another statement on the sacred cow was published in the United States: Katherine Mayo's *Mother India*. *Mother India* is perhaps the foremost example of eugenic writing about India in the English language. Its publication sparked a scandal "across three continents" that reverberated in British and American imperialist politics, Indian nationalist and women's movements, and anti-immigrant politics in the United States.[48] For ten years after its publication, "Mayo shared the world's spotlight on India only with M. K. Gandhi," who aptly dubbed her book a "drain inspector's report."[49] *Mother India* remained a key source of common-sense knowledge about India in the United States well into the 1950s, second only to Rudyard Kipling.[50]

Mayo's *Mother India* is most infamous for its disparaging portrayal of Hindus' "oversexed" reproductive culture;[51] it is not known for its treatment of the sacred cow. Still, this far-fetched American account renders explicit the homology between Indian cattle and Indian people in the British colonial discourse of the sacred cow. The four chapters devoted to the sacred cow in *Mother India* mirror both the RCAI *Report* and Mayo's own account of Indians' sexual habits, demonstrating the colonial sacred cow's affinity with expressly eugenic claims about the unfitness of Hindu culture.

Authored not by a commission of British experts speaking in the name of neutral, objective scientific knowledge but by an American journalist known for the unabashed imperialism of her views, *Mother India* made the same points as the RCAI in strident tones. The first of these chapters began:

> "Why, after so many years of British rule, is India still so poor?" the Indian agitator tirelessly repeats.
>
> If he could but take his eyes from the far horizon and direct them to things under his feet, he would find an answer on every side, crying aloud for honest thought and labor.
>
> For example, the cattle question, by itself alone, might determine India's poverty.
>
> India is being eaten up by its own cattle. And even at that the cattle are starving.[52]

Engaging the central refrain of the Indian nationalist movement, Mayo pointed the finger at Indians' own bad management as the real source of Indian poverty and championed British good government, paired with British science, as its solution. Amplifying colonial claims that at least "50 per cent" of India's enormous cattle population was "unprofitable" or "uneconomic," Mayo quickly invoked religious sentiment as the root cause: "in the Hindu mind nothing is so deep-rooted as the sanctity of the cow."[53] Like the RCAI *Report*, Mayo's account began with an oversized population of cattle of poor quality, attributing this to (Hindu) Indians' religious aversion to killing or culling. In *Mother India*, however, the description of Indian cattle reinforced the eugenic argument elaborated in the preceding chapters: "The whole pyramid of the Hindu's woes, material and spiritual—poverty, sickness, ignorance, political minority, melancholy, ineffectiveness . . . rests upon a rock-bottom physical base. This base is, simply, his manner of getting into the world and his sex-life thenceforward."⁵⁴ Mayo painted a garish portrait of Indians' customs of excessive and ungoverned fertility, reflected both in their own material conditions and in those of their livestock.

Where the RCAI spoke rather subtly of progressive deterioration in India's cattle, *Mother India* baldly asserted that "the general conditions under which Indian animals have lived and propagated might have been specially devised for breeding down to the worst possible type."⁵⁵ Like the RCAI *Report*, *Mother India* portrayed a conflict between Hindus' lofty spiritual purposes and their disastrous disregard for practical management. Echoing the RCAI's conclusions, *Mother India* decried Indians' neglectful feeding and inattentive breeding. Indian cattle were languishing because "callous" and "unintelligent" Indian cultivators failed to provide for them: "They have raised food for themselves, but they will not raise food for their mother the cow."⁵⁶ The village stud bull was an invariably "feeble" animal that "serve[d] as sire to a neighborhood herd. Straying together, starving together, young and old, better and worse, the poor creatures mingle and transmit to each other and to their young manifold flaws and diseases."⁵⁷ Mayo's account of the inferior and reproductively overworked male and the neglected and suffering female duplicated *Mother India*'s account of the effects of Hindu marriage practices—the Indian male who "from immaturity or from exhaustion . . . has small vitality to transmit" and the Hindu woman who "because of her years and upbringing and because countless generations behind her have been bred even as she, . . . is frail of body"—as anti-eugenic.⁵⁸

The duplicative logic that rendered Indians and their cattle virtually interchangeable was signaled in the book's title, which was a tearing down of the nationalist ideal of Bharat Mata or Mother India, the personified Indian nation. Mayo plunged this lofty ideal into the mire of a paradox: "even as Indian nationalists say Mother India is an ideal worth dying for . . . actual Indian mothers are dying."⁵⁹ Mayo's account of the sacred cow played to the same paradox: although venerated, mother cow was "starving"; her signature "tranquil, far-off gaze" a sign not of contentment but of "low vitality."⁶⁰ With regard to either "mother"— Mother India or Mother Cow, Bharat Mata or Gau Mata—Mayo portrayed Indian culture negatively through a eugenic lens.

Also like the RCAI *Report*, *Mother India* portrayed British government in India as engaged in "the gradual repair of *the results of centuries of neglect.*" While Indians whingingly blamed the British, Mayo described the British as striving manfully: "by selective breeding, by crossing, and by better feeding and housing, slowly and steadily the results of centuries of inbreeding, starvation, infection, and of breeding from the worst are being conquered."[61] This portrait of the progressive, improving labor of the British stood in marked contrast to Mayo's portrait of the abject incapacity of Indians: "Given men who enter the world physical bankrupts out of bankrupt stock, rear them through childhood in influences and practices that devour their vitality ... need you ... seek for other reasons why they are poor and sick and dying and ... too weak ... to hold the reins of Government?"[62] Mayo's sacred cow justified British imperialism in India in blatantly racialist terms.

Like the RCAI, Mayo posited colonial rule—under the scientific government of the British—as indispensably necessary for a people whose dysgenic traditions made them unfit to manage their own affairs. The condition of Indian cattle, which mirrored that of the Indian people and derived from the same cause, was made to illustrate the solution: better breeding. The degenerative tendencies of Indian culture required a eugenically superior solution: British rule.

Mayo's imperialist, racist argument in *Mother India* was also informed by the specter of dysgenic reproduction in the United States. Mayo put the sacred cow to work for white supremacy in U.S. immigration politics. Mayo was a staunch opponent of the 1926 Hindu Citizenship Bill, which proposed to extend citizenship rights previously denied to Indian immigrants, and she promoted *Mother India* as a source of information about "what sort of American citizen ... the British Indian [would] make."[63] Animated by fears of miscegenation, *Mother India* was also a warning to white women against Indian men. Mayo spoke for white, Anglo-Saxon America when she described Asian immigration as a threat to "our own heritage," "our government," and "the preservation of our standards ... [and] the unborn children of America, of India and of the world."[64] As a figure of unregulated Hindu fertility, the sacred cow was yoked in the United States to eugenic fears of white "race suicide." These fears were characteristic of the American politics of eugenics; yet they were not so distant from politics on the subcontinent as we might suppose.

QUESTIONING COW PROTECTION

What relation does the colonial discourse of the sacred cow bear to today's Hindu nationalist politics of cow protection, which threatens so many Indian lives and livelihoods? Whatever other work the sacred cow has been called to perform, in

India it has continued to work the field of cow protectionist politics to which it was summoned in the nineteenth century. Like the colonial sacred cow, like the conditions of rearing and breeding cattle, like bovine bodies in India, the politics of cow protection has significantly altered since that time. By way of conclusion, this section traces some important continuities in this politics.

Hindu nationalists have endorsed some of the defining characteristics of the colonial sacred cow—its identification with unchanging Hindu tradition and its association with an uncompromising interdiction on cattle killing—while emphatically rejecting others—in particular, its identification as antiscience. In this fashion, Hindu nationalism has helped secure understandings of the sacred cow as an essentially religious object of Hindu reverence that are now common sense. But the sacred cow was not a self-possessed and bounded figure, intrinsically religious or Hindu, to be rescued either by Western science or from it. In its very fibers, the sacred cow spoke in Malthusian and eugenic tones of the laws of nature. Recognizing this, it is possible to unravel the politics of cow protection—and to question the Hindu nationalist sacred cow.

Posing as defenders of Indian tradition at home and abroad, Hindu nationalist advocates of cow protection have vigorously denounced the colonial discourse that would portray Hinduism as maladaptive. But, far from rejecting the imperialist imputations of reproductive excess or irresponsible breeding and feeding that were embodied in the colonial sacred cow, Hindu nationalism hitched them to the politics of caste-Hindu supremacy.

Colonial discourse of the sacred cow didn't split hairs over India's enormous social and cultural diversity, heedlessly eliding *Indian* with *Hindu*. The political ramifications of this elision are made vividly clear in Hindu nationalism, which identifies the Indian nation as exclusively Hindu and demonizes Indian Muslims in particular. Hindu nationalism dates to the 1920s, when its defining ideology was coined, and the first Hindu nationalist organization was founded. It was also in the 1920s that Indians began to expound overtly eugenic ideas and to establish eugenic societies.[65] Indian eugenicists were preoccupied with India's poverty problem; they were also overwhelmingly middle class, upper-caste Hindu, and male.[66] Embracing notions of the "differential fertility" of inferior classes, they portrayed Muslims and lower castes as engaged in excessive, "foolish procreation . . . without regard to the natural resources of the land."[67] Compounding existing stereotypes, Indian eugenicists characterized "the working classes, lower-caste Hindus, and Muslims as sexually irresponsible subjects" and blamed their "irresponsible procreation" for their poverty, and the poverty of the nation.[68]

The politics of cow protection drew on these influences. Cow protectionist discourse characteristically disparaged consumers of beef as selfishly feeding themselves at others' expense.[69] The foundational text of cow protection made this point in quantitative terms in 1881: it contrasted the number of people who could be fed by the beef of a slaughtered cow with the number of people who could be fed by the milk and grain produced by that cow and her future offspring and concluded that beef eating caused public harm.[70] Around 1910, this argument was recast in qualitative terms: contending that cow slaughter was "seizing food" from Hindus, cow protectionists pointed the finger at Indian Muslims: "Even as everyone is raising a ruckus that everywhere whites and Anglo-Indians are slaughtering [cattle], the population of Muslims is expanding twofold by day and fourfold by night—and how many poor children of [Hindus] are dying every moment."[71] Anticipating coming developments in cow protectionist politics, it portrayed cow-slaughtering Muslims as a direct threat to the survival of individual Hindus and to Hindu racial survival.

By the 1920s, a recognizably Hindu nationalist discourse of cow protection drew together this anti-Muslim rhetoric with eugenic concerns about poverty, differential fertility, and Hindu "race suicide." Identified narrowly with antislaughter, cow protection was intermeshed with a distinctive set of campaigns to increase Hindu strength, as encapsulated in the 1926 tract *Hindu Sangathan: Saviour of the Dying Race*.[72] Cow protection's foundational logic was reworked and tied to a religious conversion campaign to reduce Muslim numbers: calculating the number of cows killed by a single Muslim in a lifetime, one publication quantified the (Hindu) food that might be "saved" by the conversion to Hinduism of "just one cow-eater."[73] Simultaneously, cow protection was linked to fears of miscegenation and efforts to manage the reproductive practice of Hindu women. Contending that Hindu widows "seek shelter under Muslim roofs ... [and] in this way, while reducing the numbers of Hindus, ... add to the numerical strength of beefeating religious societies," *Hindu Sangathan* advocated for marriage reform.[74] By ensuring that Hindu women bore Hindu children, cattle and the (present and potential) foods they produced could be preserved to fortify Hindus in body as well as in number. The Hindu nationalist discourse of cow protection pitted fears of runaway population expansion among Indian Muslims against the very possibility for Hindus' eugenic development—that is, their proper nourishment.

It portrayed Hindus as doubly endangered by the dysgenic presence of Indian Muslims, figured as both "irresponsible breeders" and "irresponsible feeders."

Since independence, Hindu nationalist proponents of cow protection have posed as defenders of Hindu religion and champions of Hindus' economic and political empowerment. Condemning both British and Indian secularist opponents of cow protection, they have appeared to uphold an essential tenet of Hindu religion and to defend the sacred cow—"the symbol of Hindus' culture"—against colonialist slander, past and present. Hindu nationalist proponents of cow protection claim to represent "those people . . . who have a cultural and religious connection" to the cow but also those "whose bodies' every particle [is] indebted to the cow."[75] This formulation renders Muslims as a dysgenic national influence whose presence is an impediment to the physical and economic development of upper-caste Hindus. When caste-Hindu society is conflated with the Indian nation, then the consumptive and reproductive habits of Muslims and others who defy cow protectionist values are rendered responsible for Hindu weakness and Indian poverty. And so Hindu nationalist cow protection puts the colonial sacred cow to work in service to Hindu supremacy.

NOTES

I am especially grateful for the collegiality, support, and comments that made this chapter possible. My thanks to Myrna Perez Sheldon, Ahmed Ragab, Terence Keel, Tisa Wenger, Suman Seth, and all the scholars of the Critical Approaches to Science and Religion workshop; Nancy Reynolds and the Wastelands collective at Washington University in St. Louis; Kristin Distil; and J. Barton Scott. My research was funded by an NEH-American Institute of Indian Studies fellowship.

1. *Oxford English Dictionary Online*, s.v. "sacred cow (n.)," accessed November 19, 2021, https://www-oed-com.libproxy.wustl.edu/view/Entry/169557?redirectedFrom=sacred+cow#eid.
2. Joanne Waghorne, "Caste, Cows and Karma," *Hinduism Today*, August/September/October/November 1996, https://www.hinduismtoday.com/magazine/august-september-october-november-1996/1996-08-caste-cows-and-karmaoa/.
3. Wendy Doniger, "A Burnt Offering," review of *The Myth of the Holy Cow*, by D. N. Jha, *The Times Literary Supplement* no. 5183, August 2, 2002, 9.
4. Indian National Congress, National Planning Committee, *Animal Husbandry and Dairying* (Bombay: Vora, 1948), 18.
5. For accounts of contemporary Indians' relationships to cattle that belie this characterization, see James Staples, *Sacred Cows and Chicken Manchurian: The Everyday Politics of Eating Meat in India* (Seattle: University of Washington Press, 2020); the essays collected

in *South Asia: Journal of South Asian Studies* 42, no. 6 (2019); and Radhika Govindrajan, *Animal Intimacies: Interspecies Relatedness in India's Central Himalayas* (Chicago: University of Chicago Press, 2018).

6. A substantial body of scholarship has built upon Talal Asad's observation that religion does not precede the secular, awaiting its judgment; the two concepts emerge and are formed together, in mutual relation. E.g., Talal Asad, "Religion and Politics: An Introduction," *Social Research* 59, no. 1 (1992): 3–16.
7. David Ludden, introduction to *Agricultural Production and Indian History*, ed. David Ludden (Delhi: Oxford University Press, 1994), 6–7.
8. Great Britain, Royal Commission on Agriculture in India, *Evidence of Officers Serving Under the Government of India*, vol. 1, pt. 1 (London: His Majesty's Stationery Office, 1927), iii (hereafter RCAI, *Evidence*).
9. Great Britain, Royal Commission on Agriculture in India, *Report* (Bombay: Government Central Press, 1928), 175, 168 (hereafter RCAI, *Report*).
10. Great Britain, Royal Commission on Agriculture in India, *Abridged Report* (Bombay: Government Central Press, 1928), 21 (hereafter RCAI, *Abridged Report*).
11. RCAI, *Abridged Report*, 20–21, my emphasis.
12. RCAI, *Evidence*, 15.
13. This was a prominent representation of India, especially Hindu India, as spiritual rather than material—that is to say out of touch with material reality and incapable of self-management or self-government. See Peter van der Veer, *The Modern Spirit of Asia: The Spiritual and the Secular in China and India* (Princeton, NJ: Princeton University Press, 2014); Ronald Inden, *Imagining India* (Bloomington: Indiana University Press, 2000); and Richard King, *Orientalism and Religion: Post-colonial Theory, India and "the Mystic East"* (London: Routledge, 1999).
14. Patrick Brantlinger, *Dark Vanishings: Discourse on the Extinction of Primitive Races, 1800–1930* (Ithaca, NY: Cornell University Press, 2003); Philippa Levine, "Anthropology, Colonialism, and Eugenics," in *The Oxford Handbook of the History of Eugenics*, ed. Alison Bashford and Philippa Levine (New York: Oxford University Press, 2010).
15. "Sacred Cow Country: Hindu Politics and Cattle in Modern India." Unpublished manuscript, last modified Sept. 14, 2022. Scrivener file.
16. Margaret E. Derry, *Masterminding Nature: The Breeding of Animals, 1750–2010* (Toronto: University of Toronto Press, 2015); Philippa Levine and Alison Bashford, "Introduction: Eugenics and the Modern World," in *The Oxford Handbook of the History of Eugenics*, ed. Alison Bashford and Philippa Levine (New York: Oxford University Press, 2010) 4–5.
17. David Arnold, *Famine: Social Crisis and Historical Change* (New York: Blackwell, 1988) 36.
18. *The Cow Question in India, with Hints on the Management of Cattle* (Madras: Christian Literature Society, 1894), 26.
19. *Cow Question*, 30, 32.
20. *Cow Question*, 34, 35.
21. This was the expanded edition; the original publication was issued in 1798.
22. Arnold, *Famine*, 35.
23. G. Keatinge, *Rural Economy in the Bombay Deccan* (New York: Longmans, Green, 1912). Keatinge was director of agriculture for the Bombay Presidency.

24. Albert Howard and Gabrielle L. C. Howard, *The Development of Indian Agriculture* (London: Oxford University Press, 1927), 19. The authors served as agricultural adviser to the states in Central India and second imperial economic botanist, respectively.
25. Howard and Howard, *Development of Indian Agriculture*, 54.
26. Keatinge, *Rural Economy*, 125, 126–127.
27. Rebecca J. H. Woods, *The Herds Shot Round the World: Native Breeds and the British Empire, 1800–1900* (Chapel Hill: University of North Carolina Press, 2017); Sarah Franklin, *Dolly Mixtures: The Remaking of Genealogy* (Durham, NC: Duke University Press, 2007).
28. RCAI, *Abridged Report*, 21. The Board of Agriculture had declared in 1925 that Indians' failure to grow more fodder must be addressed before improved breeding could be attempted. India, Board of Agriculture and Animal Husbandry, *Proceedings of the Board of Agriculture in India* (1925) (Calcutta: Office of the Superintendent of Government Printing, 1926), 10.
29. RCAI, *Abridged Report*, 21. Also C. E. Low, "The Supply of Agricultural Cattle in India." In Indian Science Congress Association, India. Department of Agriculture, and Imperial Council of Agricultural Research. *Agricultural Journal of India*, Calcutta: Government of India, Central Publication Branch for the Imperial Council of Agricultural Research, vol. 7 (1912): 331–341, p. 334. HathiTrust Digital Library, https://hdl.handle.net/2027/uc1.b2938007
30. Neeladri Bhattacharya, *The Great Agrarian Conquest: The Colonial Reshaping of a Rural World* (Ranikhet, India: Permanent Black, in association with Ashoka University, 2018); David Gilmartin, *Blood and Water: The Indus River Basin in Modern History* (Oakland: University of California Press, 2015); Elizabeth Whitcombe, *The United Provinces Under British Rule, 1860–1900*, vol. 1, *Agrarian Conditions in Northern India* (Berkeley: University of California Press, 1972).
31. E.g., Punjab (India), Department of Agriculture, *Cattle and Dairying in the Punjab* (Lahore: Printed at the "Civil & Military Press," 1910). The summary account in this section is derived from Adcock, "Sacred Cow Country."
32. This was in keeping with the priority of military over civil requirements by the government of India. James L. Hevia, *Animal Labor and Colonial Welfare* (Chicago: University of Chicago Press, 2018); Saurabh Mishra, *Beastly Encounters of the Raj: Livelihoods, Livestock, and Veterinary Health in North India, 1790–1920* (Manchester, UK: Manchester University Press, 2015); Brian P. Caton, "The Imperial Ambition of Science and Its Discontents: Animal Breeding in Nineteenth-Century Punjab," in *Shifting Ground: People, Animals, and Mobility in India's Environmental History*, ed. Mahesh Rangarajan and K. Sivaramakrishnan (New Delhi: Oxford University Press, 2014) 133–154.
33. India, Board of Agriculture and Animal Husbandry, *Proceedings of the Board of Agriculture in India* (Calcutta: Office of the Superintendent of Government Printing, 1914), HathiTrust, https://hdl.handle.net/2027/coo.31924066971635. Colonial experts were divided on breed policy.
34. I discuss this alliance and its contemporary ramifications in " 'Preserving and Improving the Breeds': Cow Protection's Animal-Husbandry Connection," *South Asia: Journal of South Asian Studies* 42, no. 6 (2019), https://doi.org/10.1080/00856401.2019.1681680.
35. David Arnold, *Colonizing the Body: State Medicine and Epidemic Disease in Nineteenth-Century India* (Berkeley: University of California Press, 1993); Sarah Hodges, *Contraception, Colonialism and Commerce: Birth Control in South India, 1920–1940* (Burlington, VT: Ashgate, 2008).

36. Rebecca Woods has discussed how colonial arguments about livestock stood in "proxy" for valuations of human beings that were too impolitic to state directly. Woods, *Herds Shot Round the World*, 111. Also Harriet Ritvo, *The Animal Estate: The English and Other Creatures in the Victorian Age* (Cambridge, MA: Harvard University Press, 1989).
37. David Arnold, "Official Attitudes to Population, Birth Control and Reproductive Health in India, 1921–1946," in *Reproductive Health in India: History, Politics, Controversies*, ed. Sarah Hodges (Hyderabad, India: Orient Longman, 2006), 47.
38. Dr. J. W. D. Megaw of the Indian Medical Service, quoted in Arnold, "Official Attitudes to Population," 30.
39. J. T. Marten, *Census of India, 1921, Part I—Report*, quoted in Arnold, "Official Attitudes to Population," 23.
40. Sarah Hodges, "Governmentality, Population and Reproductive Family in Modern India," *Economic and Political Weekly* 39, no. 11 (March 13–19, 2004), 1160.
41. Levine and Bashford, "Introduction," 4, 15.
42. Quoted in Levine and Bashford, "Introduction," 5.
43. Quoted in Levine and Bashford, "Introduction," 5.
44. Lucy Bland and Lesley A. Hall, "Eugenics in Britain: The View from the Metropole," in *The Oxford Handbook of the History of Eugenics*, ed. Alison Bashford and Philippa Levine (New York: Oxford University Press, 2010), 213–227; Sarah Hodges, "South Asia's Eugenic Past," in *The Oxford Handbook of the History of Eugenics*, ed. Alison Bashford and Philippa Levine (New York: Oxford University Press, 2010), 228–242; Pauline M. H. Mazumdar, "The Eugenists and the Residuum: The Problem of the Urban Poor," *Bulletin of the History of Medicine* 54, no. 2 (Summer 1980): 204–215.
45. Quoted in Mazumdar, "The Eugenists and the Residuum," 211.
46. Levine, "Anthropology, Colonialism, and Eugenics," 51, 52.
47. Asha Nadkarni, *Eugenic Feminism: Reproductive Nationalism in the United States and India* (Minneapolis: University of Minnesota Press, 2014), 24. My argument has not yet benefited from the important new publication, Mytheli Sreenivas *Reproductive Politics and the Making of Modern India* (Seattle: University of Washington Press, 2021).
48. Mrinalini Sinha, *Specters of Mother India: The Global Restructuring of an Empire* (Durham, NC: Duke University Press, 2006), 1.
49. Sinha, *Specters of Mother India*, 2.
50. Sinha, *Specters of Mother India*, 25.
51. Sinha, *Specters of Mother India*, 79.
52. Katherine Mayo, *Mother India* (New York: Harcourt, Brace, 1927), 223.
53. Mayo, *Mother India*, 223–224.
54. Mayo, *Mother India*, 22.
55. Mayo, *Mother India*, 230–231.
56. Mayo, *Mother India*, 229, 234.
57. Mayo, *Mother India*, 231.
58. Mayo, *Mother India*, 22, 26, 22.
59. Nadkarni, *Eugenic Feminism*, 114.
60. Mayo, *Mother India*, 226–227.
61. Mayo, *Mother India*, 237.

62. Mayo, *Mother India*, 32.
63. Nadkarni, *Eugenic Feminism*, 113; Sinha, *Specters of Mother India*, 96.
64. Nadkarni, *Eugenic Feminism*, 113.
65. Historians describe a global "peak" in eugenics in the 1920s, although eugenics language had been taking shape since the 1880s. Harald Fischer-Tine, "From Brahmacharya to 'Conscious Race Culture': Victorian Discourses of 'Science' and Hindu Traditions in Early Indian Nationalism," in *Beyond Representation: Colonial and Postcolonial Constructions of Indian Identity*, ed. Crispin Bates (New Delhi: Oxford University Press, 2006), 241–269.
66. Hodges, "South Asia's Eugenic Past," 229.
67. Sanjam Ahluwalia, *Reproductive Restraints: Birth Control in India, 1877–1947* (Urbana: University of Illinois Press, 2008), 38, 32.
68. Ahluwalia, *Reproductive Restraints*, 39–40.
69. Compare Andrew Sartori, *Bengal in Global Concept History: Culturalism in the Age of Capital* (Chicago: University of Chicago Press, 2008), ch. 6.
70. Dayananada Sarasvati, *Atha Gokarunanidhih* ([Allahabad, India]: Vaidikyantralaya, [1886]).
71. "Goraksha aur us ki virodhi," *Saddharm Pracharak* (Kangri, India) 23, no. 12 (June 28, 1911): 5.
72. Shrāddhanand Sanyasi, *Hindu Sangathan: Saviour of the Dying Race* ([Delhi?]: Swami Shraddhanand Sanyasi, 1926).
73. Sri Swami Chidanand Sannyasi, "Shuddhi sankat-mochan har," in *Rshyank*, special issue, *Arya Mitra* 30, no. 42–43–44 (March 1927): 71; the author was the head of the Bharatiya Hindu Shuddhi Sabha. And see Charu Gupta, *Sexuality, Obscenity, Community: Women, Muslims, and the Hindu Public in Colonial India* (Delhi: Permanent Black, 2001).
74. Shraddhanand, *Hindu Sangathan*, 98.
75. Lala Hardev Sahay, *Govadh Nishedh* ([India]: Lala Hardev Sahay, 1946), 10, 7. Lala Hardev Sahay was a prominent proponent of a total prohibition on cow slaughter upon independence.

CHAPTER 8

"AND GOD KNOWS BEST"

Knowledge, Expertise and Trust in the Postcolonial Web-Sphere

AHMED RAGAB

In 1945, the Egyptian Ministry of Interior sent a query to the office of the grand mufti of Egypt.[1] It was almost pilgrimage season, and the ministry, which supervised the pilgrimage proceedings, intended to require the immunization of all Egyptian pilgrims against cholera and typhoid.[2] The query was not related to the legality of vaccines in themselves, as this point had been debated before; rather, it was related to whether a cholera vaccine would invalidate the religious fast, seeing that many pilgrims were fasting. The office's response to the query referenced a previous query and a fatwa that was issued in 1932. In 1932, the Ministry of Interior had queried the mufti about whether a vaccine—that time the smallpox vaccine—violated the fast. In the earlier fatwa, the mufti explained that he consulted with a physician to find out details about the composition of vaccines and how they enter the body. When assured that vaccines do not enter the body by any of its regular orifices, he proclaimed vaccines safe to use during fasting.[3] In 1945, the mufti affirmed the 1932 fatwa but did not believe it was necessary to consult another physician. Relying on what seemed to have been established in 1932, he felt comfortable offering the same fatwa: vaccines were safe to use during fasting.[4]

The twin fatwas of 1932 and 1945 present a number of important questions. On one hand, while the Ministry of Interior had already received the 1932 fatwa, officials there believed it was necessary to acquire a new fatwa on basically the same topic, at least as judged by the mufti. On the other, the modes of legal reasoning that the muftis of 1932 and 1945 used were markedly different. In 1932, the mufti attempted to consult medical experts to help him determine the nature of the vaccine and its route of entry, whereas in 1945, he simply relied on the precedent or, rather, on the previously supplied medical information.

While the 1932 and 1945 fatwas and the attendant communications exist in the archives of Dar al-Ifta, the House of the Grand Mufti, I did not access them in the archives. Instead, I am able to look up these fatwas on the relatively new online portal of Dar al-Ifta.[5] In its current appearance, the portal was launched in 2016 and offers a series of services, only one of which is related to fatwas. The fatwas on the website are not organized chronologically but rather thematically, allowing the 1932 fatwa on smallpox vaccine to coexist alongside a 2015 query by an anonymous person about eye drops and another from 2013 about facial cream.[6] In two different places, one is invited to submit their own queries and is promised that the queries will be received by learned scholars and muftis who will respond either directly or on the website.[7] Dar al-Ifta's website is part of a larger system of websites and online portals that offer fatwas and legal opinions to questioners. This system itself is arguably a corner of a larger ecology of advice and forum websites, networks, and social media groups, all of which offer the opportunity for different people to ask for advice or to poll their friends or other contacts on their opinions on particular issues.[8]

Dar al-Ifta's website is not the most popular or the oldest website of its kind.[9] Other more prominent examples include IslamOnline, which is issued and maintained by the Qatari society al-Balāgh, and Islamweb, which is formally issued by the Qatari Ministry of Endowments and Religious Affairs. Similarly, alifta.net is the formal portal of the Saudi Permanent Committee for Islamic Research and Fatwa.[10] While many of these websites and portals, such as that of Dar al-Ifta, offer a mixture of fatwas that originate online, meaning that the queries and the answers were delivered entirely over the internet, and ones that originate in more traditional settings, such as the 1945 vaccination fatwa just mentioned, the online medium adds different dimensions to the interactions and the behaviors of the *mustaftī*: the one presenting the query, or the consumer, and the mufti, or the producer of the fatwa. These differences relate, in part, to the nature of the medium, its accessibility, and the kinds of mannerisms and behaviors that it enables and constitutes. Moreover, the medium implies and conditions particular articulations and manifestations of authority and reliability, and it offers specific spaces for performing piety by both the consumers and the producers of these religious opinions.[11]

Contemporary scholarship has shown growing interest in the modern life of the fatwa. A number of scholars have focused on the institutions that produce fatwas and how they fit within a modern legal system.[12] Others, who have paid more attention to the online spaces where fatwas are sought and produced, have focused primarily on websites directed to Muslims in the West.[13] Here, scholars have explained the role played by these websites in producing common identities and affirming the common belonging of Muslims as they aim to shape their lives along common goals and views. Less attention, however, has been paid to Arabic language websites, which are often primarily addressed to Muslims in Muslim-majority countries and are therefore embedded within a fatwa ecology that seamlessly extends across the real-virtual divide.

It is across this real-virtual divide that the smallpox fatwa lives. Issued in the real world through one of the more bureaucratized methods of seeking legal and religious opinions—namely, a petition from one government agency to another, complete with sets of signatures and stamps and a response in kind—the fatwa lives in a new environment that looks to inform contemporary Muslims seeking fatwas online on matters relating to medicine, vaccines, and fasting, among others. While this may not be a question asked again today, or at least not with the same urgency, it works to establish a frame of reference that continues to be operative in similar questions, as will be explained later.

In this chapter, I will investigate the life of this and similar fatwas. Focusing on Sunni online space, I will first look at the spaces of fatwa making online, investigating the place of online media in the economy of piety and fatwa. Here, I will also look at how fatwa-producing media and institutions perceive and aim to regulate the behavior of fatwa seeking and how such disciplining functions to regulate particular modes of pietistic behavior. Then I will look at the role played by the internet in constructing Muslim identities in contemporary securitized discourse. Throughout, this chapter looks to investigate different layers of religious and legal authority and how they manifest in the space of fatwas and legal opinions.

THE ONLINE SPACE FOR FATWAS

Recent scholarship on the use of different websites that provide fatwas, along with other Islamic websites, does not indicate a strong user loyalty. In other words, there is no evidence that users continue to visit the same website for their different needs or that they rely on a given website to the exclusion of others.[14] Moreover, and despite some claims to the opposite, the most popular Islamic websites, especially those originating in Egypt, Qatar, and Saudi Arabia, rely openly on authorities and scholars who are also affiliated with other websites. For instance, Egypt's Dar al-Ifta's website often cites the works of Saudi muftis and the work of the famous Egyptian Qatar-based scholar al-shaykh al-Qaraḍāwī, who was connected to a number of the Qatari websites, such as IslamOnline and Islamweb.[15] And despite claims of religious and doctrinal differences between Salafis and others, which became popular during the IslamOnline crisis, as will be discussed later, there are few distinguishing features in terms of content between these different websites—so much so that after a crisis that reshaped IslamOnline and ousted al-Qaraḍāwī from its helm, IslamOnline continued to cite and archive al-Qaraḍāwī's fatwas as key parts of its online environment.[16] In other words, these different websites continue to invest in figures of religious authority without deep attention to doctrinal differences among Sunni Muslims.[17]

At another level, the familiar content is presented in significantly different forms that impact users' experience and are informed by a particular site's claim to authority. A key difference between these platforms is evident in how these sites present themselves to their users. The visitor to Dar al-Ifta's original website was immediately greeted with the grand mufti's picture and a link to a welcome note attributed to him. At the top, one can access a link to a page describing the history of Dar al-Ifta and its supposed role in spreading "proper and sound" Islamic knowledge, right next to links for contacting Dar al-Ifta. These links take the user to a list of addresses of Dar al-Ifta's central location in Cairo and its satellites in the provinces of Alexandria and Asyut, along with phone and fax numbers. In May 2022, Dar al-Ifta issued a new beta version of its website, which reorganized some of the content. In the beta version, a new slideshow was added to the top of the landing page. A new page was dedicated to the Mufti's curriculum vitae, while the history of Dar al-Ifta was moved a bit deeper in the website. While Dar al-Ifta's website is similar to all the websites analyzed here in providing a space for users to submit queries, its query portal also offers a hotline number, which can be dialed from anywhere in Egypt. The rest of the website is devoted to various other activities that Dar al-Ifta is engaged in, including a program to prepare muftis; a portal to monitor manifestations of Islamophobia around the world; a series of news articles, most of which are aimed at either fighting or condemning Islamophobia and extremism; and a separate page dedicated to the grand mufti himself.[18] The website functions primarily as an extension of the physical Dar al-Ifta. It aims at facilitating access to the Dar and its various services. Moreover, it intentionally locates itself within the context of its immediate audience—Egyptians seeking the services of Dar al-Ifta as a government institution. This focus is clear when one compares the Arabic website to the English one. While the English website maintains the overall branding and the direct connection to Dar al-Ifta as a physical institution, gone are the pages dedicated to the Dar's different activities in training and education and those related to the mufti himself. Despite the fact that the English website is not as successful or popular as its originators might have hoped, the difference in orientation is clear.

The Dar al-Ifta website anchors the legitimacy of its virtual content in the nonvirtual. The website is simply another medium, more efficient perhaps than a fax or a phone, that represents the same nonvirtual institution that derives its legitimacy from its long colonial and postcolonial history and from its government mandate.[19] As a government institution, Dār al-Ifta has consistently assumed specific roles designated by the postcolonial state that aim to further legitimize the regime and to augment the integration of religious authority into the postcolonial legal regime.[20] For instance, in 2019, the website highlighted the insistent calls of Egyptian President Abdel Fattah al-Sisi to religious leaders to fight terrorism and extremism through "spreading the true teachings of Islam." The website, which was launched well before the military coup that delivered al-Sisi to power, was therefore positioned as an answer to the president's call and a response to the newly assigned mission. Al-Sisi's call continued to be referenced in the space dedicated to training muftis and scholars, and in various other sections of the website. Here, the relative imbalance between the assumed international mission called for by the president and the local orientation that dominates Dar al-Ifta's presence online offers important insights into the making of its postcolonial legitimacy as one concerned primarily with fulfilling the institution's permanent and easily recognizable roles while attempting to respond to the charges offered by the regime.

The Permanent Committee for Islamic Research and Fatwa, which was established in Saudi Arabia in 1971, is the highest fatwa-issuing body in Saudi Arabia and is one of the more respected such bodies among Sunni Muslims. Its web portal (https://al-ifta.gov.sa) is similar in many aspects to Egypt's Dar al-Ifta website. There, too, the branding is deeply connected to the institution, with its logo on the top and its major figures prominently featured. The fatwa portal (https://www.alifta.gov.sa/Ar/IftaContents/Pages/default.aspx) is organized in two different ways. One can access fatwas organized by published volume, mirroring the printed collections that the committee produces, or thematically, similar to other online portals. It also includes dedicated pages for the fatwas of the famous Saudi scholar Ibn Bāz and to the opinions and works of the Council of Senior Islamic Scholars, which is headed by the Saudi grand mufti.[21] This website is not the committee's first foray on the internet. In fact, it launched an English language website, Fatwa-Online (https://fatwa-online.com), back in 1997. That website continues to be connected to a host of other online platforms, constituting an ecosystem that aims to provide services for Muslims in the West. These websites include madeenah.com, markazquba.com, smatch.net, subulassalaam.com, and ummahserve.com.

Fatwa-Online explains its mission as follows: "English-speaking Muslims have been starved of access to officially-published fatwas which originate in the Arabic language, our aim is to make available online, these and many other fatwas in the English language for the first time." Here, "officially-published fatwas" are those issued by the committee or the council, the two bodies officially sanctioned by the Saudi state to issue religious opinions, so it is not surprising that Fatwa-Online features the works and fatwas of these scholars prominently in its attempt to provide their views to English speakers.[22]

The Arabic portal, which was first launched in 2007 with the address al-ifta.com before moving in 2022 to alifta.gov.sa, does not propose that the inaccessibility of sanctioned fatwas was a reason for its operation, in part because these fatwas continue to circulate through satellite TV channels, booklets, newspapers, and other Arabic-language media.[23] Instead, its goal resembled what Egypt's Dar al-Iftā would claim later: to provide reliable fatwas that resist extremist interpretations and spread the "true" teachings of Islam. Here, again, and despite the different platforms and histories, alifta.com/alifta.gov.sa functioned in a similar way: to extend the reach of its issuing institution. It derived its legitimacy from the legitimacy of this institution and offered only to extend its services beyond its immediate walls. Perhaps this is why the Arabic website seemed to be an afterthought. The reach of the council, as well as its different satellite locations and TV stations, is already remarkable in its accessibility to Arabic-speaking users. The newer website functions simply within this ecosystem of media tools that extend the availability of the official institution.

Contrary to the clearly branded websites of Dar al-Ifta and the Saudi Permanent Committee for Islamic Research and Fatwa, the two most popular

Islamic websites, in English and Arabic—IslamOnline and Islamweb—lack any immediate branding that links them to the nonvirtual world. IslamOnline (https://islamonline.net) is one of the most, if not the most, popular among Arabic Islamic websites. Established in 1997 by the al-Balāgh Cultural Society in Qatar, it is an Islamic nongovernmental organization dedicated to spreading Islamic teachings in the Islamic world and beyond.[24] In its first decade, the website was run by a team of editors located in Cairo under the guidance of the famous scholar al-Shaykh Yusūf al-Qaraḍāwī.[25] The website was published simultaneously in Arabic and English in line with al-Qaraḍāwī's intellectual project, which he termed *fiqh al-aqaliyyāt* (minoritarian jurisprudence).[26] The term was deployed by al-Qaraḍāwī and his students and associates to refer to the need for particular jurisprudence dedicated to Muslims living as minorities in the West. Such jurisprudence would respond to their unique needs, including their lack of proficiency in Arabic.

In 2010, the website went through a major crisis, as the funding organization in Qatar looked to move the editorial office from Cairo to Doha while laying off about three hundred editors who worked in Cairo at the time. Al-Qaraḍāwī sided with the Cairo editors and consequently lost his job in Qatar. At the time, many argued that the reshuffle was intended to replace progressive voices with conservative ones in line with the new orientations of Qatari politics. Regardless of the motives behind such change, the editorial staff in Cairo launched a new website called onislam.net, which eventually was moved to aboutislam.net, both of which featured content directed to Muslims in the West. IslamOnline, however, continued to function normally after the interruption of 2010. Despite the changes in the editorial team and the ousting of al-Qaraḍāwī, the website continues to carry his fatwas and the content produced before 2010.[27]

Unlike with the previously discussed websites, the visitor to IslamOnline today will be hard-pressed to find a direct affiliation or connection to a particular nonvirtual institution or organization. The easiest way to obtain this information is through the link to submit one's own article. After being informed of a series of general requirements, from article length to prohibition of hate speech, one is told that IslamOnline is a nonprofit subsidiary of the Al-Balāgh Cultural Society. The fatwa section on the Arabic website is organized thematically and has no mention of who issues these fatwas. Each fatwa starts with a question, presumably in the user's own words, followed by an answer, which is attributed to the "website's team," with no specifications. All communication with the website or its team is through email or input forms, where one could presumably direct queries for fatwas. In other words, IslamOnline is simply and completely online, without any obvious or clear connection to the outside world.

In the same vein, Islamweb (https://islamweb.net), also is one of the most popular Islamic websites at the moment, shows no clear or direct branding that reveals its connection to the Qatari Ministry of Endowments and Religious Affairs.[28] Inside the fatwa portal, one can find a page dedicated to information on the Fatwa Center. There one can find out that the center was created by the Qatari ministry and that it adopts a *wasaṭī* (moderate) approach, using a multiplicity of *madhhabs* (or schools of law) in an attempt to reach middle ground and uncover the areas of agreement.

Islamweb's website offers some details concerning the process of fatwa making. First, a query is received. If it is identical or similar to a previous query, the user is referred by email to the previous query. If not, it moves to the next step, where a scholar of the fatwa team studies the question and provides a fatwa. The fatwa then is reviewed by two other scholars to ensure that it is "scientifically" sound. In the final stage, the fatwa is revised linguistically before it is sent to the website editor, who copyedits the text, adds a suitable searchable title, and then publishes the fatwa. Should there be disagreements, the website editor calls a meeting among the scholars to reach an agreement before a fatwa is published. While this process includes a number of different actors, the website editor supervises and organizes it. The fatwa clears each step only with the approval of the editor, who also categorizes and connects the fatwa thematically to other fatwas. This central role played by the website editor betrays the centrality of the online content. While Islamweb belongs to a network of websites funded by the Qatari Ministry of Endowments and Religious Affairs, it does not brand itself in this way and does not base its legitimation on the authority of that institution. Moreover, the website's Fatwa Center does not exist outside the website. In short, the website is not a publishing tool of a preexisting institution but rather the institution itself.

Scholars working on information technology have shown a direct connection between a website's appearance, its usability, and the cultures of usage that it cultivates, on one hand, and the authority and reputation that it garners, on the other.[29] User-friendly websites, which boast a sleek and clean appearance, provide a feeling of reliability and are more effective in guiding users through the website in an organized and predesigned fashion. This focus on usability and appearance is key to websites like IslamOnline and Islamweb. In both cases, the websites rely solely on their online presence as a tool for establishing confidence and cultivating their user base. While the authority of the portals maintained by Dar al-Ifta and the Permanent Committee for Islamic Research and Fatwa stems from the nonvirtual institutions with which they are affiliated, IslamOnline and Islamweb refer back to none and rely entirely on their own fame, usability, and accessibility in grounding their authority and influence.

At another level, the authority-building process of these websites relies on the form through which they provide their fatwas and create the local ecologies in which these fatwas are prompted, produced, and consumed. Here, I would like to focus on three major points—the format in which the fatwas are presented, the online ecology in which they are produced, and the intellectual resources on which they rely—as each of these factors contributes to how these fully online institutions establish their authority.

A fatwa on Islamweb, for instance, is offered in a way that is similar to that used by all the other websites. A question appears first, almost always presented anonymously and in the user's own words. The fact that these questions undergo very little (if any) editing is evident in their different styles and their use of colloquial structures at times. The question is followed by an answer. The answer starts in a formulaic fashion by invoking God's name in the *basmala* and praising the prophet. Then the muftis recast the question, especially if it contains significant colloquialisms or is not particularly clear, before starting their answer. While the answer is often concise, it always contains specific reference to Quranic verses or hadith and sometimes references to major legal scholars. At times, the muftis (or the editor) refers the questioner to other related fatwas, which are then referenced through hyperlinks. On Islamweb, editors often summarize long fatwas, stating their major points in two or three lines. In all cases, an informative title, which includes either the question or both the question and the answer, is provided, and the fatwa is cross-listed across multiple places within the online archive. At the bottom of the page, suggestions for related fatwas are offered.

The fatwa, as such, lives in an online ecology that links it to many other questions and multiple references. While neither IslamOnline nor Islamweb offers connections to an archive of nonvirtual fatwas, such as those offered by Dar al-Ifta and the Permanent Committee for Islamic Research and Fatwa, they create a more effective online archive that allows users to explore a variety of questions and related issues all at once. In this online ecology, hyperlinks are used for two main reasons. First, as commonly used in online media, they provide expansions for contracted references or answers to repeated questions. Second, and more relevant for our discussion here, they offer the simulation of a discussion, allowing the user, be it the one who asked the question or others, the chance to explore further questions. Every hyperlink leads to a new question or a new conversation between the user and the mufti, and the collection of these conversations functions to build a model of pious living. This connected ecology is not unique to the fatwa section. In fact, the fatwa section, as a whole, lives in a larger ecology where fatwas are connected to news, opinion pieces, and speculative and philosophical debates, among others. In the same way that the smaller fatwa-centered ecology offers an extended space for conversation and discussion through exploring various pages, the larger ecology of the website affords the user a whole world of connections where one can encounter the news and its analysis and philosophical and literary discussions, as well as entries by parents talking about their kid's birthday and young people discussing Valentine's Day.[30]

Throughout, these different aspects of the daily life are framed within the structure of religious observance and piety. The website becomes a space to think about living a pious Muslim life that is tethered, connected, or wired in to the lives of other Muslims.[31] Scholars have shown how this connectivity and the ensuing community of users that is formed provide important identity-formation strategies for Muslims in the West.[32] In fact, as a number of scholars have observed, websites like Fatwa-Online and IslamOnline English call on their users to precisely affirm their identities and live a life that might vary from that of the communities where they live. While this dynamic of identity formation is not similarly operative in relation to users living in Muslim-majority communities, these websites, with their aim of constant connectivity, extend the level of pious performance beyond prescribed rituals to almost all aspects of life.[33]

In his *Simulacra and Simulation*, Jean Baudrillard offered a reflection on simulacra as representations of reality that lose their connection to the real, intentionally or otherwise, and that engage in creating a level of hyperreality.[34] Unlike Plato's simulacra, which distorted the real in the interest of making it appear more real, Baudrillard's simulacra are rooted in a different mode of reality, one that is not invested in the world outside the simulacra system or a system of referents without references. I argue that websites such as Islamweb and IslamOnline create and function as simulacra. In doing so, they emulate the seeming qualities of an imagined reality, as they create a parallel system of the real. In structuring fatwas, the seemingly ancient, classical, or original system is preserved. We are offered a question by a questioner or a *mustaftī* framed mostly in their own words. Because the *mustaftī* receives the answer to their question directly in most cases, the published fatwa is primarily meant for other users and visitors of the website, who may have similar questions or may be browsing the fatwa pages. As such, the user, who is not the *mustaftī* in most case, is offered a seat to witness the interaction of a *mustaftī* and a mufti, as if mid-speech. We are allowed to see only part of the details and to listen in on what may have been a longer conversation between the two. The answer also preserves the forms of oral religious communication, starting with the *basmala*, praising God and the prophet, in a manner reminiscent of religious sermons or suggestive of the proper environment of fatwa, before proceeding to detailed answers. Moreover, the sources cited are derived from a particular archive of classical Islamic texts that ground the practice in timelessness.[35] Throughout, users are invited to "ask" additional questions or request further details, which are available through hyperlinks in a manner that also recalls the practice of fatwa giving in the nonvirtual world.

However, this simulation of fatwa-giving encounters does not rely on a particular model or a specific historical example. Instead, it precisely recreates an imagined, almost mythical reality, which users are to believe was the reality of fatwa giving in the times of the *salaf* (good ancestors) and other great scholars. Moreover, the fatwa space online extends beyond reading other people's fatwas to inviting users to engage and ask their own questions. The space of the fatwa, unreal as it may be, materializes in the construction of interactive online spaces intended to provide constant connection between pious Muslims and religious authorities. This attempt to ensure that Muslims can live every aspect of their lives according to pious guidance is evident in how these websites developed tools to allow constant access. For example, both IslamOnline and Islamweb explain that their apps are meant to provide immediate access to muftis without delay to help Muslims get the information they need to live a pious life. Islamweb offers a chat function for immediate access to a mufti. At another level, the websites offer various sections that intend to answer or guide the visitor through almost all aspects of life. A section in Islamweb, for example, is dedicated to children. There parents can find games, stories, books, and activities for their children, as well as articles discussing Islamic approaches to child rearing. Another section dedicated to "contemporary issues" contains articles addressing social media, dating websites, life coaching, and other issues. Similarly, IslamOnline offers a section dedicated to book reviews, which summarizes and offers "Islamic" opinions on books ranging from management and policy to the psychology of money.

In the process, this simulacrum creates a new hyperreality that is desired or sought after but is also different from the lived reality of Muslims around the world. In this hyperreality, Muslims are indeed invited to live even the minute details of their lives according to the dictates of Islam by asking questions about everything in their lives and about any decision they might want to make. This new hyperreality is one that attempts to reproduce the idealized narrative of a classical founding community that lived under the prophet's own guidance. For example, Islamweb explains that it models its fatwa-producing method on the traditions of authorities and on the practice of generations of the pious ancestors.[36] In the same way, a mufti who can provide advice is available at all times and can be reached online or on one's smartphone. In this context, IslamOnline's and Islamweb's interest in transcending *madhhab* difference and offering what they believe to be "true teaching" of Islam contributes to this hyperreality, whereby there is an achievable consensus-based and noncontroversial religious answer for every question. Similarly, the availability of these answers without references to space or time—as fatwas from 2005 can live alongside ones issued yesterday—highlights the belief that Islamic teachings transcend place and time and are fully available to everyone everywhere.

BAD MUSLIMS AND BAD INTERNET

The internet ecology I have described extends beyond these fatwa portals to include a variety of other websites and apps, including apps that help find the *qibla* (direction for prayers), offer Quran readings, and even provide dating services for pious Muslims interested in meeting other pious Muslims. On November 16, 2020, Joseph Cox of Motherboard published an investigative report uncovering how the U.S. military bought location data from various apps, many of which were directed to and used by Muslims around the world.[37] These included Quran apps as well as dating apps, which were downloaded more than ninety-eight million times in 2019. This acquisition was consistent with long-standing practices of the U.S. military, which uses location data to guide drone strikes, for example. This particular practice and the security of location data are beyond the scope of this chapter. Yet the logic behind such acquisition points to a particular perception of Islam and Muslims and a particular role played by the internet in the production of such perceptions.

In Derrida's analysis of archives, he explained that archives offer not only a space to collect documents but also a site for commandment and commencement—a point where events and connections are created in a manner that renders further collection and additional events and connections meaningful.[38] Indeed, the collection of such data offers a commandment to collect more, organize more, and acquire more. It also highlights an archival logic of worth: this data is worthy because it is produced by Muslims. On one hand, this logic of worth is unsurprising considering the epistemic connection in securitized discourses between Muslims and terrorism. On the other, being a logic invested in the production of knowledge, not in its content, it assumes a degree of sameness and attributes this worth to the producers rather than to the content of their production. In other words, this knowledge is worthy because it is produced by Muslims, regardless of what this knowledge includes or how it could be used. Moreover, these Muslims are perceived to exhibit a degree of sameness that justifies the collection of their information in a coherent archive. The national identities, origins, views, and histories of these Muslims are negligible and beside the point precisely because of their Islam-ness—an overarching quality that negates all other modes or types of difference.

In this process, the transnational quality of these online platforms—with their ability to cross postcolonial national boundaries and to circumvent immigration or border authorities—offers a threat as well as an opportunity to the securitized narrative. On one hand, this transnational quality is reminiscent of terrorism itself as an epistemic category connected with Islam. *Domestic terrorism*, a term used to describe plots hatched locally by non-Muslims, is confined by its domesticity and nativity and is seen as tamed precisely because of these characteristics. *Islamic terrorism*, however, is perceived in securitized discourse as—by definition—foreign, antithetical to Western civilization, and transnational.[39] The nature of these websites and apps makes them therefore a threat in their aesthetic resemblance to terrorism.[40] Moreover, they animate the fantasy of isolationism that is used to further mark Muslims as noncitizens in their societies. On the other, this transnational nature offers securitized narratives an opportunity for validation. It is an additional sign of sameness and the centrality of Islam-ness in the identity of Muslims. This is to say that this transnational nature provides additional chances for further intensification of Islam as the sine qua non of Muslims.

Across much of the media coverage, ISIS is shown as using social media effectively and producing neatly cut and edited propaganda videos—abilities causing palpable alarm and concern. While campaigning for the presidency, Donald Trump argued that the United States (or the West) needs to effectively recapture its own technology: "ISIS is recruiting through the Internet. ISIS is using the Internet better than we are using the Internet, and it was our idea. What I wanted to do is I wanted to get our brilliant people from Silicon Valley and other places and figure out a way that ISIS cannot do what they're doing."[41] The same rhetoric has been deployed in various ways in relation to the Iranian nuclear program, where such technology in the hands of bad Muslims would doom the West. According to this understanding, Islam-ness, in its intensified form, has been deployed to police the practice and distribution of science. For instance, following 9/11, enrollment in flight schools by Muslims constituted a significant red flag for U.S. security officials. Muslim travelers to the United States were often asked about their training in flying, mathematics, and engineering as potential telltale signs of their terrorist abilities. The Islam-ness of those students and travelers becomes a reason to police their consumption (and production) of scientific knowledge and technological expertise.

Here, the intensification of Islam is coupled with another type of intensification—that of technology. Intensified, technology is perceived as neutral, useful, and context-free. It requires protection and presents danger when it falls into the "wrong hands." Yet the manner in which these wrong hands use technology further criminalizes these perpetrators while exonerating technology, which acquires more authority precisely because of its perceived neutrality. Technology is therefore evidence of but not a substitute for Western modernity, which becomes the only way to extricate Muslims from their overpowering identities and allow them to produce and consume science. In fact, in this narrative, without this modernization/reform/enlightenment that Islam has missed out on, science and technology can do more harm than good—to the West, that is.

CONCLUSION

In her book *Updating to Remain the Same*, Wendy Chun explained the key role played by habits and their formation in the economy of new media. Habits provided media with the necessary connection with their users, and media became important as they created habits. In this context, updating becomes a key feature of network time.⁴² Updating ensures the constant connection between media and users by enforcing the constant need for a change in order to maintain the habit-based connection. Time in this context is rather spiral. It keeps visiting the same space repeatedly while progressively changing. In the context of the Islamic websites discussed here, habits are not simply means to sustain the spiral temporality of networked time. Instead, habits are, indeed, pious practices on their own that carry meaning beyond their importance for a given website or even a particular genre of media. The habit here is one of inhabiting a different and variant pious environment that reenacts the idealized form of connection between pious Muslims and the history and religious authority that should help them create their pious life. In this context, and to follow Chun's analysis, these websites offer networks that connect pious users locally and globally, helping to collapse the spaces these users inhabit into the more focused moral life that they seek.

At another level, the interactive nature of these websites provides a different dimension whereby asking for a fatwa or even reading one involves an interaction with a larger community of largely anonymous simultaneous viewers. The connection between these viewers is forged not only at the point when they share consumption of common content but also at the point when they perceive their role in the production of such content. On one hand, fatwas are ultimately answers to questions posed by these viewers themselves. These questions recall the viewer's own problems and create further questions that are answered without being directly asked. On the other, the ecosystem that governs these fatwas invites clicking through, following links and seeking more answers and more conversations. In this framework, the animated networked archive of fatwas, opinions, news, and other related content can come alive only through the active action of the user. This action is not implied or assumed but rather is deliberate and taken with an actively cultivated agency. Pious users perform their piety precisely through this following and clicking through. Their piety relies equally on their belief in the importance of these online actions, to which they are invited by the platform that they are using. In other words, the platform does not aim to lure users into clicking habits but rather offers to facilitate the cultivation of deliberate pious acts of following, clicking, and reflecting on such pious online ecosystems.

Here, it is instructive to consider Bernie Hogan's distinction between frontstage and backstage in social media.⁴³ While frontstage offers a space for performing specific identities for the benefit of others, backstage provides a more intimate and secretive space that allows a different performance of the self. While the two remain connected, the difference is illustrative and operative in understanding differences in online behavior across different platforms. On fatwa websites, the secrecy and anonymity provided by the platform for those who ask questions or those who consume the fatwas resemble the backstage dynamics described by Hogan, while commenting and interacting represent aspects similar to frontstage. However, in this context, the connection between backstage and frontstage practices is more dynamic and integrated. In both spaces, the user strives to build a pious self through asking and listening/reading answers. Here, the pious act is not simply affirming authoritative opinions offered on these platforms. Instead, it is also the act of asking, of considering whether one needs to account for religious opinions before making decisions, and the desire to be connected to such a pious community that functions as self-fashioning strategy. In this context, the spaces between backstage and frontstage collapse precisely because of the self-reflective nature of the pious practice of fatwa seeking.

As mentioned before, these online platforms offer the opportunity to produce a transnational community that transcends the borders and boundaries of the postcolonial state. This community is built on structures of nostalgia and hypermodernism. On one hand, the online environment created by these platforms offers a nostalgic view to an imagined past, infused with a practice of piety that is rooted in self-observance and reflectiveness. On the other, this environment is facilitated through hyperreality produced online, surrounded by Google ads and placed in a browser alongside other aspects of one's online experience.

NOTES

1. The office of the grand mufti was established in Egypt in the middle of the nineteenth century, following the precedent of Istanbul. The mufti was responsible primarily for providing the state with legal and religious opinions when requested. See Jakob Skovgaard-Petersen, *Defining Islam for the Egyptian State: Muftis and Fatwas of the Dār Al-Iftā*, vol. 59 (Leiden: Brill, 1997).
2. On the hajj and its health and political implications in the early twentieth century, see Margaret Follett, " 'Such a Method of Doing Business': Local Shipping Agents, the Hajj, and the Divide Between Corporate and Colonial Priorities in Late 1870s Jeddah" (bachelor's thesis, Brown University, 2019); and Michael Christopher Low, "The Mechanics of Mecca: The Technopolitics of the Late Ottoman Hijaz and the Colonial Hajj"

(PhD thesis, Columbia University, 2015). On vaccination in the late nineteenth and early twentieth centuries, see Layla Aksakal, "The Sick Man and His Medicine: Public Health Reform in the Ottoman Empire and Egypt" (paper, Harvard Law School, 2003); Asa Briggs, "Cholera and Society in the Nineteenth Century," *Past and Present* 19, no. 1 (1961): 76–96; Liat Kozma and Diane Samuels, "Beyond Borders: The Egyptian 1947 Epidemic as a Regional and International Crisis," *British Journal of Middle Eastern Studies* 46, no. 1 (2019): 50–67; and Michel Lombard, Paul-Pierre Pastoret, and A. M. Moulin, "A Brief History of Vaccines and Vaccination," *Revue Scientifique et Technique—Office International des Epizooties* 26, no. 1 (2007).
3. https://tinyurl.com/3cdued37, accessed June 29th, 2022.
4. Perhaps since the 1989 fatwa by Ayatollah Ruhallah al-Khomeini proclaiming Salman Rushdie an apostate, the term *fatwa* has been used in Euro-American media to refer to rulings by Muslim authorities condemning others to death or justifying terrorism, etc. The reductionist securitized lens through which fatwa is often understood was further emphasized after the 9/11 terrorist attacks and the beginning of the War on Terror. It also followed the same trajectory of *madrasa*, a term that came to mean a religious Muslim school aiming at training fanatics. In this chapter, I will use the term *fatwa* in its original meaning: a legal opinion offered by a religious or secular authority—judge, jurisconsult, lawyer, etc. In the Sunni context, these opinions are almost always a response to a question and are never mandatory. Individual Muslims are expected to choose the fatwas that are more suitable for their own lives. Fatwas offered by the grand mufti of Egypt are equally optional. However, as the grand mufti is head of a government body, the Egyptian government almost always accepts his opinions. Notably, the word *madrasa* is simply the Arabic word for school, which continues to be used to refer to any school—religious or secular.
5. On Dar al-Ifta and its relation to the Egyptian state, see Skovgaard-Petersen, *Defining Islam for the Egyptian State*. See also Malika Zeghal, "Religion and Politics in Egypt: The Ulema of Al-Azhar, Radical Islam, and the State (1952–94)," *International Journal of Middle East Studies* 31, no. 3 (1999): 371–399, and "The 'Recentering' of Religious Knowledge and Discourse: The Case of Al-Azhar in Twentieth Century Egypt," *Schooling Islam: The Culture and Politics of Modern Muslim Education*, ed. Robert W. Hefner and Muhammad Qasim Zaman (Princeton, NJ: Princeton University Press, 2007).
6. https://tinyurl.com/453whcxt, accessed June 29, 2022.
7. See Mohamed Chawki, "Islam in the Digital Age: Counselling and Fatwas at the Click of a Mouse," *Journal of International Commercial Law and Technology* 5, no. 4 (2010): 165–180. On muftis, see also Muhammad Khalid Masud, Brinkley Morris Messick, and David Stephan Powers, *Islamic Legal Interpretation: Muftis and Their Fatwas* (Cambridge, MA: Harvard University Press, 1996).
8. See Derek John Illar, "Cyber Fatwās and Classical Islamic Jurisprudence," *John Marshall Journal of Computer and Information Law* 27, no. 4 (2010): 2. This online ecology extends to various aspects of online interactions, including, but not limited to, comments on videos and articles. See, for example, Liesbet van Zoonen, Farida Vis, and Sabina Mihelj, "YouTube Interactions Between Agonism, Antagonism and Dialogue: Video Responses to the Anti-Islam Film *Fitna*," *New Media and Society* 13, no. 8 (2011): 1283–1300. See also Annisa M. P. Rochadiat, Stephanie Tom Tong, and Julie M. Novak, "Online Dating and

Courtship Among Muslim American Women: Negotiating Technology, Religious Identity, and Culture," *New Media and Society* 20, no. 4 (2018): 1618–1639.

9. Barney Warf, "Islam Meets Cyberspace: Geographies of the Muslim Internet," *Arab World Geographer* 13, no. 3–4 (2010): 217–233.

10. On Saudi religious institutions and their relationship to the state, see Frank E. Vogel, *Islamic Law and the Legal System of Saudi: Studies of Saudi Arabia*, vol. 8 (Leiden: Brill, 1999); Joseph Nevo, "Religion and National Identity in Saudi Arabia," *Middle Eastern Studies* 34, no. 3 (1998): 34–53; Madawi Al-Rasheed, *Contesting the Saudi State: Islamic Voices from a New Generation*, vol. 25 (Cambridge: Cambridge University Press, 2006); Muhammad Al-Atawneh, *Wahhābī Islam Facing the Challenges of Modernity: Dār Al-Iftā in the Modern Saudi State* (Leiden: Brill, 2010); and David Commins, *Islam in Saudi Arabia* (Ithaca, NY: Cornell University Press, 2015).

11. See Jon W. Anderson, "The Internet and Islam's New Interpreters," *New Media in the Muslim World: The Emerging Public Sphere* 2 (2003): 45–60; Vit Šisler, "The Internet, New Media, and Islam: Production of Islamic Knowledge and Construction of Muslim Identity in the Digital Age" (PhD diss., Charles University, Prague, 2011).

12. See, among others, Hussein Ali Agrama, "Ethics, Tradition, Authority: Toward an Anthropology of the Fatwa," *American Ethnologist* 37, no. 1 (2010): 2–18, and *Questioning Secularism: Islam, Sovereignty, and the Rule of Law in Modern Egypt* (Chicago: University of Chicago Press, 2012).

13. See, for instance, Šisler, "The Internet, New Media, and Islam"; David Herbert, Tracey Black, and Ramy Aly, "Arguing About Religion: BBC World Service Internet Forums as Sites of Postcolonial Encounter," *Journal of Postcolonial Writing* 49, no. 5 (2013): 519–538; Stef Van den Branden and Bert Broeckaert, "Living in the Hands of God: English Sunni e-Fatwas on (Non-)Voluntary Euthanasia and Assisted Suicide," *Medicine, Health Care and Philosophy* 14, no. 1 (2011): 29–41; and Alexandre Caeiro, "The Making of the Fatwa: The Production of Islamic Legal Expertise in Europe," *Archives de Sciences Sociales Des Religions*, no. 155 (2011): 81–100.

14. See Mansur Aliyu et al., "A Preliminary Investigation of Islamic Websites' Design Features That Influence Use: A Proposed Model," *Electronic Journal of Information Systems in Developing Countries* 58, no. 1 (2013): 1–21; Waleed Mugahed Al-rahmi et al., "Information Technology Usage in the Islamic Perspective: A Systematic Literature Review," *Anthropologist* 29, no. 1 (2017): 27–41.

15. On al-Qaraḍāwī and his role, see Gary R. Bunt, "Islam Interactive: Mediterranean Islamic Expression on the World Wide Web," *Mediterranean Politics* 7, no. 3 (2003): 172.

16. Most work done on differences between traditionalist (*salafi*) and moderate (*wasati*) approaches on internet-based platforms focused on the question of Muslim minorities in the West. See, for example, Uriya Shavit, "The Wasati and Salafi Approaches to the Religious Law of Muslim Minorities," *Islamic Law and Society* 19 (2012): 416–457.

17. It is worth noting that Shia websites and platforms offer a set of different questions, which are not analyzed here. See Susanne Olsson, "Shia as Internal Others: A Salafi Rejection of the 'Rejecters,'" *Islam and Christian-Muslim Relations* 28, no. 4 (2017): 409–430.

18. On Dar al-Ifta in contemporary Egyptian politics, see Yohanan Manor, "Inculcating Islamist Ideals in Egypt," *Middle East Quarterly* (2015) https://www.meforum.org/5480

/inculcating-islamist-ideals-in-egypt, accessed June 29, 2022; Nathan J. Brown, "Contention in Religion and State in Postrevolutionary Egypt," *Social Research* 79, no. 2 (2012): 531–550; *Post-revolutionary Al-Azhar*, vol. 3 (Washington, DC: Carnegie Endowment for International Peace, 2011).

19. On the use of the internet and social media to extend nonvirtual messages, see Dale F. Eickelman and Jon W. Anderson, *New Media in the Muslim World: The Emerging Public Sphere* (Bloomington: Indiana University Press, 2003). Internet-based platforms became popular for government institutions only after the Arab Spring; see Marc Lynch, "After Egypt: The Limits and Promise of Online Challenges to the Authoritarian Arab State," *Perspectives on Politics* 9, no. 2 (2011): 301–310; Ralf Klischewski, "When Virtual Reality Meets Realpolitik: Social Media Shaping the Arab Government-Citizen Relationship," *Government Information Quarterly* 31, no. 3 (2014): 358–364; and Charles Hirschkind, "From the Blogosphere to the Street: Social Media and Egyptian Revolution," *Oriente Moderno* 91, no. 1 (2011): 61–74. However, the use of mass media in Islamic preaching has a much longer history that extends as far back as 1970s. On these media, see Armando Salvatore, "Social Differentiation, Moral Authority and Public Islam in Egypt: The Path of Mustafa Mahmud," *Anthropology Today* 16, no. 2 (2000): 12–15; Susanne Olsson, *Preaching Islamic Revival: Amr Khaled, Mass Media and Social Change in Egypt* (London: Tauris, 2015); and Yasmin Moll, "Islamic Televangelism: Religion, Media and Visuality in Contemporary Egypt," *Arab Media and Society* 10 (2010): 1–27.

20. Skovgaard-Petersen, *Defining Islam for the Egyptian State*, 59. In relation to al-Azhar, another key Islamic institution in Egypt, see Zeghal, "Religion and Politics in Egypt."

21. See, for example, https://tinyurl.com/y2fj8fre.

22. Nadav Samin, "Dynamics of Internet Use: Saudi Youth, Religious Minorities and Tribal Communities," *Middle East Journal of Culture and Communication* 1, no. 2 (2008): 206–207. On the impact of these scholars on Muslims around the world, see, for example, Vit Sisler, "Cyber Counsellors: Online Fatwas, Arbitration Tribunals and the Construction of Muslim Identity in the UK," *Information, Communication and Society* 14, no. 8 (2011): 1136–1159

23. On Fatwa-Online, see Samin, "Dynamics of Internet Use"; Rusli Rusli, "Progressive Salafism in Online Fatwa," *Al-Jami'ah: Journal of Islamic Studies* 52, no. 1 (2014): 205–229; Roxanne D. Marcotte, "Fatwa Online," in *Political Islam and Global Media: The Boundaries of Religious Identity*, ed. Noha Mellor and Khalil Rinnawi (London: Routledge, 2016); and Illar, "Cyber Fatwas and Classical Islamic Jurisprudence."

24. On dissemination of Islamic legal opinions over television, see, among others, Yael Warshel, "It's All About Tom and Jerry, Amr Khaled and Iqra, Not Hamas's Mickey Mouse: Palestinian Children's Cultural Practices Around the Television Set," *Middle East Journal of Culture and Communication* 5, no. 2 (2012): 211–245; and Osama Kanaker and Zulkiple Abd Ghani, "Animation Programs of Islamic Television: A Study of Al-Hijrah Television Channel," *Al-'Abqari: Islamic Social Sciences and Humanities* 282, no. 3646 (2016): 95–109.

25. Bettina Gräf, "Islamonline.Net: Independent, Interactive, Popular," *Arab Media and Society* 4 (2008): 1–21.

26. Ermete Mariani, "The Role of States and Markets in the Production of Islamic Knowledge On-Line: The Examples of Yūsuf Al-Qaraḍāwī and Amru Khaled," in *Religious Communities on the Internet* (Uppsala: Swedish Science Press, 2006).

27. Shavit, "The Wasati and Salafi Approaches"; Richard Papík, "The Internet, New Media, and Islam: Production of Islamic Knowledge and Construction of Muslim Identity in the Digital Age" (2010).
28. Mona Abdel-Fadil, "The Framing of the Islam Online Crisis in Arab Media," in *Media and Political Contestation in the Contemporary Arab World*, ed. Lena Jayyusi and Anne Sofie Roald (New York: Palgrave Macmillan, 2016).
29. Y. Salhein, "State-Society Relations in Digital Fatwas: A Study of Islamweb in Qatar" (master's thesis, Hamad Bin Khalifa University, Ar-Rayyan, Qatar, 2018).
30. See, for example, Rohit Khare and Adam Rifkin, "Weaving a Web of Trust," *World Wide Web Journal* 2, no. 3 (1997): 77–112; Jennifer Golbeck, "Trust on the World Wide Web: A Survey," *Foundations and Trends® in Web Science* 1, no. 2 (2008): 131–197; and Jingwei Huang and Mark S. Fox, "An Ontology of Trust: Formal Semantics and Transitivity" (paper presented at the 8th International Conference on Electronic Commerce, Fredericton, New Brunswick, Canada, 2006). In relation to internet use in the Islamic world, while most researchers focus on web commerce, they offer important insight on how users deal with and trust their interactions online. See Shrafat Ali Sair, "Examining Factors Affecting the Acceptance and Adoption of Mobile Commerce Through the Consumers' Lens in Pakistan" (thesis, National College of Business Administration and Economics, Lahore, 2019); A. A. Chaudhry, "Determinants of Users Trust for Branchless Banking in Pakistan," *Journal of Internet Banking and Commerce* 21, no. 1 (2016), https://www.icommercecentral.com/open-access/determinants-of-users-trust-for-branchless-banking-in-pakistan.php?aid=67474, accessed June 29, 2022; Maha Helal, "An Investigation of the Use of Social Media for E-Commerce Amongst Small Businesses in Saudi Arabia" (PhD thesis, University of Salford, UK, 2017); and Mohammad Alzahrani, "An Acceptance Model for Citizen Adoption of Web-Based E-Government Services in Developing Countries: An Investigation in the Saudi Arabia Context" (PhD thesis, Flinders University, Adelaide, 2014).
31. Samin, "Dynamics of Internet Use."
32. See Göran Larsson, "The Death of a Virtual Muslim Discussion Group: Issues and Methods in Analysing Religion on the Internet," *Heidelberg Journal of Religions on the Internet* 1, no. 1 (2005), https://doi.org/10.11588/rel.2005.1.382, accessed June 29, 2022. See also P. H. Cheong, "Religion and the Internet: Understanding Digital Religion, Social Media and Culture," in *Religion and American Cultures: An Encyclopedia of Traditions, Diversity and Popular Expressions*, ed. G. Laderman and L. Leon (Santa Barbara, CA: ABC-CLIO, 2014).
33. See, among others, Vit Sisler, "The Internet and the Construction of Islamic Knowledge in Europe," *Masaryk University Journal of Law and Technology* 1 (2007): 205–217; Giulia Evolvi, "Hybrid Muslim Identities in Digital Space: The Italian Blog Yalla," *Social Compass* 64, no. 2 (2017): 220–232; Stine Eckert and Kalyani Chadha, "Muslim Bloggers in Germany: An Emerging Counterpublic," *Media, Culture and Society* 35, no. 8 (2013): 926–942; Güney Dogan, "Moral Geographies and the Disciplining of Senses Among Swedish Salafis," *Comparative Islamic Studies* 8, no. 1–2 (2014); Roxanne D. Marcotte, "Gender and Sexuality Online on Australian Muslim Forums," *Contemporary Islam* 4, no. 1 (2010): 8; and Sisler, "Cyber Counsellors."
34. Azimaton Abdul Rahman, Nor Hazlina Hashim, and Hasrina Mustafa, "Muslims in Cyberspace: Exploring Factors Influencing Online Religious Engagements in Malaysia," *Media Asia* 42, no. 1–2 (2015): 61–73.

35. Jean Baudrillard, *Simulacra and Simulation* (Ann Arbor: University of Michigan Press, 1994).
36. See Illar, "Cyber Fatwas and Classical Islamic Jurisprudence."
37. See, for example, https://www.islamweb.net/ar/fatwa/الفتوى-مركز/عن-ه.
38. Jacques Derrida, *Archive Fever: A Freudian Impression* (Chicago: University of Chicago Press, 1996).
39. Joseph Cox, "How the U.S. Military Buys Location Data from Ordinary Apps," Motherboard, November 16, 2020, https://www.vice.com/en/article/jgqm5x/us-military-location-data-xmode-locate-x.
40. Much has been written on the connections between Muslims and terrorism. For a critical race theory–based analysis, see, for example, Caroline Mala Corbin, "Terrorists Are Always Muslim but Never White: At the Intersection of Critical Race Theory and Propaganda," *Fordham Law Review* 86 (2017): 455. See also Lamiyah Zulfiqar Bahrainwala, "Where Time and Style Collide: The Muslim in U.S. Discourse" (PhD diss., University of Texas, 2016); and Madeline-Sophie Abbas, "Conflating the Muslim Refugee and the Terror Suspect: Responses to the Syrian Refugee 'Crisis' in Brexit Britain," *Ethnic and Racial Studies* 42, no. 14 (2019): 2450–2469.
41. The aesthetics of terrorism were discussed in many works, most prominently in Jasbir K. Puar's *Terrorist Assemblages: Homonationalism in Queer Times* (Durham, NC: Duke University Press, 2018). On the same topic, see, among others, Marshall Battani and Michaelyn Mankel, "Terrorist Aesthetics as Ideal Types: From Spectacle to 'Vicious Lottery,'" *Contemporary Aesthetics* 15, no. 1 (2017): 5; GerShun Avilez, "The Aesthetics of Terror: Constructing 'Felt Threat' in *Those Bones Are Not My Child* and *Leaving Atlanta*," *Obsidian* 13, no. 2 (2012): 13–28; Maria Flood and Florence Martin, "The Terrorist as *ennemi intime* in French and Francophone Cinema," *Studies in French Cinema* 19, no. 3 (2019): 171–178; and Jenifer Chao, "Portraits of the Enemy: Visualizing the Taliban in a Photography Studio," *Media, War and Conflict* 12, no. 1 (2019): 30–49.
42. Team Fix, "5th Republican Debate Transcript, Annotated: Who Said What and What It Meant," *Washington Post*, December 15, 2015, https://www.washingtonpost.com/news/the-fix/wp/2015/12/15/who-said-what-and-what-it-meant-the-fifth-gop-debate-annotated/?utm_term=.ec0e781eec60.
43. Wendy Hui Kyong Chun, *Updating to Remain the Same: Habitual New Media* (Cambridge, MA: MIT Press, 2016).
44. Bernie Hogan, "The Presentation of Self in the Age of Social Media: Distinguishing Performances and Exhibitions Online," *Bulletin of Science, Technology and Society* 30, no. 6 (2010): 377–386.

PART III

NARRATIVES

AHMED RAGAB, TERENCE KEEL, AND MYRNA PEREZ SHELDON

Scholarship on science and religion has long been framed within a set of renewable narratives. From Galileo's trial before the Vatican to Scopes's role in a Dayton courtroom, stories of science and religion have been canonized, circulated, and retold. These stories continue to offer a foundation for much of the public and scholarly thinking about science and religion. Beyond these historical anecdotes, normative narratives—whether of conflict, conciliation, or independence—provide a mainframe in which investigations, discussion, and explorations are nested.

This part approaches narratives from a different viewpoint in order to focus on the political, ethical, and moral work done by storytelling. Narratives of science and religion are rooted and invested in a certain production of *science* and *religion* as concepts, categories, and social forces. These terms stand in for sets of virtues, such as objectivity, neutrality and replicability in relation to science or spirituality, divinity, and devotion in the case of religion. The contributors in this part explore how such narratives develop and what role they play in inscribing Euro-American modernity and white supremacy at the heart of normative or descriptive writings. In other words, we consider not only the content of narratives of science and religion (are they too simplistic? are they accurate?) but also the manner by which they operate to create a frame of reference.

Thus, for the purpose of this discussion, narratives stand as distinct from rhetoric. When we identify narratives, we are not merely indexing certain patterns of speech or types of argumentation. Nor do narratives simply refer to a temporality (say, of the modern) or a geography (say, of the West)—imaginary as these categories may be. Instead, they represent the epistemic infrastructure that gives these speech patterns, arguments, and logics a foundation. They endow categories with meaning through links and interruptions, concordance and distinction, and generalizations and specifications.

Ultimately, narratives are entanglements of meaning making. For instance, in the entanglement characterizing the temporality of the modern, fourteenth-century Aztec pyramids are compared not to plague-devastated London of the same period, but instead to ancient Greece in a manner that renders the Aztec always ancient and always behind. In the same way, Bronze Age African artifacts are placed alongside twenty-first-century African weaving or woodwork in a museum exhibit. Similarly, Europe extends to include Poland, Greece, and Russia but not Turkey. And the Middle East shares a geographical extent interchangeable with Islam. These narrative entanglements are not themselves arguments or descriptions; rather, they form the spaces from which arguments emerge. At the same time, the survival of these narratives is intractably connected to their seeming simplicity and the structural investment in their circulation.

The chapters in this part ask how these narratives come to be and what political, social and moral roles they play. Erika Lorraine Milam's chapter, "Secular Grace in the Age of Environmentalism," looks at narrative making at the intersection of science, religion, gender, and whiteness. Milam recounts the life and work of Jane Goodall and Dian Fossey, each of whom leveraged Western intuitions about feminine knowledge and mystical propensity to advocate for the conservation of great apes in various locations in Africa. Milam argues that it was a secular vision of grace, explicitly in contrast to traditional modes of masculine science, that enabled a feminine scientific authority in these contexts. In particular, she highlights how the use of nonsectarian spirituality allowed these women to claim intuitive knowledge about the apes that they studied, as well as the moral authority to push for the conservation of great ape habitat. Throughout the chapter, she draws attention to how these narratives of secular grace and white femininity enabled and relied on colonial power in Rwanda and Tanzania.

In "Performing Polygenism: Science, Religion, and Race in the Enlightenment," Suman Seth revisits scientific narratives of human origin in the Enlightenment period and their relation to debates around slavery. He investigates the rise of polygenism—a scientific theory advanced by Euro-American naturalists that there had been separate creations for different human groups. He considers polygenism not simply as a justification for slavery but rather as a historically significant performative heresy that also intended to reorganize knowledge and authority in the Euro-American context. In this sense, polygenism cannot be simply attributed to a political ploy to justify slavery when such justification was more soundly offered by religious narratives. Instead, Seth's retelling presents the roots of narratives of difference, which intersected but survived transatlantic slavery.

Finally, Joseph Graves analyzes narratives of human origin. In his chapter, "Out of Africa: Where Faith, Race, and Science Collide," he traces how scientists and theologians in the nineteenth and twentieth centuries were reluctant to accept Africa as the point of origin of humanity. Drawing on both a historical analysis of evolutionary debates and his own biological research, he highlights the stakes of this debate and demonstrates that the contests between science and religion were directed squarely at racist efforts to deny Africa a key role in the story of humanity.

Although these chapters draw on a critical historical sensibility, they keep a keen eye on contemporary social and political dilemmas. Yet their goal is not to simply offer historical correctives. Instead, this part looks to unveil the structures, spaces, and historical and political processes that govern the politics of science and religion.

CHAPTER 9

SECULAR GRACE IN THE AGE OF ENVIRONMENTALISM

ERIKA LORRAINE MILAM

Attention is the beginning of devotion.

—Mary Oliver[1]

Jane Goodall. Dian Fossey—these names are known to us as celebrity scientists who captured the hearts of families that watched their *National Geographic* television specials. Scenes of chimpanzees playfully flickered in their living rooms. Gorillas chewed celery and thistles. As science studies scholar Donna Haraway argued thirty years ago, Goodall and Fossey served as white, female intermediaries between the perceived civilization of their viewers and the tropical wilderness through which they trekked.[2] Having spent years living in close proximity to these elusive great apes, they translated primate social worlds for eager listeners, looping through the wild to the divine.

In their colloquial scientific publications, Goodall and Fossey came to represent a spiritual-epistemological connection women could potentially establish with wild communities of animals through their experiences studying the social behavior of chimpanzees and gorillas, respectively.[3] Often paired, thanks to the sponsorship of Louis S. B. Leakey and the National Geographic Society, Goodall and Fossey provided contestable grounds for claiming experience with a feminine sublime. I argue that their descriptions of interactions with wild primates contained moments of a spiritual-epistemological grace—in which truth was revealed to them about the inner workings of nature—enabled by their identity as white women and established tropes for each as saint and martyr.[4]

9.1 Peter Else, *Mandrillus Sphinx* (2019), acrylic on wood, reproduced with kind permission. To see this painting in full color, please see his website: https://www.peterelse.com/.

When insights into nature's order from Goodall or Fossey were framed as "revealed" truth, as unearned, this obscured the substantial work required to generate the knowledge to which they laid claim.[5] The ecofeminist and science studies literature of the 1980s amplified the idea that a feminine way of knowing could act counter to that of mainstream science, still conceptualized as male.[6] Recall the incredible success enjoyed by psychoanalyst Clarissa Pinkola Estés with *Women Who Run with the Wolves*, as well as by novelists like Barbara Kingsolver, Gloria Naylor, and Alice Walker, and poets like Mary Oliver, all of whom invested women with special ways of knowing and navigating the natural world.[7] Consumers of these writings assumed a feminine connection not just to nature but also to the ability of the supranatural to transcend culture, allowing for synergistic book sales at the border of magical realism and ecofeminism.[8] In this context, a naturalized environmental grace thus acted to both empower the capacity of Goodall and Fossey to speak on behalf of the animals they studied and undercut their authority as careful, disinterested scientists.[9]

Who then was being redeemed and from what? In granting white women the capacity to heal the estrangement of the capitalist West from nature, narratives of their heroic journeys reinforced assumptions that the technologies of Western culture were destructive, masculine in design, and in need of repair, and that the locations to which Goodall and Fossey traveled (and from which they returned) were transcendent spaces removed from historical time and geographic location, such that foreigners could enter and find a connection to a naturalized divine that was unavailable to people who called Tanzania or Rwanda home.[10] Local residents were often framed as threatening the integrity of the land and its animal inhabitants.[11] Despite the widespread nature of this rhetoric, we must not lose sight of the fact that marginalized and colonized peoples have long traditions of reflecting on and engaging with nature.[12]

This chapter explores tropes of Goodall as saint and Fossey as martyr, drawing on both their own writings and the reception of their research in the popular media and in the history of science. Their roles as oracles of nature were both of their own crafting and, once they gathered steam, beyond their direct control.[13] By weaving the spirituality of ecofeminism, feminist epistemology, and women in primatology into a single conversation about the fabric of public culture in the last quarter of the twentieth century, I suggest that moments of environmental grace in this period were central to the presentation of women as authorities on nature and especially primates in the wild. When gendered, a feminine environmental grace paralleled the spontaneous epiphanies of brilliance attributed to men, albeit stripped of individual agency.[14] By removing evidence of their hard work and scientific analysis from their narratives, these moments amplified the ability of Goodall and Fossey to defend the animals they studied and warn of coming environmental harm.

SAINT: JANE GOODALL AND THE CHIMPANZEES OF GOMBE

"Jane paved the way for all of us," Biruté Galdikas wrote in 1995.[15] She continued by quoting Roger Fouts's description of Goodall: "Jane's is a humble science. She asks the animals to tell her about themselves." By characterizing Goodall's science as one of collaboration rather than control, Galdikas repeated well-worn criticisms of Goodall's approach as outside of mainstream research in animal behavior but transformed these objections into salutary field practice. In Stephen Jay Gould's introduction to a 1988 republication of Goodall's first memoir, *In the Shadow of Man*, he did much the same: "When we read about a woman who gives funny names to chimpanzees and then follows them into the bush, meticulously recording their every grunt and groom, . . . [we] wonder if she represents forefront science or a dying gasp from the old world of romantic exploration." But, he added, "Jane Goodall's work with chimpanzees represents one of the Western world's great scientific achievements."[16]

Time alone in the wilderness was key to Goodall's descriptions of her early years at Gombe in the 1960s. She wrote about how she initially felt lonely when her mother returned to England after five months in Tanzania but quickly "accepted aloneness as a way of life."[17] Each day she followed the chimpanzees; each night she worked on her notes. The Tanzanian camp staff made sure she was stocked with food and supplies so she could devote her days and nights to understanding and documenting chimpanzee behavior. Looking back on these months of felt solitude—staff were present in camp—Goodall wondered if she "might have become a rather peculiar person" if she had maintained this schedule for over a year.[18] She repeated this account years later, although in the new version, the awakening of her senses became a dawning awareness of the world she was to inhabit in the coming years of research, a world replete with spiritual meaning. Goodall wrote that after her mother left, she "was able to penetrate farther and farther into a magic world that no human had explored before—the world of the wild chimpanzees."[19] She wrote, "The longer I spent on my own, the more I became one with the magic forest world that was now my home. Inanimate objects developed their own identities and, like my favorite saint, Francis of Assisi, I named them and greeted them as friends."[20] Goodall emphasized that as she grew "closer to animals and nature," she also became closer to herself and "more in tune with the spiritual power I felt all around." She wrote, "For those who have experienced the joy of being alone with nature, there is really no need for me to say much more; for those who have not, no words of mine can ever describe the powerful, almost mystical knowledge of beauty and eternity that come, suddenly, and all unexpected. The beauty was there but moments of true awareness were rare."[21]

In *Shadow of Man*, Goodall posited that scientists in their long quest for truth were never able to "provide a platform for man's ancient belief in God and the spirit." And yet the two pursuits seemed interwoven to her, and she asked, "Who in the silence of the night or alone in the sunrise has not experienced—just once perhaps—a flash of knowledge 'that passeth all understanding'? And for those of us who believe in the immortality of the spirit, how much richer life must be."[22] Published in 1971, *Shadow of Man* recounted Goodall's first years in Gombe at a distance of a decade. She wrote as a young woman, married and recently a mother, having just earned her PhD in zoology at Cambridge with Robert Hinde. In retrospect, Goodall has intimated that "snide remarks in the 1960s and 1970s about the '*Geographic* cover girl'... rankled more than I had admitted."[23] In 1986, she published her extensive analysis of chimpanzee behavior to date, *The Chimpanzees of Gombe*, and the book's positive reception among scientists boosted her confidence.[24] By the time she wrote *Reason for Hope* in 1999, much had changed. She had divorced amicably and remarried, and then her second husband, Derek Bryceson, passed away in 1985 after a difficult struggle with cancer. In this retold account of her first years in Gombe, the transformative potential of the forest, the silence, and the sunrise meant far more than "a flash of knowledge."

Aloneness was not just an obstacle to be overcome in Goodall's writings; it was accompanied by the sensual pleasure of freedom and independence rarely available to women in the world back home—both were elements of ecofeminism as it developed in the 1980s.[25] Recalling her first years in the field, Goodall wrote that she used to dread the early morning climb to "the Peak," her name for the rocky outcrop from which she could survey chimpanzee activity in Gombe. She had to leave the warmth of her bed when it was still dark and plunge through cold, wet grass to get there. She soon took to bundling her clothes in a plastic bag and carrying them with her. "There was no one to see my ascent, and it was dark anyway," she wrote. "Then, when I knew there were dry clothes to put on at my destination, the shock of the cold grass against my naked skin was a sensual pleasure. For the first few days my body was criss-crossed by scratches from the tooth-edged grass, but after that my skin hardened."[26] Goodall emerged from this intimate experience more confident and physically stronger. Rare moments of true insight into the forest world came during these times when she felt alone. In *Reason for Hope*, this story dropped out, but she did note that "the forest, and the spiritual power that was so real in it, had given me the 'peace that passeth understanding.'"[27]

Such fleeting moments of wilderness-inspired ecstasy cemented Goodall's belief in God. Her trust in the power of nature had started out strong.[28] She wrote of her belief when she began her study that wild animals would never hurt her for they would sense that she meant them no harm.[29] Leakey, she added, encouraged her in this but also made sure she took no unnecessary risks like coming between a mother and her young or confronting a wounded animal. Even those hazards, she had reckoned, were far less risky than living in a city. Then in 1974, she entered the Cathedral of Notre Dame, and as she stared at the Rose Window, the organist filled the enormous space with the opening chords of Bach's Toccata and Fugue in D Minor. Goodall felt as though the music entered and possessed her whole self: "That moment, a suddenly captured moment of eternity, was perhaps the closest I have ever come to experiencing ecstasy, the ecstasy of the mystic."[30] Yet it took her time, she wrote, to learn how to keep the transformative potential of the forest with her after these moments passed.[31]

Twelve years later, back in Gombe following her husband's death and seeking to reground herself in what she knew best, Goodall experienced another such moment of "heightened awareness." I quote the passage at length to give you a sense of how she narrates this experience:

> It is hard, impossible really—to put into words the moment of truth that suddenly came upon me then. . . . It seemed to me, as I struggled afterward to recall the experience, that *self* was utterly absent; I and the chimpanzees, the earth and trees and air, seemed to merge, to become one with the spirit power of life itself. . . .
>
> Suddenly a distant chorus of pant-hoots elicited a reply from Fifi. As though wakening from some vivid dream I was back in the everyday world,

cold, yet intensely alive. When the chimpanzees left, I stayed in that place—it seemed a most sacred place—scribbling some notes, trying to describe what, so briefly, I had experienced. I had not been visited by the angels or other heavenly beings that characterize the visions of the great mystics or the saints, yet for all that I believe it truly was a mystical experience.[32]

She compared this moment of true understanding with flashes of scientific insight that take place only with the mind, calling her own experience (by way of contrast) "a flash of 'outsight'" in which she "had known timelessness and quiet ecstasy, sensed a truth of which mainstream science is merely a small fraction." She continued, "And I knew the revelation would be with me for the rest of my life, imperfectly remembered yet always within. A source of strength on which I could draw when life seemed harsh or cruel or desperate."[33] What a pity, she concluded, that so many people are convinced "that science and religion are mutually exclusive."[34]

Goodall's conversion to a life of devotion—to the spiritual and physical care of nature—took place around the same time as the second of these experiences. In October 1986, she embarked on what she would later describe as a "spiritual journey." The call to the hard road of environmentalism tested her strength and her courage; yet "like Saint Paul," she wrote, "I found myself unable to 'kick against the pricks.'"[35] An airplane flight over Gombe National Park in Tanzania served as her road to Damascus. She looked down and saw clearly for the first time how the forest habitat of the chimpanzees had shrunk to the restrictive edges of the park itself. This journey followed a transformative conference at which she realized "the extent to which the chimpanzees across Africa were vanishing." Looking back, she recalled, "When I arrived in Chicago I was a research scientist. . . . When I left I was already, in my heart, committed to conservation and education."[36] To secure a future for chimpanzees, Goodall decided to craft a new life spreading compassionate awareness of the plight of nonhuman species and a new spiritual connection with all living creatures.

After Goodall's conversion to conservation, she devoted her life to spreading the gospel of hope, "hope for the future of life on earth."[37] Before COVID, she was ever on the move. In the introduction to *Hope*, Goodall enumerated a typical seven-week lecture tour in the United States: 27 cities, 32 airplanes, 71 lectures, 170 media interviews, 32,500 people in audiences. Two or three times a year she allowed herself a "roosting period"—no more than three weeks at home, largely devoted to writing. She visited Gombe "only a few times a year." She wondered, looking back, if she would make the same choices knowing the grueling path ahead and answered yes. "I didn't have to make the choice, for my life, it almost seems, was taken over by a force far too strong to fight against."[38]

Goodall's earnest message, recounted in *Reason for Hope: A Spiritual Journey* and repeated to reverent audiences around the world, relies on four fundamental bases: "(1) the human brain; (2) the resilience of nature; (3) the energy and enthusiasm that is found or can be kindled among young people worldwide; and (4) the indomitable human spirit."[39] For her audiences, however, Goodall herself is their reason for hope. On Thursday, September 29, 2016, I attended a talk she gave to the Philomathean Society at the University of Pennsylvania. The excitement in the audience was palpable before she began, and those gathered fell into a quiet hush when she began speaking in a clear, calm voice. They were intent on catching every word. The Jane Goodall Institute, for which she so diligently raises money, works with women in small communities in Africa who live adjacent to protected areas, providing them with a livelihood premised on maintaining and regenerating the endangered forests. Goodall speaks, too, about the horrors to which animals around the world are subjected in the name of science, agriculture, zoos, and human health.[40]

Since the 1980s, Goodall has been transformed into a revered symbol of care for the environment, humanity, and our intertwined futures. She framed this understanding in terms of a nondenominational universal "Spirit within" each person that would also enable a "reunion with the Spiritual Power that we call God, or Allah, the Tao, Brahma, the Creator of whatever our personal belief prescribes."[41] The task she set before each of her readers was to hasten their own moral evolution without investment in any particular religious practice.[42] She wrote, "We will have to evolve, all of us, from ordinary, everyday human beings—into saints! Ordinary people, like you and me will have to become saints, or at least mini-saints."[43] For those who are atheists, she concluded that this "does not make any difference. . . . A life lived in the service of humanity, a love of and respect for all living things—those attributes are the essence of saintlike behavior."[44] At a minimum, it is clear that these are the goals she has set for herself. "I really believe that the essence of the message I share," Goodall insists, "comes to me from outside; as if I am an eolian harp with strings vibrating to an invisible wind."[45]

MARTYR: DIAN FOSSEY AND THE MOUNTAIN GORILLAS OF KARISOKE

In her accounts of her time in Rwanda in *Gorillas in the Mist*, Fossey melded seamlessly her desire to study the social relationships of mountain gorillas and her conviction that she needed to actively defend the fate of those animals she devoted her life to understanding. She had established the Karisoke Research Centre the afternoon of September 24, 1967. It was situated in the saddle terrain between two mountains from which she took the name: Karisimbi and Visoke.[46] In her description of the early days of her research, the game at lower altitudes had begun to grow more meagre, and as a result, the "antipoacher work became a greater part of Karisoke's day-to-day fight for survival of the gorillas, as well as for other animals that occupied the saddle area of the forest."[47] Fossey recounted with joy the feeling of finding traplines and cutting them down before they ensnared their intended victims. She developed expertise in finding and destroying spring traps, pit traps, neck rope nooses, and, rarely, tree-trunk stockades. Gorillas were not usually the targets of these snares, she noted, but when they accidentally became entangled, the effort to free themselves left wounds that became infected and often led to their death.[48] She lamented, as had Leakey, that the mountain gorilla might go extinct in the same century it had been scientifically identified.[49] Knowledge of the gorillas and environmental action were thus intertwined throughout her narrative.

Karisoke was Fossey's second attempt to establish a research center for mountain gorillas. The first site, located in Zaire (now the Democratic Republic of the Congo), lasted for less than a year. She was arrested and spent two weeks confined in Rumangabo, the park and military headquarters for the Kivu Province where she had set up camp.[50] After realizing she had been "earmarked for the general," Fossey wrote, she managed to engineer her escape with promises of being a willing captive if only she could regain her cash stored just over the border in Uganda. She managed to escape with not only her life but also her car, data, photographic equipment, and two chickens. After a further series of interrogations in Uganda, she was told she "would be shot on sight" if she tried to reenter Zaire. Undeterred, when she met with Leakey to discuss the options, they quickly decided she should return to the Virunga Mountains but stay on the Rwandan side of the border. "There still were gorillas to find and mountains to climb. It was like being reborn."[51] According to journalist Alex Shoumatoff, "Dian Fossey spent eighteen years on and off among the mountain gorillas of Rwanda. She was to them what Jane Goodall is to the chimpanzees of Tanzania: she devoted her life to them and made us aware of their existence."[52]

In *Gorillas in the Mist*, Fossey wrote that she had longed to go to Africa for many years "because of what that continent offered in its wilderness and great diversity of free-living animals."[53] After she established her home among the mountain gorillas in Karisoke, she reflected that when she thought of Africa, she recalled not visions of open savannahs but "the montane rain forest of the Virungas—cold and misty, with an average rainfall of seventy-two inches."[54] She remembered vividly her first encounter with gorillas: "Sound preceded sight. Odor preceded sound in the form of an overwhelming musky, barnyard, human-like scent."[55] Several groups of gorillas moved in different areas of the mountains and over time she came to know all the group members who traversed her area.

Fossey described two moments central to her commitment to protect the gorillas at all costs. The first of these happened early in her time in Rwanda. In 1968, she booked a flight over the eight peaks that constitute the Virungas: two are active volcanoes and located in the former Zaire; the other six lie dormant and are shared by the former Zaire, Rwanda, and Uganda. She described the flight as an "ethereal experience"—her words echo Goodall's own conversion.[56] "From the air," she recalled, "it was possible to comprehend just how much of the park had been usurped for cultivation. . . . The devastated forest was pocked with smoke from burning *Hagenia* trees where small land plots were being cleared for pyrethrum cultivation. The pillage extended up to 8860 feet on Visoke and 9680 feet on Karisimbi."[57] She lamented that the western slope of Visoke, leading toward Karisimbi and Mikeno, constituted the "heartland, or the core of the Virunga Volcanoes . . . [and] will probably become the last stronghold for the mountain gorilla."[58]

The second was devastating. Digit, one of the gorillas that Fossey had come to know well, was killed defending the members of his group. His body had five wounds, "any one of which could have been fatal," from spears thrown into both the front and the back of his body.[59] Poachers had decapitated him and cut his hands off. Fossey had first met Digit in 1967 when he was a "bright-eyed, inquisitive ball of fluff," and she named him Digit because one of his middle fingers appeared to have once been broken, and although it healed, it remained twisted.[60]

A decade later he served a crucial role as sentry to his group. The other individuals escaped unharmed, thanks to Digit's ability to hold off six antelope poachers and their dogs—his head and hands appeared to be collected as an afterthought.[61] "There are times when one cannot accept fate for fear of shattering one's being," she wrote. "From that moment on, I came to live within an insulated part of myself."[62] She and Ian Redmond, a researcher who spent eighteen months at Karisoke and was especially interested in the parasites carried by gorillas, debated whether or not to tell the press of Digit's death.[63] Fossey's "greatest fear was that the world would climb evangelistically onto a 'save the gorilla' bandwagon upon hearing of Digit's death. Was Digit going to be the first sacrificial victim from the study groups if monetary rewards were to follow the news of his death?"[64] In the end, she launched the Digit Fund so that his death was not in vain.[65] Six months later two more gorillas from the same group were shot defending their young. According to an article published in *Vanity Fair* after her death, "The mountain gorilla proved to be as good a fund-raising animal as the panda or the whale."[66] Fossey used revenue generated by the Digit Fund and the outcry over these additional deaths to establish independent antipoacher patrols.[67]

Fossey counted the members of these patrols, especially cattle-herder Mutarutkwa, as among her "truest friends" and considered their actions part of her "active conservation."[68] She reserved special attention for the problem of conservation in the so-called third world. Most conservation efforts in Africa she decried for ignoring the social and cultural complexities of each country's bureaucracies, the basic needs of their citizens, and the widespread corruption among local officials.[69] She argued that, under such conditions, active conservation must precede theoretical conservation: ensuring the availability of food for the gorillas, cutting traps set for other animals, maintaining stringent surveillance of park boundaries, and strictly enforcing penalties (including imprisonment) against all poachers. Fossey repeated many stereotypes about newly independent African countries, but her anger was directed equally at the naïveté of American and European nonprofits.[70] In order for conservation efforts to work, she argued, they must enroll local populations to make tourism more profitable than poaching. Her frustrations with conservation practices in the 1970s fit with transformations afoot more broadly in the field. Her methods did not.

In the first article she wrote for *National Geographic*—"Making Friends with Mountain Gorillas"—Fossey emphasized her hard-won familiarity with individual gorillas.[71] Concerned that the gorillas would greet with suspicion any "alien object that only sat and stared," she instead chose to act like a gorilla herself. "I imitated their feeding and grooming, and later, when I was surer what they meant, I copied their vocalizations, including some startling deep belching noises."[72] The article ended by sounding a grave note regarding the gorillas' danger of future extinction, but mostly it contained stories about the individuals she had come to know during the previous six years of research.

The first version of the article had looked quite different. An editor at the magazine worried that it would compare unfavorably with Goodall's essays because it rambled: "Her house catches fire. She kidnaps a native boy. The boy's father steals her dog. She 'holds for ransom' another native's cattle (to get her dog back). She dresses up in a Halloween mask to spook the natives; they're disturbing her gorillas."[73] Although Fossey revised the article to include more of her "scientific findings," she later incorporated these stories in her episodic monograph, *Gorillas in the Mist*.

If solitary connection with nature had proved crucial to Goodall's salvation, the rhetoric contributed equally to Fossey's downfall. Rather than concluding with a note of hope, Fossey's book and her story end in distress:

> For Digit, Uncle Bert, Macho, Lee, N'Gee, and so many other gorillas, I sorrow that I was too late to change the quixotic ways of many Europeans and Africans who, in hoping for a brighter dawn tomorrow, have yet to realize that avoidance of very basic conservation issues may ultimately push Beethoven, Icarus, Nunkie, their mates, and their progeny into the mountain mist of times past.[74]

When one of the gorilla groups she followed began to forage at the edge of the park boundary, she related, villagers took note of the seasonal change. They would yell "*Ngagi! Ngagi!*" (Gorilla! Gorilla!) as the gorillas appeared on the bluff overlooking their fields. Fossey followed the group, climbing onto the bluff herself, and was surprised one year to hear a new series of shouts: "'*Nyiramachabelli! Nyiramachabelli!*' they cried, meaning 'The old lady who lives in the forest without a man.'" She added, "Although my new name was pleasantly lyrical, I would have to admit that I did not like its implications."[75] Fossey embraced the name Nyiramachabelli, and both it and her implied Africanized identity were often repeated in stories after her death.

Fossey was murdered in 1985. She was found in her cabin the morning of December 26 by men she had employed for the antipoacher patrols. Her body had been repeated slashed by a machete, the weapon left at the scene. Like Goodall, Fossey had moved to Africa to study the behavior of great apes, although she started six years later. She, too, had become an ardent advocate for conservation. When she had published *Gorillas in the Mist* in 1983, only 242 gorillas remained in the Virunga Mountains—almost half the number estimated to have lived in the area in the early 1960s. One journalist wrote in *Entertainment Weekly*, "To the millions who learned of her either through her book, *Gorillas in the Mist*, or the movie of the same name, Dian Fossey became the very model of secular sainthood: the Mother Teresa of the mountain ape."[76] Her murder turned her into a martyr.

Like most martyrs, Fossey's legacy is deeply contested: Did she die in valiant defense of her principles, or had she descended into irrecoverable madness?[77] Her death prompted a flurry of media attention. Beloved nature writer Farley Mowat penned a biography, *Woman in the Mists*, interweaving passages from her journals with his own narrative glue, which Michael Apted, in turn, adapted for the screen.[78] Sigourney Weaver played Fossey in the film and found herself sympathetic to the woman: Fossey, she said, "felt the urgency of the gorilla situation the way no one else did. . . . She felt she had no support, so she resorted to violence."[79] The film creates a vivid scene of Weaver in Fossey's Halloween mask, indicating Fossey's darker side and leaving to viewer imagination how far she really went in her antipoacher activities.

By lore, Fossey cultivated enemies through the violence of her "active conservation" methods. Ian Redmond, the scientist with whom she collaborated the longest, compared her to the "front-line members of Greenpeace" for her willingness "to take direct action to protect animals at risk (even without the legal authority to do so)."[80] One account of her life contended, "Fossey was less a scientist than a crusader, a woman single-minded in her campaign to save the mountain gorillas."[81] According to a later exposé, penned by Harold Hayes, Fossey "had people beaten, burned their property, killed their cattle and abducted their children" and was known to have antagonized everyone, including government officials in Rwanda, her fellow researchers, and graduate students eager to contribute to her research. The more Hayes wrote, the more explicit his condemnations became: "Fossey had shot at her enemies, kidnapped their children, whipped them about the genitals, smeared them with ape dung, killed their cattle, burned their property, discredited their work, and sent them to jail."[82] These accusations came from taking her correspondence seriously: many of these are actions she told her friends she committed. Not all of them believed her.

Having learned to live with the gorillas, Fossey died alongside them.[83] She was buried in the gorilla graveyard where she had lovingly interred Digit, as well as other gorillas who had died during her time in Rwanda. She had told a Rwandan journalist that she hoped her gravestone would include a single word: Nyiramachabelli.[84] The permanent marker bore a longer inscription: "No one loved gorillas more. Rest in peace dear friend, eternally protected in this sacred ground for you are home where you belong."

After Fossey's death, long-time collaborator Redmond (by then living in Bristol, England) told a journalist, "Dian as an individual was in many ways like the gorillas . . . in that if you are easily put off by bluff charges, screaming and shouting, then you probably think that the gorillas are monsters. But if you are prepared to sidestep the bluff charges and temper and shouting and get to know the person within . . . then you'll find that Dian, like the gorilla, was a gentle, loving person."[85] Others were less kind. Bill Weber had been frustrated by Fossey's lack of interest in studying gorilla behavior toward the end of her life. He asked, "Why did she hardly ever go out to the gorillas if they were her life-motivating force?"[86]

If not her passion in life, then Fossey's gruesome murder called attention to the plight of the mountain gorillas. Mowat's *Woman in the Mists* emphasized her commitment, above all, to conservation, which is often how she is now remembered. One reviewer of his book wrote, "It is difficult to pin down her major contributions to science, partly because her publications are not equal to her reputation, and partly because her research did not keep pace with her efforts at conservation."[87] More recently, scholars have called for a reckoning of Fossey's legacy, noting her deeply colonialist habits in both the imperious way she treated her staff and her willingness to engage in abusive conservation methods outside the law.[88]

Yet Fossey's legacy continues. According to David Attenborough, looking back on these events, whatever her personal faults or scientific shortcomings had been, "she undoubtedly saved the species."[89] Novelist Margaret Atwood awarded her a "Saint Day" in *The Year of the Flood*, in which she was "consecrated to interspecies empathy" and remembered for being martyred while defending the Virunga gorillas.[90] In 2018, Portia de Rossi gave her wife, Ellen DeGeneres, a birthday present of a charitable donation sufficient in size to build the Dian Fossey Gorilla Fund a permanent structure in Rwanda. The highly publicized gift and their subsequent pilgrimage to see the gorillas in person brought Ellen to tears—Fossey had been a childhood hero.[91] In her words, "Fifty years after

Dian started her work, I'm honored to help carry on her legacy. Together we can help save these beautiful creatures."[92] A recent census of mountain gorillas in the Virungas has suggested their numbers now exceed 1,000.[93] That makes it the only population of wild apes known to be increasing.

CONCLUSION

In the 1980s and '90s, as Goodall and Fossey wrote accounts of their years studying great apes in the wild, concern for the environment and its wild inhabitants resonated with very different kinds of spiritual communities. Together their early narratives reinforced the notion that wild primates lived in paradise, in endangered Edens threatened by the ever-encroaching worlds of the Fallen, which readers could access for themselves through the written accounts of lives spent studying the behavior of the inhabitants.[94]

Since the 1980s, this perspective has been criticized by both conservationists and conservatives. Scholars in the environmental humanities lamented in retrospect that highlighting some natures as worthy of scientific and conservationist attention, as Edenic, implied other natures already tainted by human presence were irredeemable and further erased the laborious interventions required in conservation and restoration.[95] At the same time, conservatives decried environmentalism as a new form of pseudoreligion, with Rachel Carson as its first patron saint.[96] The well-known writer of scientific and medical fiction Michael Crichton has served as a vocal critic of leftist politics, decrying environmentalism as "the religion of choice for urban atheists."[97] He cynically dismisses environmentalism as mere religious rhetoric, empty of true belief. Crichton's claims resonated with many of the narrative tropes that writers like Goodall and Fossey employed in their writing in previous decades. The difference, of course, is that Crichton identified the salvation of most conservationists through the symbolic consumption of organic food, whereas Goodall and Fossey called for a global transformation in humanity's economic relationship with the natural environment.

By deconstructing the religious overtones of their public images, I do not intend to amplify depictions of Goodall and Fossey as saints of modern environmentalism. Instead, I hope that by exploring the religious language in which their public personas are framed, we can open up new questions about the reasons millions of readers have devoted their time and money to a handful of key figures, like Goodall and Fossey, who are invested with the capacity for oracular truth telling about environmental causes.[98] As Haraway articulated in *Primate Visions*—thereby importing these religious tropes into academic discussions of primatology—the power of such stories lay in their capacity to rewrite the Edenic narrative, with women as purveyors of normative truth, and, I would add, to replace the stigma of the apple with the capacity for cultural redemption.[99] Haraway made race central to her argument too, noting that racialized tropes of so-called natives were intimately bound to the rhetoric of Africa as part of either a preindustrial, prelapsarian paradise that served as fertile ground for cultural and biological anthropologists keen on elaborating an alternative to the warlike present or a reenvisioned developing world beset by modernizing capitalist forces, often in the form of poachers and corrupt government officials, that threatened Eden's very existence. Either way, primatologists represented for Haraway voyagers from afar, a visual link to the civilization of their readers that required the women's isolation within the wilderness of their surroundings. In short, for those scientists who would speak on behalf of animals, their status as white and Western secured a ready audience.

These are powerful narratives to be sure. They succeeded in turning Jane Goodall and Dian Fossey into household names. In the political climate of the late Cold War, their stories of self-sacrifice resonated with feminists who sought to establish that by ignoring women as creators of knowledge, science was suffering from observational and theoretical biases. Yet their very popularity threated their status as scientists.[100] This locked Goodall and Fossey on one side of a perceived popular/professional binary, with less public-facing primatologists securely on the other.[101] In the colloquial scientific press of the late Cold War, Goodall's career remains unfreighted by controversy, while Fossey was transformed into the "avenging angel" Nyiramachabelli. Each has been hailed as an oracle warning of the likely extinction of the animals to which their public identity was intimately linked. In turn, readers have accorded them devotion.

Although ecofeminist literature of the era reinforced the possibility of environmental grace as a path to truth for women, this at once empowered these primatologists' capacity to speak with fervor on behalf of the animals they studied. Revealed truth came unexpectedly in moments of solitary observation, enabled by their status as women seeking new understandings of our closest living nonhuman relatives—an environmental grace that moved inward to illuminate humanity and upward to illuminate God, as well as outward to illuminate nature.

NOTES

For stimulating and thoughtful discussion, I thank the participants of the Symposium on Critical Approaches to Science and Religion, organized by Myrna Perez Sheldon, Ahmed Ragab, and Terence Keel, as well as at Suman Seth's graduate seminar at Cornell University, and the interdisciplinary community of historians, philosophers, and biologists at the John J. Reilly Center for Science, Technology, and Values at the University of Notre Dame. At Princeton, Spence Weinreich, Jonathan Baldoza, and Catherine Clune-Taylor offered crucial feedback, along with the participants of our HOS Program Seminar. To Myrna, I owe a sincere debt for engaging in multiple enriching conversations, for inviting me to participate in the symposium, and for providing generous guidance throughout.

1. Mary Oliver, *Upstream* (New York: Penguin, 2016).
2. Donna Haraway, *Primate Visions: Gender, Race, and Nature in the World of Modern Science* (New York: Routledge, 1989).
3. Jane Goodall, *In the Shadow of Man* (1971; repr. New York: Mariner, 2010); Jane van Lawick-Goodall, *My Friends the Wild Chimpanzees* (Washington, DC: National Geographic Society, 1967); Dian Fossey, *Gorillas in the Mist* (1983; repr. New York: Mariner, 2000), "Making Friends with Mountain Gorillas," *National Geographic* 137 (1970): 48–67, "More Years with the Mountain Gorillas," *National Geographic* 140 (1971): 574–585, and "The Imperiled Mountain Gorilla," *National Geographic* 159 (1981): 501–523.
4. By invoking a naturalized grace, I mean something close to Kai Wiegandt's analysis of secular grace as "creature-feeling" in J. M. Coetzee's fiction as a way of escaping the interpretive baggage of "secular" that has dogged scholarly conversations that pit religion and science against each other: Kai Wiegandt, "The Creature-Feeling as Secular Grace: On the Religious in J. M. Coetzee's Fiction," *Literature and Theology* 32, no. 1 (2017): 69–86. See also Bernard Lightman, ed., *Rethinking History, Science, and Religion: An Exploration of Conflict and the Complexity Principle* (Pittsburgh, PA: Pittsburgh University Press, 2019). I hope it also allows me to sidestep philosophical debates over the possibility of grace within a secularized understanding of the world, as tackled by Dana Freibach-Heifetz and Barbara Harshav, *Secular Grace* (Leiden: Brill Rodopi, 2017).
5. Their dismissal as scientists is often attributed to the emotional valence of their writings and empathetic (rather than objective) interactions with their research subjects as individuals; see Amanda Rees's lucid explanation in "Reflections on the Field: Primatology, Popular Science and the Politics of Personhood," *Social Studies of Science* 37, no. 6 (2007): 881–907. Their value as *oracles*, speaking for the animals, could be another way through this material.
6. Carolyn Merchant, *Death of Nature: Women, Ecology, and the Scientific Revolution* (San Francisco: Harper & Row, 1980); Evelyn Fox Keller, *A Feeling for the Organism: The Life and Work of Barbara McClintock* (San Francisco: Freeman, 1983); Riane Eisler, *The Chalice and the Blade: Our History, Our Future* (San Francisco: Harper & Row, 1987); Londa Schiebinger, *The Mind Has No Sex? Women in the Origins of Modern Science* (Cambridge, MA: Harvard University Press, 1989); Helen Longino, *Science as Social Knowledge: Values and Objectivity in Scientific Inquiry* (Princeton, NJ: Princeton University Press, 1990); Lorraine

Code, *What Can She Know? Feminist Theory and the Construction of Knowledge* (Ithaca, NY: Cornell University Press, 1991); Sandra Harding, *Whose Science? Whose Knowledge? Thinking from Women's Lives* (Ithaca, NY: Cornell University Press, 1991).

7. Clarissa Pinkola Estés, *Women Who Run with the Wolves: Myths and Stories of the Wild Woman Archetype* (New York: Ballantine, 1992); Barbara Kingsolver, *The Bean Trees* (New York: Harper & Row, 1988); Gloria Naylor, *Mama Day* (Boston: Ticknor & Fields, 1988); Alice Walker, *The Temple of My Familiar* (New York: Harcourt, 1989).

8. On the connection of primatologists to "wild" women tropes, see Rebecca Bishop, "Writing the Body Wild: Primatological Narrative and Spaces of Animality," *Culture, Theory and Critique* 49, no. 2 (2008): 133–148.

9. On long-standing connections between ecological and spiritual convictions, see Nicolas Howe, *Landscapes of the Secular: Law, Religion, and American Sacred Space* (Chicago: University of Chicago Press, 2016); Mark R. Stoll, *Inherit the Holy Mountain: Religion and the Rise of American Environmentalism* (New York: Oxford University Press, 2014); and Thomas Dunlap, *Faith in Nature: Environmentalism as Religious Quest* (Seattle: University of Washington Press, 2004).

10. On tropes of the Edenic tropics, see Candace Slater, *Entangled Edens: Visions of the Amazon* (Berkeley: University of California Press, 2002), esp. "Warrior Women, Virgin Forests, and Green Hells," 81–102; Tony Bennett, *Pasts Beyond Memory: Evolution, Museums, Colonialism* (New York: Routledge, 2004); Elizabeth DeLoughrey and George B. Handley, eds., *Postcolonial Ecologies: Literatures of the Environment* (New York: Oxford University Press, 2011); and Elizabeth Hennessy, *On the Backs of Tortoises: Darwin, the Galapagos, and the Fate of an Evolutionary Eden* (New Haven, CT: Yale University Press, 2019).

11. On the racialization of access to wilderness in the United States that shaped the assumptions of *National Geographic* readers, see Carolyn Finney, *Black Faces, White Spaces: Reimagining the Relationship of African Americans to the Great Outdoors* (Chapel Hill: University of North Carolina Press, 2014); Karl Jacoby, *Crimes Against Nature: Squatters, Poachers, Thieves and the Hidden History of American Conservation* (Berkeley: University of California Press, 2001); and Mark David Spence, *Dispossessing the Wilderness: Indian Removal and the Making of the National Parks* (New York: Oxford University Press, 1999).

12. E.g., Camille Dungy, ed., *Black Nature: Four Centuries of African American Nature Poetry* (Athens: University of Georgia Press, 2009); Rob Nixon, *Slow Violence and Environmentalism of the Poor* (Cambridge, MA: Harvard University Press, 2013); and Jacob Dlamini, *Safari Nation: A Social History of the Kruger National Park* (Athens: Ohio University Press, 2019).

13. Kathryn Lofton's *Oprah: The Gospel of an Icon* (Berkeley: University of California, 2011) was especially helpful in framing the celebrity and globalization of inspiration proffered by Goodall. Like that book, this chapter is not an exposé but an analysis of what Goodall and Fossey have come to represent.

14. See, e.g., Sarah-Jane Leslie et al., "Expectations of Brilliance Underlie Gender Distributions Across Academic Disciplines," *Science* 347, no. 6219 (2015): 262–265.

15. Biruté M. F. Galdikas, *Reflections of Eden: My Years with the Orangutans of Borneo* (Boston: Little, Brown, 1995), 386.

16. Stephen Jay Gould, introduction to *In the Shadow of Man*, revised ed., by Jane Goodall (Boston: Mariner, 1988), v.

17. Goodall, *Shadow*, 50.
18. Goodall, *Shadow*, 50.
19. Jane Goodall with Phillip Berman, *Reason for Hope: A Spiritual Journey* (New York: Grand Central, 2000), 71.
20. Goodall, *Hope*, 73. Robert Kohler describes zoologist Adriana Kortlandt's disparagement of "Goodall's field practice as the 'Saint Francis of Assisi' approach" in *Inside Science: Stories from the Field in Human and Animal Science* (Chicago: University of Chicago Press, 2019), 142.
21. Goodall, *Hope*, 72.
22. Goodall, *Shadow*, 246.
23. Goodall, *Hope*, 205
24. Jane Goodall, *The Chimpanzees of Gombe* (Cambridge, MA: Harvard University Press, 1986). On self-confidence, see Goodall, *Hope*, 205.
25. Gregg Mitman, "Life in the Field: The Sensuous Body as Popular Naturalists Guide," in *Primate Encounters: Models of Science, Gender, and Society*, ed. Shirley C. Strum and Linda M. Fedigan (Chicago: University of Chicago Press, 2000): 421–435.
26. Goodall, *Shadow*, 54.
27. Goodall, *Hope*, 181; Phil. 4:7 (King James Version): "And the peace of God, which passeth all understanding, shall keep your hearts and minds through Christ Jesus."
28. Jane Goodall, "Jane Goodall," in *Courage of Conviction*, ed. Phillip L. Berman (New York: Dodd, Mead, 1985), 74–93. About her childhood upbringing Goodall wrote, "We didn't talk much about religion in my family when I was growing up but we went to church fairly often, our values were Christian values, and the rules we had to obey were based on the Ten Commandments" (76). She writes about the role the writings of Pierre Lecomte du Noüy played in convincing her that humans are in an ongoing process of "moral evolution" (77); likely she read his *Human Destiny* (London: Longmans, Green, 1947).
29. Goodall, *Hope*, 63.
30. Goodall, *Hope*, xiii, 92.
31. Goodall, *Hope*, 84.
32. Goodall, *Hope*, 173–174.
33. Goodall, *Hope*, 175.
34. Goodall, *Hope*, 175.
35. Goodall, *Hope*, 206.
36. Goodall, *Hope*, 206. Thanks to Spence Weinreich for pointing to the parallels between conservation education and evangelization in this context.
37. Goodall, *Hope*, xv. There are two aspects of conversion that I find fascinating in these narratives: the stories of personal transformation and subsequent attempts to convince others to follow a similar path. Following Kenneth Mills and Anthony Grafton, I am interested in "the significant transformations which were fervently sought and sometimes achieved by individuals and groups *within* or just apart from dominant social and religious communities" rather than tracing conversion as a form of religious or cultural conquest. Kenneth Mills and Anthony Grafton, eds., "Introduction," in *Conversion: Old Worlds and New* (Rochester, NY: University of Rochester Press, 2003), xi.
38. Goodall, *Hope*, 208.

39. Goodall, *Hope*, 233.
40. One can draw easy parallels between Goodall's public image and Lofton's account of Oprah as a cultural, spiritual icon and an embodied mechanism of personal transformation. Lofton, *Oprah*.
41. Goodall, *Hope*, 199.
42. On the physicality of devotional practice without religion, see Amanda Lucia, *White Utopias: The Religious Exoticism of Transformational Festivals* (Berkeley: University of California Press, 2020). I am grateful to Myrna Perez Sheldon for pointing me to the guru-like reverence with which Goodall's fans can treat the physical traces of her presence in the world.
43. Goodall, *Hope*, 200.
44. Goodall, *Hope*, 202.
45. Goodall, *Hope*, 265–266.
46. Fossey, *Mist*, 25.
47. Fossey, *Mist*, 28.
48. Fossey, *Mist*, 30.
49. Fossey, *Mist*, xvii.
50. This account of her imprisonment and escape is taken from Fossey, *Mist*, 14–18.
51. Fossey, *Mist*, 17.
52. Alex Shoumatoff, "The Fatal Obsession of Dian Fossey," *Vanity Fair*, September 1986, https://www.vanityfair.com/style/1986/09/fatal-obsession-198609.
53. Fossey, *Mist*, 1.
54. Fossey, *Mist*, 14.
55. Fossey, *Mist*, 3.
56. Fossey, *Mist*, 36.
57. Fossey, *Mist*, 38.
58. Fossey, *Mist*, 38. As you will have noticed from these page references, Fossey established the plight of the mountain gorillas early in the book and repeated the message often.
59. Fossey, *Mist*, 259.
60. Fossey, *Mist*, 167.
61. On the trade in relics, see Patrick J. Geary, "Sacred Commodities: The Circulation of Medieval Relics," in *The Social Life of Things: Commodities in Cultural Perspective*, ed. Arjun Appadurai (Cambridge: Cambridge University Press, 1986), 169–191.
62. Fossey, *Mist*, 206.
63. As Redmond was interested in parasites, twelve of the gorillas who died when he was in residence were autopsied and checked for any internal and external parasites. The results of these autopsies are discussed in the book and are included, too, in an appendix at the end. In contrast to Goodall's memoirs, there is significantly more attention to death in *Gorillas in the Mist*.
64. Fossey, *Mist*, 207.
65. Fossey, *Mist*, 209.
66. Shoumatoff, "The Fatal Obsession of Dian Fossey."
67. Fossey, *Mist*, 222.
68. Fossey, *Mist*, 165–166.
69. Fossey, *Mist*, 241–242.

70. Fossey, *Mist*, 241.
71. Fossey, "Making Friends."
72. Fossey, "Making Friends," 51. On the necessity of reverse habituation—that is, changing oneself to match a primate community's "way of being in the world" in order to gain their familiarity and trust—when observing animals in the wild, see Barbara Smuts, "Encounters with Animal Minds," *Journal of Consciousness Studies* 8, no. 5–7 (2001): 285.
73. R. L. Conly, as quoted in Nina Strochlic, "The Renegade Scientist Who Taught Us to Love Gorillas," *Lost and Found: National Geographic*, August 31, 2017, https://news.nationalgeographic.com/2017/08/Dian-Fossey-national-geographic-archives/. On myths of so-called natives being scared by audiovisual tricks of Western media, see Brian Larkin, *Sound and Noise: Media, Infrastructure, and Urban Culture in Nigeria* (Durham, NC: Duke University Press, 2008).
74. Fossey, *Mist*, 242.
75. Fossey, *Mist*, 91.
76. Gene Lyons, "The Dark Romance of Dian Fossey," *Entertainment Weekly*, June 22, 1990. On martyrdom, see also Marguerite S. Schaffer, "A Transnational Wildlife Drama," *American Quarterly* 67, no. 2 (2015): 317–352.
77. As one example, Isaac and the martyrs of Córdoba, AD 711, see Kenneth Baxter Wolfe, *Christian Martyrs in Muslim Spain* (Cambridge: Cambridge University Press, 1988).
78. Farley Mowat, *Woman in the Mists: The Story of Dian Fossey and the Mountain Gorillas of Africa* (New York: Warner, 1987); Michael Apted, dir., *Gorillas in the Mist* (Universal City, CA: Universal Pictures, 1988), 129 min.
79. Nina J. Easton, "Film Makers in the Mist: Bringing Dian Fossey Story to Screen Proves to Be a True High Adventure," *Los Angeles Times*, September 16, 1988, https://www.latimes.com/archives/la-xpm-1988-09-16-ca-2300-story.html. Weaver had read Fossey's book before she was contacted about acting in the film. She remembers the experience of interacting with the gorillas as transformative. "You are kind of with your family only they are just that much more wonderful than we are," she told Ellen DeGeneres on *The Ellen Show*, March 23, 2018.
80. Ian Redmond, "Dr. Dian Fossey 1932–1985," *Oryx* 20, no. 4 (1986): 270. A scientific collaborator later published an essay in which she described Fossey as adopting a "Joan of Arc" mentality and mentioned that Fossey initially insisted on carrying a Beretta in case of encounters with poachers. Kelly Stewart, "The Gun," in *I've Been Gone Far Too Long: Field Trip Fiascoes and Expedition Disasters*, ed. Monique Borgerhoff-Mulder and Wendy Logsdon (Oakland, CA: RDR Books, 1996), 5–18.
81. Easton, "Film Makers in the Mist."
82. Harold Hayes, "The Dark Romance of Dian Fossey: Caring for Gorillas More Than People Was Fatal," *Life Magazine* 9, no. 11 (1986): 65; later developed into *The Dark Romance of Dian Fossey* (New York: Simon & Schuster, 1990).
83. Many theories have circulated about her murder: poachers, a disgruntled official who paid off someone she knew, even one of her scientific collaborators. When the latter caught wind that Rwandan police suspected him, he fled the research site and the country. Tried in his absence, he was sentenced to death by firing squad, although no one in Canada or the United States believes he was responsible. For example, see the list of possibilities in Shoumatoff, "The Fatal Obsession of Dian Fossey."

84. Sean Kelly, "American Naturalist Buried Among Her Beloved Gorillas," AP News, January 1, 1986, https://www.apnews.com/17968da883148c270711aaf5d2fa27e2.
85. As quoted in Shoumatoff, "The Fatal Obsession of Dian Fossey."
86. As quoted in Shoumatoff, "The Fatal Obsession of Dian Fossey."
87. David Chiszar, "Nyiramchabelli, the Lone Woman of the Forest: 'A Parable of That Magnificent Condescension,'" *Zoo Biology* 7 (1988): 381–383.
88. Michelle Rodrigues, "It's Time to Stop Lionizing Dian Fossey as a Conservation Hero," Lady Science, September 20, 2019, https://www.ladyscience.com/ideas/time-to-stop-lionizing-dian-fossey-conservation.
89. Zara Hayes, dir., *Dian Fossey: Secrets in the Mist* (Washington, DC: National Geographic Films, 2017), 3-episode miniseries, 141 min.
90. Margaret Atwood, *The Year of the Flood* (New York: Doubleday, 2009), 311.
91. *The Ellen Show*, February 1, 2018. The official YouTube video has been viewed over fifteen million times: https://www.youtube.com/watch?v=XrTDwpLECtA.
92. The Ellen DeGeneres Wildlife Fund, accessed December 29, 2019, https://theellenfund.org.
93. Max-Planck-Gesellschaft, "Number of Wild Mountain Gorillas Exceeds 1000," May 31, 2018, https://www.mpg.de/12057180/number-of-wild-mountain-gorillas-exceeds-1-0001.
94. Myrna Perez Sheldon and Naomi Oreskes, for example, have shown how the creation care movement emphasized stewardship of land as part of evangelical Christianity. Myrna Perez Sheldon and Naomi Oreskes, "The Religious Politics of Scientific Doubt: Evangelical Christians and Environmentalism in the United States," in *Wiley-Blackwell Companion to Religion and Ecology*, ed. John Hart (Hoboken, NJ: Wiley Blackwell, 2017), 348–367. On the long tradition of Christian faith through nature, see Thomas Dunlap, *Faith in Nature: Environmentalism as Religious Quest* (Seattle: University of Washington Press, 2004).
95. On the redemptive possibilities of ecological restoration, see Elizabeth Hennessy, "The Politics of a Natural Laboratory: Claiming Territory and Managing Life in the Galápagos Islands," *Social Studies of Science* 48, no. 4 (2018): 483–506; and Laura Martin, *Wild by Design: The Rise of Ecological Restoration* (Cambridge, MA: Harvard University Press, 2022).
96. Mark Hamilton Lytle, "Rachel Carson: Saint or Sinner?," *Human Ecology Review* 23, no. 2 (2017): 55–64.
97. Michael Crichton, "The Greatest Challenge Facing Mankind: Remarks to the Commonwealth Club of San Francisco," September 15, 2003, www.cs.cum.edu/~kw/crichton.html; Joanna Radin, "Alternative Facts and State of Fear: Reality and STS in an Age of Climate Fictions," *Minerva* 57 (2019): 411–431.
98. Neil Gaiman, *American Gods* (London: William Morrow, 2001).
99. Haraway, *Primate Visions*, especially "Apes in Eden, Apes in Space: Mothering as a Scientist for National Geographic" (133–185) and "Women's Place Is in the Jungle" (279–303).
100. The majority of reviews were positive, but see Peter Rodman, "Flawed Vision: Deconstruction of Primatology and Primatologists," *Current Anthropology* 31, no. 4 (1990): 484–486.
101. Shirley C. Strum and Linda Marie Fedigan, eds., *Primate Encounters: Models of Science, Gender, and Society* (Chicago: University of Chicago Press, 2000).

CHAPTER 10

PERFORMING POLYGENISM

Science, Religion, and Race in the Enlightenment

SUMAN SETH

What became known in the nineteenth century as polygenism—the theory that there had been separate creations for different human peoples—is an old heresy. The orthodox explanation for the origins of nations looked to the book of Genesis. Humans made up a family, all of us having descended from Adam and Eve. If more specificity was required, peoples found in each of the three regions of the medieval world were related to one of Noah's three sons (Ham, Shem, and Japhet), who had spread out from the settling point of the Ark on Mount Ararat. Hints of alternatives to these logics may be found even in early Christian texts,[1] yet it would be the discovery of the so-called New World that posed the most profound and pressing questions. From whence had these "new" peoples arisen? If one assumed an original unity and eventual diffusion, how had they crossed oceans so long ago, and why were they never mentioned in otherwise authoritative and seemingly exhaustive ancient accounts?

For the incendiary medical reformer who styled himself as Paracelsus, the puzzle presented by Native Americans was too much to handle through established means. Given biblical chronologies, he argued, Adam's children would not have had time to make their way from western Asia to the lands they now inhabited. "Some hidden countries," he wrote, "have not been populated by Adam's children, but through another creature, created like men outside of Adam's creation . . . one would well consider that these people are from a different Adam."[2] Paracelsus was not alone in his heterodoxy. Giordano Bruno is perhaps best known today for his support of the Copernican theory, yet his denial of the logics of Genesis presented a far greater problem. There were three original patriarchs, he asserted, only one of whom was Adam, who was the father solely of the Jewish peoples and not of the Ethiopians or Americans. A profound iconoclast who questioned multiple other doctrines, including the divinity of Christ and the doctrine of eternal damnation, Bruno would be burned at the stake by the Inquisition in 1600.

An expansive survey of polygenetic ideas is found in David Livingstone's detailed, complex, and precise *Adam's Ancestors: Race, Religion, and the Politics of Human Origins* (2008). Here, I am interested in a more narrow question: What was the role of scientific argumentation in defending and supporting polygenist claims during the Enlightenment? For Livingstone, science in this period should be seen as a kind of support for a more fundamental set of political reasonings and stakes. "Claims about comparative anatomy, cultural anthropology, natural history, regional climatology, and colonial history," he writes, "were all folded into the competing narratives of humanity's genesis that both monogenists and polygenists offered. Yet, underlying these scientific preoccupations were fundamentally political fixations: the elaboration of moral maps of the globe, the ethics of plantation economies, and the management of domestic polities."[3] I offer in this chapter a different take, not because I believe that polygenism was somehow apolitical but because I do not believe that scientific discourse, even that involved in debates about race, functioned in such superstructural ways during the eighteenth century. It is true that we can find more apparently instrumentalist arguments in the *nineteenth* century, where forms of inequality were, indeed, buttressed and defended on the naturalistic grounds that different races were of different species. A myriad of examples may be adduced, but the most straightforward is probably the sort of advocacy against abolition to be found in the work of the American polygenist and physician Josiah Nott.[4] In 1850, for example, he gave a lecture to the Southern Rights Association in Mobile, Alabama, on "The Natural History of Mankind, Viewed in Connection with Negro Slavery." In an essay in the book *Types of Mankind*, which he edited with the equally vicious George Gliddon, Nott laid out what he thought "natural history" could teach—namely, that Caucasians were natural rulers of other races and that "no philanthropy, no legislation, no missionary labours can change this law."[5] Polygenist science, for the American school of anthropology, could serve very explicitly as a tool for the defense of the South's peculiar institution. Having established the "fact" of multiple creations, nineteenth-century polygenists then deployed that fact in further arguments as to the current and ideal future place of nonwhite races.

Yet even in this case we should, as Terence Keel has persuasively argued, evince skepticism at the thought of nineteenth-century polygenism as merely proslavery politics in another guise, as a kind of scientific window dressing. Were Nott's aim simply to garner support for slavery, after all, there were easier and more likely methods than calling widely accepted foundational biblical narratives into question. Instead, Keel argues, it is more reasonable to "take Nott's science at face value and admit that although polygenism was used to support a proslavery agenda, Nott was driven to his position by what he understood to be scientific interests."[6] Polygenism could serve proslavery ends, but it was not reducible to them, even in the nineteenth century. Why this methodological point matters for both the history and the current state of understandings of race is a point to which I will return in the conclusion.

If we should be leery of instrumentalist explanations of the work done by polygenist arguments in the nineteenth century, we should be even more cautious about such claims in the eighteenth century, when direct links between antiabolitionist and polygenist claims are even harder to find. In seeking to understand what work scientific polygenism actually did in the eighteenth century, I would suggest that polygenist positions largely functioned negatively rather than positively. They worked as a weapon against established opinions and, most importantly, as a kind of signal of the author's willingness to buck contemporary norms: we might, were we inclined to anachronism, regard them as a species of "trolling." More formally, most evocations of polygenism from the end of the seventeenth century to at least the time of the French Revolution should be understood primarily as *performances of antiorthodoxy*. They did not do positive work, serving as the basis for novel causal arguments or conclusions. In almost every one of the examples I discuss here, statements about polygenism constituted the *end* of a discussion rather than the beginning—or even a step along the way.

In what follows, then, I am as interested in the kinds of arguments that were *not* made using scientific evidence as I am in those that were. I begin, in section I, with a quick survey of four texts—written by the French theologian Isaac La Peyrère (1596–1676); an author known only as L. P., who published in 1695; the French savant Voltaire (1694–1778); and the Scottish philosopher David Hume (1711–1776). An analysis of the first should make clear the kinds of purposes to which a theological polygenism could be put in the seventeenth century. The early modern polygenism of Paracelsus, Bruno, and La Peyrère, we will see, was intended not to destroy biblical narratives but to improve them. La Peyrère sought to replace flawed readings of the story told in Genesis with his own and thereby buttress a messianic vision for the future. The three subsequent examples demonstrate the essentially negative functions of naturalistic arguments in the Enlightenment. Polygenism, we will see, has little purpose in these latter texts beyond its iconoclasm.

In section II, I focus on the writings of the naval surgeon John Atkins. Here, we can be even more specific about the kinds of arguments Atkins would *not* make. Polygenism is used neither as a defense of slavery—Atkins was critical of the institution—nor as an explanation for his medical claims about the seeming idiosyncrasy of African diseases. In later editions of his works, in which he would retract his polygenism, what took its place was a different kind of performative subversion.

In section III, I examine the notorious arguments made by Edward Long concerning the humanity of Africans and their descendants. In his works, we begin to see the outlines of a crucial change in the purposes to which polygenism was put. He remained performatively antiorthodox, clearly reveling in repeated calumnies concerning amorous relationships between orangutans and African women. Yet there was now a value beyond its own utterance for claims about multiple human creations. Those who were writing before the 1760s, when the abolitionist movement found its first major proponents in Anthony Benezet and others, had little need to provide new defenses for slavery. As an institution, it was not under broad-scale attack. Writing in 1772 and 1774, however, Long clearly intuited the powerful support that monogenism could provide for notions of Christian brotherhood and care. His polygenism, then, can be understood as an attempt to blunt the force of a monogenetic critique of the slave system he sought so ardently to retain. Yet, and this is crucial, Long still did not use his polygenetic conceptions as a *positive* resource. As I have argued in detail elsewhere, Long's claims in his massive *History of Jamaica* cannot be read—as many have—as a defense of slavery on natural historical grounds. His defenses of slavery were, as was common at the time, legal rather than scientific. Moreover, while naturalism established, for Long, that Blacks and whites were different species (within the same genus), he would deploy antimaterialist arguments to defend his claims concerning the former's natural intellectual inferiority. While we may thus see in Long the outlines of positions that would become common in the nineteenth century, much would still be inchoate or even impossible.

Polygenism was a position at once political and theological, social and scientific. As a result, the examples given here provide a valuable mechanism through which to deconstruct common arguments concerning the relationship between science and religion in the eighteenth century. We are well past the point where scholars would contend that the Enlightenment saw the widescale secularization or naturalization of thought. Naturalistic ideas in the period remained thoroughly imbricated with theological conceptions. Yet relative weights did shift. Across multiple and varied contexts, one finds at once the separation of science and religion into increasingly distinct arenas, as well as—and relatedly—the rising importance of the former relative to the latter. One might be tempted, as a result, to see the move toward defending polygenism on naturalistic rather than theological grounds as an effect of a more profound reworking of the basis of sociopolitical power. Yet such an argument would unjustly separate entangled effects and causes. One might as easily suggest that Long, faced with the authority of biblical supports for monogenism *and* with critiques of slavery in the hands of opponents, turned to science defensively to support the legalistic arguments he already deployed. Certainly, that turn to naturalistic polygenism was not well-received. As Seymour Drescher has noted, Long's arguments concerning inherent Negro inferiority were, for decades, more likely to be cited in public by abolitionists delighted to find a straw man come to life than by their antagonists and were "usually unused or repudiated by antiabolitionist writers themselves."[7] There was nothing inevitable, in other words, about the deployment of scientific as opposed to religious claims in support of racist ends in the Age of Reason. The Atlantic slave trade, then, should be seen as one more locus for the simultaneous emergence of—and conflict between—distinct scientific and religious epistemologies.

I. NATURALISM AND ICONOCLASM

According to Isaac La Peyrère's own account, he found proof of long-held suspicions regarding the existence of men before Adam not in travelers' accounts of the New World but in his encounter with verses 12–14 in St. Paul's Epistle to the Romans. "As by one man sin entered into the world and by sin, death: so likewise death had power over all men, because in him all men sinned. For till the time of the Law sin was in the world, but sin was not imputed, when the Law was not." La Peyrère took this to mean that while the law entered the world only with Adam, sin had existed beforehand. Or, in Richard Popkin's pithy gloss: "If law began with Adam, then there must have been a lawless world before Adam, containing people."[8] From that exegetical clue, a great deal would flow, particularly concerning Adam's position as patriarch for only *some* people. In La Peyrère's account, the Bible was a description of history and destiny that applied only to the Jews. To them alone had the word of God been revealed. This meant that one need not consider the history of other peoples, for they were not part of the divine plan. Nor did one need to worry any longer about how to square the narrative of creation or the story of the deluge with new discoveries of men in far-flung locations. Ancient pagans had not derived from the same stock as the chosen people; inhabitants of the New World had not descended from the sons of Noah.

Far from being an isolated result, the insight rooted in Paul's epistle became the key to finding and interpreting polygenetic clues throughout the Bible. "La Peyrère argued at great length and in great detail," Popkin notes, "that item after item in Genesis proves that it is 'clearer than the Sun, That the men of the first creation were created long before Adam.'"[9] He added further that Eve was not the first woman; she was only the first in a Jewish line. While the line from Adam and Eve, however, was no longer part of a universal genealogy and only Jews were the subject of a divine history, their messianic salvation would include all others, pre- and post-Adam alike.[10] From a new understanding of the beginning of humankind, La Peyrère was led to a novel conception of its end as well. Perhaps seeing only this optimistic vision, La Peyrère was apparently shocked by the scandal that followed. His *Prae-Adamitae* (*Men Before Adam*) appeared in 1655 to wide condemnation. In February of the following year, he was jailed; in June, he agreed to repent, apologize to the pope, and convert from Calvinism to Catholicism.

La Peyrère was a believer. Many of those who would follow in his polygenetic footsteps, however, would see the doctrine of separate creations as a means of mocking—rather than salvaging—biblical exposition. Thus, one finds L. P. (presumably a nod to La Peyrère) offering a defense of rationalism in the face of biblical literalism in 1695. His work combined two essays, the first of which ridiculed the notion that the Noachian flood, if it had occurred, had spanned the globe. Why drown the earth and all its innocent creations, L. P. asked, when most of the world was uninhabited? Why destroy all on account of "a few Wanton and Luxurious *Asiaticks*, who might have been drown'd by a *Topical Flood*, or by a particular Deluge, without involving all the Bowels of the whole Mass, and the remote Creatures upon the face of the Earth, in the Ruin"?[11] The second essay then heaped the scorn of a natural historian on the idea that the world had been populated from a single spot, with animals and men traveling vast distances across many different climes to reach the locations in which they were now found. "How the Animals, that cannot endure the extremity of Cold," he scoffed, "should climb over inaccessible Mountains or Ice and Snow for many Thousands of Miles together, is hardly explicable to any thinking man."[12] Nor could one imagine that they had been carried by men. Why would one do this with dangerous creatures and leave behind so many "mild and useful ones"?[13] The logics that applied to animals would seem to apply to men as well. In any case, however, there was reason beyond this to doubt that all men were related. As others had before him, L. P. noted the problems with climatic explanations for differences in skin color between whites and Blacks. One found different-colored people in similar climates; whites did not seem to darken appreciably in warm countries, and Blacks did not lighten in cooler ones. He declared of Negroes, then, that "their Colour and Wool are Innate, or Seminal from their first beginning, and seems to be a Specifick Character, which neither the Sun, nor any curse from *Cham* could imprint upon them."[14] Arguments derived from "*Eastern* Rubbish, or Rabbinical Weeds" could supply no reasonable answer to the questions raised by natural history.[15]

As we shall see later, the tone in which John Atkins would express his musings about the several origins of mankind—matter of fact, if somewhat apologetic—would be the exception among polygenists in the first half of the eighteenth century. One might, indeed, say that the doctrine of special creations was a sarcastic position in this period. It was nowhere more so than in the writings of Voltaire, who turned La Peyrère's philo-Semitism on its head. Europeans, in Voltaire's view, had preceded Jews on the earth. The Jewish patriarch was merely a pale imitation of the original, and Jews would remain a race apart, however long they lived in Western lands.[16] Writing of Blacks in *The Philosophy of History*, he noted that it had now become common practice for the "curious traveler" passing through Leiden to make a stop to view the *reticulum mucosum* (now known as the Malpighian layer) of a dissected Negro.[17] This membrane, he added, is itself black and "communicates to negroes that inherent blackness, which they do not lose." The philosopher declared that "none but the blind" could make of "the whites, the negroes, the Albinoes, the Hottentots, the Laplanders, the Chinese, [and] the Americans" anything but distinct races, and if one were to ask about the origins of such races, the answer was equally obvious: "the same providence which placed men in Norway, planted some also in America and under the Antarctic circle, in the same manner as it planted trees and made grass to grow there."[18]

David Hume concurred with Voltaire in 1753, arguing that all races other than whites were inferior and that the differences among such races were innate, nature having made an "original distinction betwixt these breeds of men." His tone might be captured by reproducing his comments on a counterexample to his claim that no Negro had or would ever distinguish themselves through their intelligence. There were those in Jamaica, he noted, who spoke of one as a man of considerable learning. Hume dismissed the thought: "'tis likely," he wrote, "he is admired for very slender accomplishments like a parrot, who speaks a few words plainly."[19]

Almost a century separated Hume's and La Peyrère's forms of polygenism. In that period, close, literalist readings of biblical passages had gone from being solutions to being problems. By 1750, one can see that naturalism was a key part of the polygenist arsenal—but largely only as a means of rendering monogenism an absurdity. Voltaire did not need polygenism to support his anti-Semitism—and one may note that Jews do not appear as one of his separately created races. Hume's footnote, Colin Kidd has argued, "cut across the specific argument" of the essay to which it was appended "and stands in contradiction to one of the basic premises of his philosophy."[20] The entirety of L. P's argument relies on a destruction of the literalist position: polygenism is the result, but the doctrine of special creations is not then used to make another claim. "I see no way to solve this new face of Nature by old Arguments . . . unless some *New Philosopher* starts up with a fresh system," L. P. wrote. "In the meantime let them all be *Aborigines*."[21] Where La Peyrère, a theological polygenist, got quickly to his premise and moved from it to a reimagining of the history and future of humanity, scientific polygenists took the doctrine of special creations as their end point, not as a new beginning. The purpose of a naturalistic polygenism was to destroy an old position, not to defend or motivate a new one.

II. JOHN ATKINS: POLYGENISM, MEDICINE, AND MATERIALISM

Born in 1685, John Atkins began his medical training as a surgeon's apprentice and then joined the British navy, serving in multiple locations. In 1721, he sailed for the Guinea coast on one of two ships sent to suppress piracy there. After meeting with moderate success but suffering significant casualties, the ships sailed to Brazil and the West Indies before returning to England in 1723. Failing to find another position at sea, Atkins turned to writing. In 1724, he published *A Treatise on the Following Chirurgical Subjects*, the fifth chapter of which was concerned with "some *African* Distempers." The *Treatise* was restructured and republished in 1734 as *The Navy-Surgeon*, with the section on African diseases moved to an appendix. Atkins expanded on this material the following year in an account of his journeys in the early 1720s titled *A Voyage to Guinea, Brasil, and the West-Indies; In His Majesty's Ships, the Swallow and Weymouth*. Second but unaltered editions of both texts appeared in 1737, and a substantially modified third edition of *The Navy Surgeon* was published in 1742.[22]

Atkins's first effort at explaining the origins of different races appeared in his *Treatise* in the context of his discussion of the diseases characteristic of native inhabitants of the Guinea coast. Having discussed the "distempered" skin that resulted from a disease known as the croakra, he then turned to the cause of skin color. The natural cause for the dark skin and wooly hair of the inhabitants of much of Africa, he declared, "must ever perplex Philosophers to assign."[23] Malpighi, it was true, had located the cause in the color of the subcuticular mucus, but that had merely pushed the question back a step: Why was the mucus differently colored in different peoples? One could explain tanning under a hot sun by suggesting that the lightly colored mucus of fine European skin was warmed until its thinner parts were eliminated, leaving "the Remainer dark, as the clearest Liquors, they say, will leave some sediment."[24] But Atkins doubted that the darkness of Africans could be explained through this mechanism, offering five reasons for thinking that such mechanical explanations would be inadequate. First came the argument by analogy with animals. He noted that the sun did not seem to have the same effects on other animals in Guinea as it was assumed to have on humans. If the sun were responsible for producing the wooly hair characteristic of Africans, why did local sheep "have hair contrary to that closer Contexture of the Skin, which is supposed to contribute to the Production of Wool in the humane Species"? The second argument involved a comparison with Europeans. Why did white skin never turn black, and why did the skin of mulattos remain intermediary in color between the two extremes? Third, Atkins observed that one could not explain African skin color entirely by pointing to the tint of a subcuticular mucus, for the color of black skin changed as the higher layers of skin altered. Black feet and hands gradually whitened, Atkins claimed, "by Friction and constant use," and the cuticle paled after the skin was burnt or scalded. Fourth, one might compare Africans to Americans, the latter of whom lived under an equally burning sun but were not as black. Fifth and finally, not all Africans—despite their residence under the same sun—were the same color. "I saw one . . . who was woolly and, in every Respect else, a Negro . . . but in Colour," he reported. A conclusion seemed to follow by necessity. "From the whole," he asserted, without signaling any awareness of the radicalism of the claim, "I imagine that white and Black must have descended of different Protoplasts [i.e., progenitors], and that there is no other Way of accounting for it."[25] He would emphasize the point in his *Voyage to Guinea, Brasil, and the West-Indies* (1735). Describing there the physical differences between Guineans and "the rest of Mankind," he declared such differences to be so vast that "tho' it be a little Heterodox, I am persuaded the black and white Race have, *ab origine*, sprung from different colored first parents."[26]

In neither text did the discussion go beyond this polygenetic conclusion. Atkins did not, for example, use his conclusion to motivate a defense of slavery. In fact, he was opposed to the practice, in spite of the low opinion he had of African culture in general.

> When the Nakedness, Poverty and Ignorance of these Species of Men are considered, it would incline one to think it a bettering their Condition, to transport them to the worst of Christian Slavery; but as we find them little mended in those respects at the *West-Indies*, their Patrons respecting them only as Beasts of Burden; there is rather Inhumanity in removing them from their Countries and Families; here they get Ease with their spare Diet; the Woods, the Fruits, the Rivers, and Forests, with what they produce, is equally the property of all.[27]

"To remove *Negroes* then from their Homes and Friends," he would write later in the same text, "where they are at ease, to a strange Country, People, and Language, must be highly offending against the Laws of natural Justice and Humanity; and especially when this change is to hard Labour, corporal Punishment, and for *Masters* they wish at the D___l."[28]

Nor—and this is perhaps more surprising given that he was writing as a surgeon—did Atkins seek to connect his polygenetic arguments to his medical explanations. Despite writing about the diseases "peculiar" to the Guinea coast, he did not suggest that Africans suffered from specific diseases *because* they had been formed in a separate creation. In fact, the connections between his polygenetic and medical arguments were so slight that in the third edition of *The Navy Surgeon* (1742), he retracted the views he had earlier deemed "a little Heterodox" without changing his etiological explanations in the slightest. In then considering the differences between black and white bodies, he did not back off his criticisms of explanations that relied on the action of the sun. All five objections to climatic causality were reproduced verbatim. But Atkins now excised his conclusion: there was no mention of different protoplasts. Instead, he offered a different explanation—and an environmental one at that. His "Guess on this abstruse question" was now that differences in skin color and hair type were due to differences in the soil where generations had been bred.[29] Expanding his geographical perspective, Atkins acknowledged that it was not only Blacks and whites who seemed utterly different. Brazilians were a different hue than North American natives, East Indians were different than each of these, "and all as distinct as black and white, which happens I imagine chiefly, on account of the Soil each is bred and nurtured in." The soil changed the taste of plants when they were transplanted to different countries, and "Men by a like Analogy, change considerably be removing, even in their own Lives, and were they to marry and abide there."[30]

Understanding Atkins's avowal of polygenism in terms of a performance of antiorthodoxy helps to explain why he would include such a heterodox position in a text ostensibly on naval medicine without connecting the position to medicine at all. It also helps to explain why he would then disavow the position—presumably in the face of critique that made the isolated performance seem vastly less appealing. Finally, we might also be able to come to grips with what Atkins added to the new edition of his text as a kind of replacement for his other heterodoxy, for the third edition of *The Navy Surgeon* was, most generally, a very different book from its earlier versions. All of those had pitched themselves as practical works, guides to the day-to-day tasks of a surgeon at sea and in foreign lands. The 1742 edition, by contrast, was considerably more scholarly. Loaded with many more quotations and references to classical and modern works, the book now opened not by getting straight into a discussion of fractures but by considering the general question of the nature of sensation. Atkins added a "dissertation" on cold and hot mineral springs, and the introduction now included a disquisition on the life and changing fortunes of medical men in the navy, covering many of the same issues that would reach a much larger audience with the publication of Tobias Smollett's *The Adventures of Roderick Random* in 1748. Most crucially for our current purposes, the third and earlier editions ended with an appendix describing Atkins's "Physical Observations on the Coast of

Guiney." Yet, while previous editions had concluded by listing and discussing the four diseases peculiar to the inhabitants—the "sleepy Distemper," the "*Croakra* or itch, the "*Yaws* or Pox," and the "*Chicoes* or Worms"—the third added a long disquisition on the "Difficulty of Accounting for their Generation" of the last.

"There is not a more abstruse Subject in Physick," Atkins wrote, "than this Generation of Worms in human Bodies, whether in the Stomach or Guts. In the Tongue, Gums, Nose, Brain, Bladder, Liver, Heart, Lungs, Blood, at all which Places, our Philosophical Transactions relate: or this Sort called Chico's, in the Legs or Feet in the *West-Indies*, and other external Parts at *Africa*." He was something of a master at understatement. At issue here was what he termed "Equivocal Generation, (a Generation without Seed or Parents)," a phrase often used synonymously in the seventeenth and eighteenth centuries with the more familiar spontaneous generation.[31] Just as calling polygenism "a little Heterodox" vastly underplayed how fraught the position was, so, too, did calling the origin of worms in the body "abstruse" fail to capture the contentiousness of the debate around this issue in the eighteenth century. Atkins cited the naturalist John Ray on the question, and Ray would make entirely clear how high the stakes could be. In his *The Wisdom of God Manifested in the Works of the Creation* (1691), Ray weighed in on a matter "which is of some moment"

> because it takes away some concessions of Naturalists, that give countenance to the Atheists fictitious and ridiculous Account of the first production of Mankind and other Animals; *viz.* that all sorts of Insects, yea and some Quadrupeds too, as Frogs and Mice, are produced Spontaneously. My Observation and Affirmation is, that there is no such thing in Nature, as Aequivocal or Spontaneous Generation, but that all Animals, as well small as great, not excluding the vilest and most contemptible Insect, are generated by Animal Parents of the same *Species* with themselves. . . . No instance against this Opinion doth so much puzzle me, as Worms bred in the Intestines of Man and other Animals. But seeing the round Worms do manifestly generate, and probably the other kinds too; it's likely they come originally from Seed, which how it was brought into the Guts, may afterwards possibly be discovered.[32]

The fact that the acceptance of spontaneous generation was directly connected with materialism and even atheism did not, it would seem, lead Atkins to draw back from it. He declared it a "poor Solution" to suggest, as Ray had, that worms within our bodies merely hatched from eggs that we had swallowed: First, the worms that emerged from the body seemed to have no parallel outside of it; they were "nothing like what we can see without us." And, second, seeds and eggs could not simply hatch anywhere. Atkins's objection was quotidian: we regularly swallow bird eggs that had been fertilized and even live fish "without any Danger of Breeding, though we are sure they carry the Principles and Powers of Animation with them." His conclusion then followed simply: "The Generation therefore of all internal Worms, at least, tempts one to think, must be *ex novo*."[33] As with his polygenism, Atkins had found his way to a heterodox and controversial position. And just as was true there, this antiorthodoxy mattered very little to the ostensible subject at hand. The question of equivocal generation had been provoked by his discussion of the worms that plagued Africans on the Guinea coast. Having digressed about their possible origins, however, the treatment he offered was essentially identical to that found in earlier editions.

One might conclude, then, that the exposition of antiorthodoxy was itself the point. In the third edition of his text, Atkins thus exchanged one form of intellectual radicalism for another, although it is important to note that the two were not equivalent. Polygenism was a very uncommon position, one usually met in Britain with a scathing response. As positive reactions to the self-regenerating polyp discovered by the Swiss natural historian Abraham Trembley demonstrate, appropriately measured enunciations of spontaneous or equivocal generation did not provoke the same ire.[34] Atkins, in other words, had located a safer form of antiorthodoxy to espouse. To make sure of that safety, he ended the new edition of his text with a devout paragraph in defense of the faith, "advising all (as far as I can influence) who wish for Immortality, to cast their Sheat-Anchor on Revelation," to follow the precepts of the Gospels, and to "cherish, promote, and even suffer for ... one Faith, one Baptism, and one Lord Jesus Christ."[35] He would seem to have hit on a satisfactory formula. The fourth edition of *The Naval Surgeon* appeared in 1758, the year after Atkins died. It contained no further retractions or omissions of his contentious positions.

III. EDWARD LONG: PERFORMANCE AND DEFENSE

Edward Long was perhaps the most odious racist of the Enlightenment, a period in which he might have been expected to face stiff competition. It was, after all, an age in which the German philosopher Immanuel Kant could casually note that "all negroes stink";[36] when a Jamaican governor could argue that the reason for the declining slave population on the island "is the Practice of the Wenches in procuring Abortions. As they lie with both colours, and do not know which the child may prove of, to disoblige neither, they stifle it at birth";[37] and when "Hotantots" could be described as "the very reverse of humankind."[38] And yet both the volume and the detail of Long's vicious calumnies—his claim, for example, that a "Hottentott" woman should find it no dishonor to be wedded to a simian husband—stand out.[39] "Few eighteenth-century authors," David Brion Davis noted in 1966, "could equal Edward Long in gross racial prejudice."[40]

That posited, there is one charge often laid at Long's feet for which he is not accountable. It has commonly been suggested that he deployed the polygenist position proffered in meticulous detail in his *History of Jamaica* (1774) as a defense of slavery. His "polygenist apologia," writes Livingstone, for example, "was part of a strategy to justify the enslaving of Africans."[41] Yet this claim is incorrect. Long did, indeed, seek to justify slavery, and he did, indeed, vigorously defend polygenism. Yet he did not put those two positions together in his text. Doing so would, in principle, have been easy. We can reconstruct the potential argument today as a near syllogism: since Africans were not of the same species as Europeans and it was unproblematic to own any species other than our own, it was perfectly justified to own Africans. Yet that is *not* Long's argument, and he worked hard to avoid having such a position imputed to him. In explaining why he did not make such an argument, this section aims to elucidate some of the *limits* of polygenism as political discourse.

"An Opinion . . . That the White and the Negroe Had Not One Common Origin"

The central issue in Long's discussion of race—as in those of his contemporaries interested in the question—was the relative effect of climate and innate, heritable racial characteristics. To be sure, no polygenist denied that climate would have some effect: Long noted of white Jamaican Creoles that the shape of their eye sockets had changed to protect them in the strong sunlight of the West Indies. "Although descended from British ancestors, they are stamped with these characteristic deviations."[42] Yet climate, for the polygenist, could explain only so much. Generalizing from his Jamaican experiences to the broadest question of human malleability, Long denied that any external effects could explain the most pertinent differences between Blacks and whites. "I do not indeed suppose, that, by living in Guiney," he wrote, that white people "would exchange hair for wool, or a white cuticle for a black: change of complexion must be referred to some other cause." And what was true of skin color and hair was true of many other features, both physical and intellectual. Innate differences from other races that characterized Africans abounded: the shapes of the features of Black people's faces, the color of the lice on their bodies, their "bestial or fetid smell," the failure of Africans to make "any progress in civility or science," and the miserable state of their houses and roads. They were, as a people, "brutish, ignorant, idle, crafty, treacherous, bloody, thievish, mistrustful and superstitious"—characteristics that were not to be explained, as Hippocrates or Montesquieu might have, in terms of climate or custom.[43] Common climates, then, failed to eradicate essential differences between whites and Blacks. And diverse climates, such as those found across Africa and the Americas, failed to produce racial diversity. America, for example, lay between 65 degrees north of the equator and 55 degrees south, with climates that encompassed those of all the known continents. Yet the complexion of its inhabitants was everywhere the same, "only with more or less of a metalline luster."[44] Distinctions between whites and Negroes would appear, then, to be innate and permanent.

To these anticlimatological arguments, which spoke to *racial* difference—difference, that is, that need not rise to the species level—Long then added one from his observations—such as they were—of interracial coupling, which suggested that the difference was even more profound. Mulattoes, he argued, were infertile with each other but not necessarily with individuals from the races of their parents. "The subject," Long concluded, "is really curious . . . it tends, among other evidences, to establish an opinion, which several have entertained, that the White and the Negroe had not one common origin. . . . For my own part, I think there are extremely potent reasons for believing, that the White and the Negroe are two distinct species."[45] Different species, but—and this was perhaps Long's most idiosyncratic argument—the same human genus. In Long's taxonomy, Negroes were indeed men, as were whites, mulattoes, and—most peculiarly—orangutans.

The amount of space devoted to demonstrating the humanity of orangutans in *The History of Jamaica* would seem to provide the most obvious evidence for Long's performative antiorthodoxy. While many authors in the seventeenth and early eighteenth centuries had floated the idea that apes and Africans were closely related, the most authoritative position was that of the Comte de Buffon, who had rejected the connection.[46] Long insisted that the matter was still undecided. "That the oran-outang and some races of black men are very nearly allied," he wrote, "is, I think, more than probable; Mr. Buffon supports his conclusions, tending to the contrary, by no decisive proofs."[47] The question was of obvious scientific interest, yet the fact that Long found the need to mention no fewer than four times in the space of fourteen pages that orangutans were believed to mate with humans—that "they sometimes endeavor to surprise and carry off Negroe women into their woody retreats, in order to enjoy them," for example—surely suggests a desire to dwell at the very edge of good taste.[48]

That said, the pages Long dedicated to the topic are not entirely without significance as part of his overall argument. His central aims in discussing the tripartite division of humankind were, first, to establish a ranking and, second, to insist that essential differences existed without physical causes. That is, the physical differences between orangutans and "other" men were minor, intellectual differences were merely of degree, and these intellectual differences could not be explained by differences in the structure of the body. The brains of these apes and men were, Long insisted, "absolutely the same in texture, disposition, and proportion, and yet *he does not think*." This was therefore proof "that mere matter alone, though perfectly organized, cannot produce thought, nor speech, the index of thought, unless it be animated with a *superior principle*."[49] And that superior principle, he claimed, was differently distributed within the human genus from the time of the original creation, so that Europeans received the most, Negroes a middling amount, and orangs next to none.[50]

This, it might be argued, was Atkins in reverse. Atkins had balked at his own polygenism and had retracted it in later editions of *The Navy Surgeon*, replacing it with his flirtation with materialism. Long doubled down on polygenism but seemed to regard materialism as a bridge too far. Though physical differences between Blacks and whites existed to such an obvious degree that one could declare the two groups to be of different species on their basis, intellectual and moral differences could not be explained simply by the physical structure of the brain. Naturalism had its limits. Both physical evidence and claims about the deity's selective doling out of the superior intellectual ability would be joined to make the case for the African's original and natural inferiority. And yet, despite his willingness to make arguments that many would find absurd on their face, Long did not—at any point in this discussion—defend the enslavement of the

"Negroe race" on the basis of their inherent inferiority. Indeed, as we will see, he would, in fact, posit essential equivalence between European and African laws, and the people bound by them, in order to make his case.

"An African Is as Much Bound by This Supreme Power, as the English Labourer"

Long first published on the legality of slavery in 1772, in a short pamphlet issued in the aftermath of the decision made in the so-called Somerset case. That decision was read by Long as asserting "that the laws of *Great Britain* do not authorize a master to reclaim his fugitive slave, confine, or transport him out of the kingdom. In other words; that a Negroe slave, coming from the colonies into *Great Britain*, becomes *ipso facto*, Free." Lawyers, he noted, had succeeded in "washing the black-a-moor white."[51] Here, I will focus on the arguments he laid out in his much longer *History of Jamaica*, where he made three broad claims defending the slave trade. First, such trade had been carried on by Africans themselves for "some thousand years." Although Europeans obviously profited by the trade in men, they had hardly invented it. Second, European participation in the trade, far from ruining African lives, had improved them. Portuguese slavers, he suggested (while relying on the most critical accounts of African societies), may have thought it a "meritorious act" to send slaves to work in their mines, thus saving them from the death, torture, cannibalism, and human sacrifice that characterized day to day life on the Guinea coast—and "mak[ing] their private gain compatible with the suggestions of humanity or religion."[52] Regardless of motive, however, economics alone had lessened the savagery of African customs. Criminals and those captured in war were now worth far more alive than dead. It was those who sought to abolish the trade, Long argued with cynical cunning, who were without sufficient humanity, for to do so "is therefore no other than to resign them up to those diabolical butcheries, cruelty, and carnage, which ravaged their provinces before the European commerce with them began."[53]

Long's humanitarian defense of slavery stemmed from more than the desire not to immediately cede the moral register to his opponents. His moralizing also drew the reader's attention to the third and most significant part of his argument: the character of those sold as slaves and, more particularly, the idea of slavery as punishment for legal infraction. Long suggested that those slaves sold to Europeans came from four distinct groups: war captives; those sold "by brutal parents, or husbands"; native slaves sold for some crime; and the free born, punished with slavery for a particularly egregious offense.[54] Ignoring the second category and minimizing the impact of the first, he concentrated his attention on the two kinds of criminals. Citing a figure he had only recently read, he claimed that it was "well-known" that 99 percent of slaves shipped from Africa were felons, their sentences having been commuted from death to exile and servitude.[55] Accepting this fact as true licensed a comparison both easy and powerful, for the British state had few qualms about the use of transportation as a punishment for criminals —and fewer still about the use of such criminals as an indentured workforce. Long had only to forget that he had recently condemned African laws as "ridiculous" and to argue instead for an equivalence between the rights of African and European states to dispose of their unwanted as they saw fit.

> It may be said of our English transported felons, as of the Negroe criminals, that neither of them go into a voluntary banishment; but it must be allowed, that the Africans may with equal justice sell their convicts, as the English sell theirs; and equally well vest a legal right to their service in the purchasers.[56]

Of course, Long was well aware of the limits of his analogy. African states may well have had the right to sell both criminals and their labor to Europeans in Africa. But by what right could one keep an individual convicted of no crime by British law in servitude in Britain or its colonies? The right, he answered, was created by an implicit contract. Those captured in war, he argued, submitted voluntarily to slavery, for they had known the consequences of their actions in going to war and had sought to reduce their opponent to the same state. Other slaves accepted the right of their owner to sell them "as part of the law or usage of his society," and that right, Long asserted, was transferred across the seas. "Surely a voyage from Afric to any other country, where this claim of property is continued, cannot dissolve the bargain."[57]

Long's "surely" in the preceding sentence was disingenuous. Abolitionists argued precisely that the claim of property over another human being—whatever its presumed legality in other nations—was anathema in Britain. Long, however, believed that abolitionists overstated what was often portrayed as an innate English opposition to any loss of liberty. The fact of the matter was that unfreedom was a basic part of social life. It was hardly restricted to Africans and their progeny and could be defended only by the exertion of force. It was force alone that kept the convict transported to America from evading his sentence, the inhabitants of debtor's prisons in confinement, the sailor and soldier in service, and the laborer in his place.

> A labourer in England never consented to the laws which impose restrictions upon him; but there is in every government a certain supreme controuling power, included in the social compact, having the energy of law, or published and declared as the law of the land; by which every member of the community, high and low, rich and poor, is respectively bound: it is in truth an association of the opulent and the good, for better preserving their acquisitions, against the poor and the wicked. For want, complicated with misery and vice, generally seeks relief by plundering from those who are better provided. An African is as much bound by this supreme power, as the English labourer.[58]

I have devoted significant space to Long's positions not because they were terribly novel—they were fairly standard for the eighteenth century—but rather because I want to make clear the assumptions upon which they were predicated. All of these arguments presupposed not only the humanity of the African slave—a humanity that licensed a direct comparison to the English laborer—but also a civilized humanity, one above a state of nature. Long's arguments relied crucially on the assumption that Africans could make laws, that they understood that they were bound by such laws, and that these laws were reasonable enough that other nations should respect them as well. "No one doubts," he insisted, "but that every contract made in Afric for the purchase of a slave, is there understood by the three parties, the buyer, the seller, and the person sold, to be perfectly firm and valid; the one knows what he buys, the other what he sells, and the third, that his services are thus become translated to his new owner."[59] To argue that Africans were little more than beasts would be to undercut such an argument, for beasts could neither formulate nor be held to contracts.[60]

It might, of course, be suggested that Long wished to have the reader make a connection he was not willing to make himself. At the very least, we might suggest that while he did not suggest that Africans should be enslaved *because* they were inferior, he did intend for their claimed inferiority to blunt some of the sympathy that abolitionists sought to evoke for their plight. Yet even this had its dangers. In the eighteenth century, the claim that those who owned slaves justified their actions on the grounds that their property were little more than animals was one made powerfully by abolitionists *against* their opponents. This was vehemently denied by those in the proslavery camp, who sought instead to cast themselves as men dedicated to the well-being of those who faced a much worse fate in their native lands—see Long's claims discussed earlier—. Thus, while Granville Sharpe compared the treatment of the enslaved and "the merciless usage practiced in England over post-horses, sand-asses, &c," Long insisted that he had "never known, and rarely heard, of any cruelty either practiced or tolerated" by Creole owners over their human property.[61] "In his habitation, cloathing, subsistence, and possessions," Long wrote in what should now be a familiar comparison, the Jamaican slave "is far happier and better provided for than most of the poor labourers, and meaner class, in Britain."[62]

I would thus suggest that we see Long's polygenism in two ways. First, it was primarily a defense intended to blunt an abolitionist attack that used scientific monogenism to further motivate a moral critique of slavery. One might avoid the full force of Christ's admonition to care for "one of the least of these brothers of mine" by denying the existence of such a brotherhood between masters and slaves. But a more positive connection was a far more dangerous move, one effectively denied to even the most vicious proponent of the slave trade. Animality could not openly be used as a defense of unfreedom in the eighteenth century. Second, particularly in Long's peculiar and largely unnecessary flirtation with the notion that orangutans might learn to use cutlery and were already engaged in amorous intercourse with African women, we should see both his profoundly unpleasant racism and his desire to thumb his nose at propriety and orthodoxy. This was a willful and, one imagines, rather gleeful performance, one we would recognize today as a form of trolling. And as with many trolls today, one suspects that the dismay and distaste with which his views were greeted were reward enough.

CONCLUSION

That polygenism could, in principle, be used for productive purposes seems clear from a reading of La Peyrère's *Prae-Adamitae*. There one finds a new social order predicated on a vision of Jews as the sole subjects, if not the sole inheritors, of a contract with God, made through Adam as their progenitor. The same cannot be said of the four examples of scientific polygenism analyzed in sections I and II of this chapter. L. P's pamphlets aim to eviscerate those who would read the Bible literally and who might seek thereby to limit rationalism in what had not yet been named an enlightened age. Why, beyond critique, it should matter that men in different places should all be regarded as "aborigines" is not made clear. A desire to mock biblical literalism and church authority was clearly what motivated Voltaire's attack on monogenetic orthodoxy—that and a performative desire to scandalize, as seen in his movement from natural historical arguments concerning the origins of races to his insistence that the existence of satyrs might be possible, formed as the progeny of animals and men. (Herodotus, he informs the reader by way of evidence, notes the case of a woman in Egypt "who publicly copulated with a he-goat.")[63] We might reasonably conclude this is not a tract that we are meant to take entirely seriously. Johann Blumenbach, at the very least, would describe the Frenchman as "witty, but badly instructed in physiology."[64] Hume's polygenism not only did no productive work but also undercut his larger claims, while in Atkins's case, we find his polygenetic arguments used neither to defend the slave trade nor to explain his medical analyses. Having decided to remove views for which he presumably received sizable criticisms, he replaced them with safer heterodoxies that also did not further argumentative work.

I have already suggested that we should not be surprised that those writing before 1760 did not seek to use naturalistic arguments to defend slavery. Such arguments had been largely unnecessary, for slavery was part of an accepted social fabric, particularly in locations far from European metropoles. Arguments established since antiquity—many of which Long would merely rehearse—justified the practice in the eyes of the majority. In the 1760s, however, Anthony Benezet—the so-called father of Atlantic abolitionism—helped to start a movement that eventually culminated in the abolition of the British slave trade in 1807. In the 1770s, when monogenism, in a newly naturalized form, could be invoked as one of the powerful moral critiques of the human trade, Long had a motivation beyond his own performative antiorthodoxy to deny the brotherhood of man that previous polygenists had not. However, as we have seen, polygenism could only blunt an opponent's attack; it could not be used to justify slavery in its own right.

Let us end by reflecting on what the history of polygenism in the eighteenth century might tell us, first, about the entangled histories of science and religion and, second, about the history of race. Once again, Nott's nineteenth-century position provides a valuable point of comparison. There was a good deal, after all, that drew Nott to polygenism. The idea that Blacks and whites could be regarded as separate species provided a rational defense of slavery, of course. It also provided a religious skeptic, as Keel terms him, with a cudgel to bludgeon the devout.⁶⁵ Although Nott would write in print about the need to assign religion and science to separate spheres—"to cut loose the natural history of mankind from the Bible and to place each upon its own foundation, where it may remain without collision or molestation"—he was less circumspect in private, relishing the dismay that his words produced in the "godly" and the fact that they would "stir up hell in the christians."⁶⁶ As I have emphasized here, however, there was no necessity, at least in the eighteenth century, for polygenism and proslavery to go hand in hand. Further—and this is perhaps more surprising given the religious basis of the monogenist alternative—eighteenth-century polygenism did not require the rejection of faith. Although both Long and Atkins, for example, performed their heterodoxies with regard to a literalist reading of the Bible, neither took it as their task to reduce the church to rubble. Atkins, after all, eventually removed his musings about the multiple origins of humanity from later editions of his texts, while Long retained an aversion to materialism that would have seemed almost quaint by mid-nineteenth-century standards.

In noting that the conflict thesis does not adequately explain eighteenth-century polygenism, I do not mean simply to argue either that religion and science were really compatible (although that may have often been the case) or that things were more complicated than that (although this is always true). Instead, rather than seeing debates over slavery and human origins as a context or locus where religion and science clashed or mutually adapted, we should see such debates as a site where religion and science emerged as partially separate spheres, separate enough that later thinkers on precisely the same topic could posit their mutual exclusion.

As for polygenism and the history of race, it may seem—particularly for those most interested in racial politics today—at best obtuse and at worst exonerative to focus on a period when polygenism was not deployed directly to buttress proslavery positions. The very terms *monogenism* and *polygenism* were, after all, coined by American defenders of slavery in the mid-nineteenth century, and much of our historical interest in the doctrine of special creations derives originally from its later connections with the explicitly racist "American school" of ethnology. I will leave to one side the historian's most obvious rebuttal to this kind of criticism—understanding the past on its own terms has value beyond its direct relation to the present—and turn instead to why understanding eighteenth-century polygenism should matter, even to such an imagined presentist reader. The issue comes down to the question of how we, as analysts, should treat racisms as intellectual positions. Should we treat them as largely derivative of more fundamental political positions, or should we understand them as sets of claims with logics, histories, and motivations of their own, whatever the political purposes for which they might then be used? This chapter should make clear that my sympathies lie with the latter camp. That racisms are odious should not stop us from treating them as serious intellectual positions. There is too much arrogance involved in assuming that those who disagree with us do so only out of ignorance or cynicism. If we begin with the premise that forms of racial thinking are to be treated as we would treat any other intellectual position, it follows that we should understand that they might serve multiple purposes, only some of which relate to explicit "capital P" Political positions. Systematically exploring racisms as modes of thought does not mean justifying them; it means taking them as seriously as the threats they pose. Those who would make racists out to be either fools or cynics do so at their—and our—peril.

NOTES

1. David N. Livingstone, *Adam's Ancestors: Race, Religion, and the Politics of Human Origins* (Baltimore: Johns Hopkins University Press, 2008), 6–8.
2. Paracelsus, quoted in Justin E. H. Smith, *Nature, Human Nature, and Human Difference: Race in Early Modern Philosophy* (Princeton, NJ: Princeton University Press, 2015), 94.
3. Livingstone, *Adam's Ancestors*, 79.
4. Reginald Horsman, *Josiah Nott of Mobile: Southerner, Physician, and Racial Theorist* (Baton Rouge: Louisiana State University Press, 1987).
5. Josiah Nott, quoted in George J. Frederickson, *The Black Image in the White Mind: The Debate on Afro-American Character and Destiny, 1817–1914* (Middletown, CT: Wesleyan University Press, 1971), 79. See more generally Livingstone, *Adam's Ancestors*, ch. 3.
6. Terence Keel, *Divine Variations: How Christian Thought Became Racial Science* (Stanford, CA: Stanford University Press, 2018), 79.
7. Seymour Drescher, *The Mighty Experiment: Free Labor Versus Slavery in British Emancipation* (Oxford: Oxford University Press, 2002), 76.
8. Richard H. Popkin, *Isaac La Peyrère (1596–1676): His Life, Work, and Influence* (Leiden: Brill, 1987), 44.
9. Popkin, *Isaac La Peyrère*, 47.
10. Popkin, *Isaac La Peyrère*, 69–79.
11. L. P., *Two Essays Sent in a Letter from Oxford to a Nobleman in London.* (London: R. Baldwin, 1695), 14.
12. P., *Two Essays*, 20.
13. P., *Two Essays*, 22.
14. P., *Two Essays*, 27.

15. P., *Two Essays*, 23.
16. Popkin, *La Peyrère*, 133–134.
17. Voltaire, *The Philosophy of History* (London: I. Allcock, 1766), 6.
18. Voltaire, *The Philosophy of History*, 6, 9. Material on the races of men first appeared in 1734.
19. The remarks appeared originally in a footnote in the 1753 edition of a work first published in 1748. David Hume, "Of National Characters," in *Hume: Essays, Moral, Political, and Literary*, ed. E. F. Miller (Indianapolis, IN: Liberty Fund, 1987), fn. 10, p. 208; for a discussion of variant readings, see 629–630.
20. Colin Kidd, *The Forging of Races: Race and Scripture in the Protestant Atlantic World, 1600–2000* (Cambridge: Cambridge University Press, 2006), 93–94.
21. P., *Two Essays*, 23.
22. For biographical materials, see N. M. (Norman Moore, MD), "Atkins, John (1685–1757)," *Dictionary of National Biography*, ed. Leslie Stephen (London: Smith, Elder, 1885), 2:220; and F. Tubbs, "John Atkins: An Eighteenth-Century Naval Surgeon," *British Medical Bulletin* 5 (August 1947): 83–84.
23. John Atkins, *A Treatise on the Following Chirurgical Subjects:* . . . (London: T. Warner, 1729), 203.
24. Atkins, *A Treatise*, 204.
25. Atkins, *A Treatise*, 204–205.
26. John Atkins, *A Voyage to Guinea, Brasil, and the West-Indies; in His Majesty's Ships, the Swallow and Weymouth* (London: Caesar Ward and Richard Chandler, 1735), 39.
27. Atkins, *A Voyage*, 61–62.
28. Atkins, *A Voyage*, 178. One of the only places where polygenism was explicitly invoked in support of another contention, rather than simply for its own sake, involved the deployment of the doctrine of separate creations as a *defense* of Africans against those who would criticize their lack of true religion. Having never received the benefit of revelation, how could they be judged for their heathenism? "They are set down as from the Clouds, without Guide, Letters, or any means of Cultivation to their better Part, but what immediately strike their Senses from beholding this Universe and the Beings contained in it; their Deductions from whence, as to a Deity devoid of Matter, is next to impossible, therefore we say mean and pitiful" (82).
29. John Atkins, *The Navy Surgeon; or, Practical System of Surgery with a Dissertation on Cold and Hot Mineral Springs; and Physical Observations on the Coast of Guiney* (London: J. Hodges, 1742), 368.
30. Atkins, *Navy Surgeon*, 368–369.
31. Atkins, *Navy Surgeon*, 372.
32. John Ray, *The Wisdom of God Manifested in the Works of the Creation Being the Substance of Some Common Places Delivered in the Chappel of Trinity-College, in Cambridge* (London: Samuel Smith, n.d.), 221–222.
33. Atkins, *Navy Surgeon*, 372.
34. On Trembley's discovery of a self-regenerating polyp in 1742 (the same year that Atkins published the third edition of *The Navy Surgeon*), see Emma Spary, "Political, Natural, and Bodily Economies," in *Cultures of Natural History*, ed. N. Jardine, J. A. Secord, and E. C. Spary (Cambridge: Cambridge University Press, 1996), 181–184.
35. Atkins, *Navy Surgeon*, 378.

36. Immanuel Kant, "Von Den Verschiedenen Racen Der Menschen," in *Der Philosoph Für Die Welt*, ed. J. J. Engel, 1777, 125–164.
37. Governor Edward Trelawney, quoted in Richard B. Sheridan, *Doctors and Slaves: A Medical and Demographic History of Slavery in the British West Indies, 1680–1834* (Cambridge: Cambridge University Press, 1985), 224.
38. J. Ovington, "A Voyage to Suratt in the Year 1689," in *India in the Seventeenth Century: Being an Account of the Two Voyages to India by Ovington and Thevenot. To Which Is Added the Indian Travels of Careri*, ed. J. P. Guha, vol. 1 (New Delhi: Associated Publishing House, 1976), 218. The text was first published in 1696.
39. Edward Long, *The History of Jamaica*, vol. 2 (London: T. Lowndes, 1774), 364–365.
40. David Brion Davis, *The Problem of Slavery in Western Culture* (Ithaca, NY: Cornell University Press, 1966), 459.
41. Livingstone, *Adam's Ancestors*, 68.
42. Long, *History*, 2:262. The material on Long in this and the next subsection is taken from Seth, *Difference and Disease*.
43. Long, *History*, 2:352–354.
44. Long, *History*, 2:375.
45. Long, *History*, 2:336.
46. Winthrop D. Jordan, *White Over Black: American Attitudes Toward the Negro, 1550–1812* (Chapel Hill: University of North Carolina Press, 1968), 28–32. See, more generally, Phillip R. Sloan, "The Idea of Racial Degeneracy in Buffon's *Histoire Naturelle*," *Studies in Eighteenth-Century Culture* 3 (1973): 293–321; and Claude-Olivier Doron, "Race and Genealogy: Buffon and the Formation of the Concept of 'Race,' " *Humana.Mente: Journal of Philosophical Studies* 22 (2012): 75–109.
47. Long, *History*, 2:365.
48. Long, *History*, 2:360. Atkins took up the same theme in considerably less detail. See Atkins, *Voyage*, 108.
49. Long, *History*, 2:363.
50. Long, *History*, 2:371.
51. Edward Long, *Candid Reflections Upon the Judgement Lately Awarded by the Court of King's Bench in Westminster Hall, on What Is Commonly Called the Negroe-Cause, by a Planter* (London: T. Lowndes, 1772), 56, iii.
52. Long, *History*, 2:387.
53. Long, *History*, 2:391–392.
54. Long, *History*, 2:388.
55. Long, *History*, 2:391. The figure is given in An African Merchant, *A Treatise Upon the Trade from Great Britain to Africa. Humbly Recommended to the Attention of Government* (London: R. Baldwin, 1772), 12.
56. Long, *History*, 2:390.
57. Long, *History*, 2:394.
58. Long, *History*, 2:392–393.
59. Long, *History*, 2:393.
60. In fact, one of the few passages to make the claim that enslaved Africans deserved servitude under British rule on the basis of their similarity to animals makes the argument on

the basis that such a bestial nature is not innate. "Among men of so savage a disposition," the passage reads, "as that they scarcely differ from the wild beasts of the wood in the ferocity of their manners, we must not think of introducing those polished rules and refinements, which have drawn their origin and force from the gradual civilization of other nations that once were barbarous. Such men must be managed at first as if they were beasts; they must be tamed, before they can be treated like men." One will note that this passage assumes, first, that the British themselves were once barbarous and, second, that Africans can and will over time cease to be like beasts (hence, their bestial nature is not fixed or racially innate) and will become like civilized men. Edward Long, *The History of Jamaica*, 3 vols. (London: T. Lowndes, 1774), 2:401.
61. Long, *History*, 2:269.
62. Long, *History*, 2:402.
63. Voltaire, *The Philosophy of History*.
64. Johann Blumenbach, excerpted and translated in Robert Bernasconi and Tommy L. Lott, eds., *The Idea of Race* (Indianapolis, IN: Hackett, 2000), 32.
65. Terence D. Keel, "Religion, Polygenism, and the Early Science of Human Origins," *History of the Human Sciences* 26, no. 2 (2013): 17.
66. Josiah Nott, quoted in William Stanton, *The Leopard's Spots: Scientific Attitudes Towards Race in America, 1815–59* (Chicago: University of Chicago Press, 1960), 122.

CHAPTER 11

OUT OF AFRICA

Where Faith, Race, and Science Collide

JOSEPH L. GRAVES JR.

WHAT DO NONSCIENTISTS BELIEVE ABOUT HUMAN ORIGINS?

Recent surveys show that the United States is still unique among modern industrialized nations for its ongoing rejection of the evidence for evolution. Gallup Poll surveys taken from 1981 to 2017 illustrate this via this question:

> Which is closest to your view: humans developed from less advanced forms of life, but God guided the process; or humans developed from less advanced forms of life, but God had no part in it; or God created humans in their current form.

There is a major problem with the formulation of these options, as the term *developed* is substituted for *evolved*. The former term is more Lamarckian than Darwinian and confuses development that occurs as an individual grows from a fertilized egg to an adult with evolution and speciation that occur in populations across multiple generations. However, given this confusion, the data show that for the "God guided the process" result between 1981 and 2017, the responses ranged from a high of 40 percent in 1997 to a low of 31 percent in 2014 but bounced back to 38 percent in 2017; for "God had no part in [the process]," the low was 9 percent in 1981, and the high of 19 percent was achieved in 2017; for the special creationist response, "God created humans in their current form," the high was 47 percent in 1999, and the low of 38 percent was achieved in 2017.[1]

Christian, Jewish, and Islamic denominations differ in their acceptance of organic evolution. This discussion will focus on the views of Christian denominations. In general, doctrinally conservative Christians reject evolution.[2] For example, the Catholic Church accepts evolution as compatible with their faith, while the Southern Baptist Convention and the National Baptist Convention (with a predominately African American membership) both reject evolution as compatible with their faith.[3] Notably, there is variation within the individuals who subscribe to the major denominations concerning their acceptance of evolution. For example, among doctrinally conservative Protestants surveyed from 1994 to 2004, 76 percent felt the statement that humans developed from earlier species of animals was definitely false or probably false, while 24 percent felt it was probably true or true. Respective values were 66 percent and 35 percent for Black Protestants, 45 percent and 55 percent for mainline Protestants, and 42 percent and 58 percent for Roman Catholics.[4] Thus, while a denomination's official position may be to accept or reject evolutionary science, individuals within that denomination tend to make up their own minds concerning evolution.

In 2017, the Gallup Poll suggested that Americans were equally split on whether God created humans in their present form or guided their "development" (read "evolution"). If God created humans in their present form, a position predominantly held by doctrinally conservative, biblical literalist Christians,[5] how does one explain the diversity of modern humans? The doctrinally conservative Answers in Genesis website uses a strange combination of biblical literalism and modern genetics to explain the origin of human biological diversity.[6] The explanation begins with classical monogenism: all humans are the descendants of Adam and Eve. For its scriptural authority, it utilizes both the Noachian flood (eight people came from the Ark, although it conveniently does not mention Noah's curse on Canaan) and the Tower of Babel (God scatters the people across the earth). For its genetics, it claims that Adam and Eve most likely had light brown skin and genes for all the other skin color variants, as well as genes for all the other physical trait variations that we see in modern humans.

This genetic explanation is, of course, scientifically impossible. The principle of biparental inheritance means that Adam and Eve would have had two possible alleles (variants of genes) at each of the genetic loci responsible for skin color. Skin color is a somewhat simple trait, as normal pigmentation variation is influenced by genetic variation in small set of genes.[7] Genome-wide association studies identify less than twenty single nucleotide polymorphisms that determine skin color.[8] The problem, of course, is that some of these variants (particularly the ones that confer light pigmentation) are younger than others, meaning that Adam and Eve could not have had all the variants at the moment of creation. This is also true for the other genetic loci that code for the physical variation that we see in human populations across the globe.[9] Many of these mutations occurred after populations left Africa, meaning that each set of continental populations has genetic variants that are unique to its specific continent (and region within that continent).[10] This is the reason that genetic ancestry tests can provide estimates of an individual's continental (and regional) ancestry. Despite the proliferation of modern genomic methods that have provided great insight into human origins and ancient migrations, it is still likely that the one-third of Americans who accept biblical literalism have been exposed to monogenist arguments of this type concerning the origin of human biological variation.[11]

We have some data that bears on the question of the public acceptance of the modern science of human origins. In 2014, Duke University's Genomics, Race, and Identity Center (GRID) conducted a nationwide poll 1,655 adults.[12] The poll consisted of statements concerning genetics and race concepts, with the responses arranged on a four-point Likert scale (strongly disagree, disagree, agree, strongly agree). The survey responses showed great variation along the Likert scale, indicating that there is still a deep confusion concerning the relationship of genetics to socially defined race. Furthermore, socially defined racial groups often displayed different score distributions to specific items on the test. An example was the response to this statement: "Everyone's ancestors originally came from Africa." The respondent answers were significantly influenced by their own self-identified race, region, age, and religiosity. Whites, Hispanics, and Blacks disagreed or strongly disagreed with this statement at probabilities of $p = 0.673$, 0.609, and 0.538, respectively (or 67.3 percent, 60.9 percent, and 53.8 percent of respondents by group disagreed or strongly disagreed, respectively). People living in the South were more likely to disagree, at $p = 0.683$, compared to 0.550 for individuals from states outside the South. The disagreement with this statement increased with the age of the respondent, with $p = 0.586$ for people in their twenties, compared to $p = 0.643$ for individuals sixty years of age and older. Finally, religiosity was associated with how an individual within a self-identified race responded to this question. The impact of religiosity was different by socially defined race. Among Blacks, not religious, weakly religious, religious, and strongly religious individuals disagreed/strongly disagreed, at $p = 0.611$, 0.545,

0.465, and 0.862, respectively. For Hispanics, these values were $p = 0.734$, 0.387, 0.783, and 0.508, respectively. Finally, for whites, these values were $p = 0.642$, 0.643, 0.759, and 0.767, respectively.

These results indicate that significant numbers of those surveyed were influenced by their socially defined race, region, age, or religiosity when it came to acceptance of the scientific fact of the African origin of our species. The differences between Hispanic (0.508) and Black/white (0.862/0.767) strongly religious individuals may be accounted for by religious denomination, as the majority of Hispanics in the United States are Catholic, a denomination whose official stance is the acceptance of evolution. A variety of studies have shown that whites, and particularly evangelical Christians, are among the most racist individuals within U.S. society.[13] Furthermore, the overall probability of rejection of the out-of-Africa science may be related to the fact that, in the main, this topic is not covered in high school biology textbooks.[14] I have argued that this results from two concerns related to the high school curriculum: an unwillingness to seriously address race, racism, and injustice and a fear of the general subject of evolution.[15]

HOW WE GOT HERE

Upon recognizing the implications of Charles Darwin's descent with modification for humans, Sir Charles Lyell mused, "Go back umpteen generations and would blacks and whites find a common ancestor? Itself the descendent of an ape? The idea would give shock to . . . nearly all men. No university would sanction it; w. ensure the expulsion of a professor already installed."[16] Lyell was the leading man of British science, and in 1856, he was deeply concerned about how Darwin's notion of common descent would be received by the Victorian gentry. The issue of race was coming to a boil in the mid-nineteenth century, and the English physician Robert Knox had predicted a coming global race war. Knox was a polygenist who believed that the socially defined races were not biological races but separate species.[17] In *The Races of Men*, published in 1850, he claimed that in history "race is everything."[18] The chief theorist of polygenism, the Swiss-born biologist Louis Agassiz (1807–1873), produced the rationale for the existence of separate human species via his zones of creation theory,[19] which proposed that God had created animals and plants in harmonious zones. Thus, to Agassiz, adaptation was best explained by divine plan. Tropical creation included tropical plants, animals, and human species.

Darwin understood that Agassiz would be a primary opponent to evolution by the means of natural selection. He particularly understood that Agassiz would be most riled by the idea that natural selection and descent with modification would imply that all humans shared common ancestry. On this point, Darwin felt that Agassiz "would throw a boulder at me, & many others would pelt me."[20] Indeed, all the natural scientists up to Darwin generally agreed with the inferiority of the Negro, whether they were monogenists, polygenists, or Lamarckian transformationists.[21] The monogenists, however, were not uniform in their views. Some held that biblical narratives of Creation were correct. This meant that Adam and Eve were the first humans, that God destroyed humanity in the Great Flood, and that the earth was repopulated by eight human beings (Noah and his wife, his three sons, and their wives) after the flood subsided. Some believed that Africans and other dark-skinned races resulted from the mark of Cain, while others thought that the inferiority of the Negro was generated by the curse on Ham's son Canaan. These people held this belief despite the complete absence of scriptural support and the confusion of medieval scholars concerning which races of man were descended from which sons of Noah.[22]

Creationism and American Slavery

A particularly vehement defense of the inferiority of the Negro as a justification for slavery from a monogenist perspective was authored by Josiah Priest in 1852. He was more a popular author than a scholar, but his *Bible Defense of Slavery* was premised on the reality of the curse of Ham, which caused his son Canaan to turn black.[23] In contrast, "rational" monogenists ascribed the inferiority of Africans to the biological phenomenon of degeneration. They differed from traditional monogenists in that they did not attempt to explain nature solely by means of scripture. Thus, Johann Friedrich Blumenbach (sometimes referred to as the father of physical anthropology) judged some but not all features of Africans inferior to those of Europeans due to degeneracy.[24] His theory of degeneracy held that while all humans were the descendants of Adam and Eve, some had moved into unhospitable climates, and this resulted in the degradation of some (or all) of their physical traits (including intelligence).

A different monogenist explanation of Negro inferiority was related to the ideas of Jean Baptiste Lamarck (1744–1829). The transformationists held that a small number of primordial germs or monads resulted from spontaneous generation and passed through successive transformations or divergences. Thus, in Lamarckian evolution the number of monads never changed, but the species that developed from them did over successive generations. Humans were the result of the slow transformation of apes. While all modern humans were descended within the human monad, the rates at which these transformations occurred differed, accounting for the races of humanity. This view would later resurface in the form of Ernst Haeckel's "ontogeny recapitulates phylogeny" explanation for the inferiority of non-Aryan races.[25]

The polygenists, in contrast, held that the races of humanity were, in fact, different species. However, they also differed in their beliefs as to the origins of these separate species. Religious polygenists supported the pre-Adamite race theory, in which God created several different Adams and Europeans were the descendants of the Adam and Eve described in Genesis. This idea goes back at least to Paracelsus (1493–1541), a Swiss physician, alchemist, and chemist. In 1591, the Italian philosopher Giordano Bruno (1548–1600) also championed this idea. Bruno, while considered a hero of science for his willingness to die in defense of the Copernican theory, was also put to death for the heresy of the pre-Adamite races.[26] Pre-Adamism makes an important appearance in American thought as a supposition in Thomas Jefferson's *Notes on Virginia*, first published in 1781. Later Dr. Charles Caldwell challenged monogenism with his book *Thoughts on the Original Unity of the Human Race*, published in 1830. He began his book with an explanation that his opposition to the unity of mankind was not an opposition to Christianity. Rather, his main purpose was to take issue with the monogenism of Samuel Stanhope Smith (1751–1819), a Presbyterian minister who founded Hampden-Sydney College and later became president of Princeton University. Smith believed that climate was the cause of color complexion changes in human beings; thus, Africans were dark-skinned because they lived in the tropics. Smith's mechanism for this change was more akin to Lamarckian inheritance of an acquired characteristic. Caldwell's main complaint with Smith was his environmentalism, particularly bringing forth numerous arguments eviscerating the idea that skin color was determined by climate.[27]

However, it was Thomas Cooper (1759–1839), the second president of South Carolina College (now the University of South Carolina), who recognized polygenism's significance as an ideological prop for Negro slavery. He was described as an ostentatious champion of free speech who taught the youth of the slaveholder class a virulent and uncompromising proslavery ideology founded on the separate creation and inferiority of Africans.[28] He found it necessary to dispense with the idea of "self-evident" truths in the Declaration of Independence, indicating that there was no reality to the idea that men were born equal. In 1823, he claimed, "People of color are, in every part of the United States, considered, not merely by the populace, but by the law as a permanently degraded people; not participating as by right, of the civil privileges they possess, as a right and grant, as a matter of favour conceded by the law and revocable by the law."[29] It is Cooper's reasoning that Roger B. Taney utilized in the Dred Scott decision of 1857.

Another school of polygenists resisted the idea that the Bible should or could provide sufficient ideological support for slavery. For example, Josiah Nott, who was a devoted student of Thomas Cooper, said, "Just get the dam'd stupid crowd safely around Moses and the difficulty will end."[30]

Agassiz, the chief scientific theorist of the polygenists, did not start out with polygenist views concerning the human species. In 1845, the year before he came to the United States, he had formulated the idea of zoological provinces of special creation, mentioned earlier in relation to Darwin.[31] Agassiz recognized that there were different races of humans living in the various climatic zones (Negroes in the tropics, Caucasians in the temperate zones, and Eskimos in the Artic), but he did not think that these races were different species. Indeed, Charles Bachman, South Carolina slaveholder, parson, and monogenist, cited Agassiz in opposition to the writings of Josiah Nott and Dr. Samuel Morton.[32]

Agassiz's conversion to polygenism began during his American tour: first, as part of his reasoning that his zones of creation theory should logically include humans as well and, second, as a result of his aversion to Negroes, which primarily drove his rethinking on the unity of mankind.[33] He traveled to the United States in 1846, and one of his stops was Morton's human skull collection in Philadelphia. The doctor was an Episcopalian and up to this point considered America's greatest man of science.[34] He had already published *Crania Americana* in 1839 and followed that with *Crania Aegyptica* in 1844. In the first of these volumes, he had claimed that the ancient Egyptians were Caucasians, and the latter study was designed to validate that claim.[35]

During the winter of 1847, Agassiz visited the Medical School in Charleston, South Carolina. In December, he discussed race with the members of the Literary and Philosophical Society of Charleston. This society included John Bachman and Thomas Smyth (both monogenists), who were unconvinced by Agassiz's polygenism. Smyth wrote *The Unity of the Human Races Proved to Be the Doctrine of Scripture, Reason, and Science*, published in 1851. Bachman would in 1855 write a response to Agassiz's zones of creation theory titled *An Examination of Professor Agassiz's Sketch of the Natural Provinces of the Animal World and Their Relation to the Different Types of Man*.

Undaunted by the rejections of his polygenist ideas by Smyth and Bachman, in 1850 Agassiz attended a lecture delivered by Nott, a South Carolina physician and slaveholder, on the genealogy of the human race at a meeting of the American Association for the Advancement of Science in Charleston. In this lecture, Nott argued that there is not a "particle of proof" that the different human races shared the same family tree. Agassiz supported Nott's contention by offering up his own version of the pre-Adamite races theory. Agassiz's remarks on this question were read by Darwin, who wrote to his cousin William Darwin Fox about "Agassiz's lectures in the U.S. in which he has been maintaining the doctrine of several species—much I daresay, to the comfort of the slave-holding Southerns."[36] After the meeting, Agassiz toured some South Carolina plantations at the invitation of Dr. Robert W. Gibbes and soon became convinced that he could identify the tribal/ethnic origin (Ebo, Foulah, Gullah) of the slaves he observed. Later Gibbes arranged for daguerreotypes to be made of the enslaved people Agassiz had observed, and these portraits were arranged to display the individuals as typical "zoological" specimens.[37]

By the 1860s, Agassiz was the most influential biologist in the United States, having already established the Museum of Comparative Zoology at Harvard University.[38] He was in many ways the first public scientist, and he intentionally infused his scientific thinking into the nation's most thorny problem, race. Ironically, he was antislavery, but like many persons of European descent, such as Abraham Lincoln and Ulysses S. Grant, he also felt that Africans were inferior to Europeans. Agassiz supported the political equality of the American Negro but stressed that this should be granted with a clear recognition of the permanent physical, social, and mental differences that made the Negro inferior to the white. His public stance as one of the country's most respected scientists, as well as his antislavery stance, facilitated his becoming a "scientific" adviser to the Freedman's Inquiry Commission. This commission was brought into existence as a result of Lincoln's Emancipation Proclamation of January 1, 1863, and was chaired by Samuel Gridley Howe.[39] An additional inconsistency in Agassiz's polygenism is revealed by his role as adviser to Howe. Agassiz was preoccupied with the prevention of miscegenation in the South. He saw this and a potential Negro takeover of these states as great concerns for the commission. He feared that once the war was concluded, whites would flee these states due to their aversion to Negroes.

EVOLUTIONARY MONOGENISM AND POLYGENISM

Asa Gray, Fisher Professor of Natural History at Harvard, had been corresponding with Darwin since 1855. By 1857, Darwin had sent Gray the outline of his ideas concerning the origin of species. Gray was a botanist whose examination of plant collections from Japan already had him thinking along the lines of common descent. Indeed, plant biogeography would be one of the primary lines of evidence supporting *The Origin of Species*.[40] Agassiz was coming into the height of his scientific fame at Harvard, outshining his botanical colleague Gray. In 1857, Agassiz published *An Essay on Classification*, which outlined an entire system of nature as the result of a divine plan of special creation and included his zones of creation thinking. Gray realized that he would have to take the initiative to bring about a debate on the origin of species, and by December 1858, he was prepared to take on Agassiz with a specific case.[41] Agassiz's zones of creation theory predicted that specific species of plants and animals were created for specific regions of the world. Thus, Gray reasoned that his studies showing that the flora of the eastern United States and Japan had a great number of congeneric, closely related species, with many of them being identical, would be a strong blow to this idea.

It was Darwin who struck the final blow to creationist polygenism. By 1870, he had become the leading figure of English science.[42] With his prestige assured, he was ready to address human evolution and the concept of biological races. Chief among his conclusions was that it was important to address the race concept in humans in the same way that it would be addressed in any other animal.[43] Using that approach, Darwin concluded that all humans were members of one species. He dismissed the polygenist idea by relying on the interfertility of the hybrids. This meant that all human races were capable of reproducing with each other with no loss of viability in the resulting offspring. He cited data supplied by the South Carolina slaveholder John Bachman, whose records indicated no loss of viability in "mulatto" slaves; as well as other data concerning the fertility of English/Australoid mating. Thus, Bachman's data demonstrated that these crosses between "races" did not result in the patterns that would be expected of crosses between species. Darwin concluded that the races of man were more like each other than not; that humans were protean or polymorphic, meaning that many of the traits used to classify them were unimportant to natural selection; and, finally, that the number of races identified by the naturalists of his day was arbitrary.[44] Yet the influence of evolutionary reasoning on human evolution was no immediate palliative against racist ideology. For example, in 1869 John Jeffries utilized an evolutionary scenario to explain that the acute facial angle of Negroes was a primitive condition from which the superior right angle of Europeans was derived.[45]

ENTER THE NEANDERTHALS

Three years before Jeffries's evolutionary explanation of the human races, the anatomist Paul Broca declared that the discovery of the remains of ancient Neanderthals provided the first anatomical evidence for the Darwinian idea applied to humans. Neanderthal fossils had been discovered in Europe as early as 1830 and then again in 1848. However, these were dismissed by the scientific community as "deformed individuals" and by the religious community as the remains of individuals (or species) destroyed by the flood.[46] In 1863, Thomas H. Huxley (known as "Darwin's bulldog") published the first paper suggesting that these fossils were the remains of ancient humans.[47] Some scholars in the early 1900s believed that these were the remains of direct ancestors of modern humans, while others thought they were a side branch. The discovery of the Piltdown Man (1912) played an important role in focusing the search for human fossils in Europe.[48] By this time, fossils of human-like apes were being discovered in other regions of the world, including Java, China, and South Africa. However, the brain cases of these specimens were more ape-like in size, whereas the Piltdown specimen had a brain case more similar to that of modern humans. For many years, this specimen was an anomaly that serious paleontologists struggled to reconcile with the rest of the fossil evidence concerning human origins. However, in 1953, it was finally revealed as a fraud.

Yet Darwin had predicted that the origin of anatomically modern humans must be in Africa.[49] This followed from descent with modification; since the most closely related nonhuman species (the great apes, gorillas and chimpanzees) lived in Africa, the common ancestor of all of these species must have lived there also. In 1924, Raymond Dart discovered the humanoid fossil Taung boy in South Africa and named it *Australopithecus africanus*.[50] His find was met with considerable skepticism, in part due to the established belief within anthropology that the ancestor of modern humans must have lived in Europe (e.g., Neanderthals) and in part due to the fact that European (English) anthropology looked down on scientists from the Commonwealth (Dart was a professor at the University of Witwatersrand in South Africa). Also, anthropology had still not fully accepted the implications of descent with modification, which predicted that the evolution of humans (or any species) was best visualized as a bush, as opposed to a straight tree trunk. In descent with modification, many species were predicted to be evolutionary dead-ends; thus, while leaving fossils, they might not have left progeny that existed in the modern world. Therefore, many of the humanoid fossil finds represented extinct lineages of hominids (e.g., the genus *Australopithecus*).

MULTIREGIONALISM

In the 1940s, a partial reconciliation of the diverse humanoid fossil finds appeared in the form of Franz Weidenreich's multiregional theory of human origins.[51] According to this theory, the modern human biological races were descended from an ancient common ancestor (*Homo erectus*), and each race of humans evolved within its current geographical range. *Homo erectus* fossils have been found all over the world (sub-Saharan Africa, China, Java, Malaysia, Saudi Arabia, England, France, Germany, and other European sites). These fossils differ from modern humans in a number of ways, including having cranial bones twice as thick and a mean cranial capacity of about 833 cubic centimeters (the mean for modern humans is 1,350 cc).[52] Today we know that the African fossils are the oldest (dating back 1.8–1.5 million years, compared to 1.6 million–53,000 years for those in other parts of the world).[53] However, the type of radiometric dating tools needed to determine that were unavailable to Weidenreich. His model of human evolution allowed for some gene flow between the regions, but the primary evolutionary events (adaptations) occurred within a region. He also posited that the rate of evolution from *H. erectus* to *H. sapiens* most likely differed from region to region. Typical of the white supremacists/Eurocentrists that dominated physical anthropology in this period, Weidenreich saw Europeans as more advanced than other groups of modern humans:

> In any case, there is no reason to doubt the possibility that the human stem produced more advanced types under favorable circumstance at a certain period and in one place on earth, while it remained stationary in another. Even our days offer examples for those occurrences: seen from a morphological point of view, the Vedda and the Australian bushman are less advanced human forms than the white man; that is, they have preserved more of their simian stigmata. Whether they have "entered" into evolution at a later time than the whites, or their evolution "rested" or was "retarded" while the whites went on, we do not know, but it's irrelevant with regards to their relation to other races of modern mankind.[54]

In 1962, Carleton Coon would incorporate Weidenreich's multiregional model into his own ideas, which more explicitly relied on the different rates of evolution by region. Coon supported the idea that Caucasoids (Europeans) and Mongoloids (East Asians) developed into *H. sapiens* before Negroids (sub-Saharan Africans) and Australoids (native Australians).[55] His claims were immediately picked up to justify ongoing racial segregation in the United States and were popularized by the arch segregationist Carleton Putnam (who was also Coon's cousin). Coon could not claim objective neutrality in the controversy generated by his ideas, as he was shown to be in correspondence with his cousin over the relevance of his "science" for the segregationist agenda. He even argued vehemently with evolutionary geneticists such as Theodosius Dobzhansky concerning the validity and application of this theory.[56]

In the 1970s, Milford Wolpoff and his colleagues revised the multiregional hypothesis. For Wolpoff, gene flow between the regions undergoing evolution toward *H. sapiens* played a more significant role.[57] In his model, *H. erectus* populations in various regions of the world maintained sufficient gene flow such that the entire species evolved from *H. erectus* to *H. sapiens* simultaneously. This statement summarized his view of multiregional evolution: "It evokes diffusion across a network of genic exchanges, a mechanism that is the opposite of independent evolution, to account for the shared pattern of evolution across the human range combined with the presence of some regional continuities in various areas."[58]

OUT OF AFRICA

Wolpoff and his collaborators were the last serious adherents of multiregionalism. The theory was first postulated in a time period when scientists did not have the tools to accurately test its predictions. In the main, it was an evolutionary theory of human origins that relied primarily on the interpretation of fossil evidence. The problem with the type of physical (phenotype) evidence that can derived from fossils is that phenotypes are always influenced by environments. Thus, it is difficult to accurately assess the extent to which environmental and genetic influences each contributed to the differences in the physical attributes of contemporary fossils. Furthermore, fossil finds of ancient hominids are rare, so the sample sizes required to make accurate inferences between supposed species have always been limited. As a result, the coming of new technologies—particularly genomic sequencing of living humans and eventually of ancient fossilized specimens—sounded the death knell for multiregionalism between 1990 and 2010.

One of the first major genomic findings that gave powerful support to the out-of-Africa replacement model was the "mitochondrial Eve" paper of Rebecca Cann, Mark Stoneking, and Allan Wilson, published in 1987.[59] The significance of the controversy surrounding this report is illustrated by the fact that it earned the cover of *Newsweek* magazine on January 26, 1987. Adam and Eve are depicted in the Garden of Eden, with both displaying African American features and 1980s hair styles (see figure 11.1). It is interesting to note that some special creationists felt compelled to attack the science behind the mitochondrial Eve hypothesis.[60] Brad Harrub and Bert Thompson, for example, rejected the claim of a mitochondrial Eve, but it is hard to understand their motivation. This idea would be consistent with the narrative of Genesis that all humans are descended from two humans, Adam and Eve. Thus, the authors might not have been objecting to the notion of the common descent of all modern humans. Rather, they might have been objecting to the age of the common descent (~200,000 years). That objection would be expected of young-earth creationists, who believe that the earth is ~6,000 years old. However, my suspicion is that it is far more likely these authors really objected to the idea that Eve was a sub-Saharan African.

The scientific objections to the mitochondrial Eve hypothesis revolved mainly around the extreme coalescence the model predicted.[61] Coalescence of phylogenetic trees results from the fact that new mutations have occurred within specific

11.1 The search for Adam and Eve, *Newsweek*, January 11, 1988.

populations at some time in their past. The new mutations are propagated as genetic codes that are replicated in individuals through time. Coalescent analysis allows us to reverse the process to work backward from present populations to see their genetic past.[62] The mitochondrial Eve study predicted that all modern humans were descended from one to four sub-Saharan African females within the last 200,000 years. Francisco Ayala argued that an effective population estimate of anatomically modern humans was no less than 10,000 individuals in that time period.[63] Ayala's estimate of a much greater human population size within the last 200,000 years challenged the prediction of a much smaller population size made by the African Eve hypothesis.

There were other methodological concerns with the original Eve study, but over the last thirty years, an avalanche of additional evidence (fossil, mtDNA, Y-chromosome, and nuclear DNA) has validated the out-of-Africa replacement model.[64] This was possible because the multiregional and out-of-Africa replacement models make some strikingly different predictions concerning what the genetic diversity of modern humans should look like. Table 11.1 summarizes the different predictions. With respect to the five predictions, multiregionalism can be "saved" only by special circumstances. This, of course, puts this hypothesis on similar grounds as the geocentric solar system ideas struggling to retain their legitimacy in the sixteenth century. Geocentric models could still predict the positions of the planets—but only so long as epicycles were applied to the other planets' motions. For multiregionalism, the deep African origin of *H. erectus/H. ergaster* and gene flow are its epicycles. Applying them to each prediction simply weakens the theory. However, these epicycles cannot explain the patterns of linkage disequilibrium found in modern humans. If multiregionalism were true, linkage disequilibrium would be roughly equal in all regions. It clearly is not. By 2010, the multiregional hypothesis was completely discredited, replaced by the out-of-Africa replacement theory by serial founder events.[65] However, in some ways, the new revelations resulting from the sequencing of ancient DNA are resurrecting some of the ideas of multiregionalism. David Reich and others view the peopling of the world as beginning with *H. sapiens* migrating out of Africa within the last 100,000 years—but with significant regional adaptation and hybridization of archaic human species (such as the Denisovans and Neanderthals).[66] For Reich and his colleagues, modern human populations are the product of gene flow between adapted regional populations, akin to biological races.

TABLE 11.1 Genetic Predictions of African Replacement Versus Multiregional Evolution

Criterion	African replacement	Multiregional evolution	Issue
Location of ancestral neutral alleles*	Africa	Random	African origin of *Homo erectus/ergaster* would favor Africa
African versus non-African divergence time	200,000 years or less	~1.8–1.0 million	Gene flow could reduce apparent divergence
Genetic diversity	Greater genetic diversity in Africa	Diversity roughly equally distributed	African origin of *H. erectus/ergaster*, gene flow, and selection could favor Africa
Sets of neutral alleles	Alleles in non-African gene pools are subsets of African	No region's gene pool is a subset of another	African origin of *H. erectus/ergaster* could mean that the genetic variation of all other regions are a subset of Africa
Linkage disequilibrium**	Lowest in Africans and should increase with distance from Africa	Should be roughly equivalent in all regions	Cannot be explained by gene flow or African origin of *H. erectus/ergaster*

Source: Modified from table 20.1, in J. C. Herron and S. Freeman, *Evolutionary Analysis*, 5th ed. (Boston: Pearson, 2014).
*Neutral alleles are genetic variants that either do not change the resulting coding of an amino acid (triplet codes are redundant) or occur within a noncoding aspect of the genome.
**Linkage disequilibrium refers to deviation from expected frequencies of alleles found in proximity to each other on a given chromosome. If natural selection favors one allele at a given locus, then all alleles at nearby loci will be increased in frequency, even if they are not favored by natural selection. These loci become genetic markers delineating portions of the chromosome. Through time, genetic recombination will reduce the size of the linkage disequilibrium blocks within the genome. The more time that passes, the smaller the blocks. Linkage disequilibrium declines with distance from sub-Saharan Africa. S. A. Tishkoff et al., "Global Patterns of Linkage Disequilibrium at the CD4 Locus and Modern Human Origins," *Science* 271, no. 5254 (1996): 1380–1387.

CONCLUSION: "ITSELF THE DESCENDENT OF AN APE..."

We are now in the beginning the third decade of the twenty-first century. Scientific knowledge has advanced at an unparalleled pace. There are now more scientists alive than all those who existed in the history of previous civilization. We know more about the universe, including living things, than ever before. Specifically, in the last decade our knowledge of human genetic diversity and its evolutionary history has expanded by leaps and bounds. Yet in the United States, the majority of its people still do not understand the origin or the biological history of our species. This is particularly problematic, as American social life is still dominated by the socially defined race concept. The majority of Americans still believe the features and the conditions of life that these socially defined groups experience result from innate biological differences among them.

This is a particularly vexing problem for faith-based communities. For example, white evangelical Christians have long been associated with racist ideology.[67] This association first began with chattel slavery, which required that "love thy neighbor" be tossed out the window when it came to persons of African descent. Theological justification followed in the form of the lack of a scriptural condemnation of slavery, but this soon required even more powerful theories, such as the nonhuman status of Africans due to monogenist degeneration (from the curse of Ham) and pre-Adamist polygeny. Ironically, the destruction of polygeny did not come at the hands of theologians but rather was due to the progress of biological science. This created a greater issue for Christianity, as the acceptance of natural selection for some meant the intellectual requirement to reject God. Thus, while the original acceptance of Darwin in the United States was somewhat tame, by the 1920s social forces had morphed this into an out-and-out confrontation between good (Christianity) and evil (science).[68]

This came to a head in 1925 in the famous Scopes Monkey Trial (*Tennessee v. Scopes*). Yet this characterization of *Scopes* (religion versus science) really obscures the importance of these events. The defense of evolution in *Scopes* was, in reality, part of a wider social agenda to defend eugenics.[69] The American Association for the Advancement of Science (AAAS), through its magazine *Science* and its Committee on Race and Eugenics, was a major force in the confrontation. The members of the committee were the notorious racists Edward Grant Conlin (chair, embryologist, Princeton University), Henry Fairfield Osborn (director of the Museum of Natural History in New York), and Charles Davenport (director of the Station for Experimental Evolution, Long Island, New York, and director of the Eugenics Records Office). Clarence Darrow and the American Civil Liberties Union sought out the advice of the AAAS committee. The agenda of the committee was to defeat religious fundamentalism so that "evolutionary principles" (meaning eugenics) could be used to guide American social policy. The committee (and *Science* magazine) fought back against William Jennings Bryan (and his conservative Protestant base) by arguing that progressive religion was already backing evolutionary science and that modernist clergy also endorsed applied eugenics. To justify this, these clergy were appealing to the authority of Christianity and professional science and often reading Judeo-Christian texts through the lens of eugenics and racism.[70] This again underscores the alliance between various sectors of Christianity and supporters of evolutionary science in the service of white supremacy in U.S. history.

Today the degree to which racist thinking is associated with accepting or rejecting the modern science of human evolution is an unanswered question. On one hand, some white supremacists are attempting to claim human evolution as a validation of their racial supremacy. A minority of vocal scholars still seem to adhere to biological determinist forms of racism. Yet the majority of professional biologists and anthropologists have repeatedly demonstrated the absurdity of racist claims concerning human variation.[71] Thus, racist strategies completely relying on modern biology should be doomed to fail on intellectual grounds. Yet given the general scientific illiteracy around genetics as associated with human variation, these programs still seem to gain traction. On the other, neither should racist strategies relying on faith succeed. All the major Christian denominations have now publicly denounced racism. For example, the Episcopal Church of America has declared racism to be a sin.[72] Racism no longer has an official home within the faith community, other than in the most extreme white nationalist fringe churches or other racist faith positions. For example, two rabbis in Israel recently utilized racist thinking to justify the domination of the Palestinian Arabs. When asked about their views, one replied to the television reporter: "Yes, we're racists. We believe in racism.... There are races in the world and peoples have genetic traits, and that requires us to try to help them.... The Jews are a more successful race."[73]

This leads to an additional unanswered question: To what degree is the Christian/or other religions belief in the unity of humankind driving white racists away from Christianity or other religions?[74] The rejection of Christianity and the substitution of Aryan legends of Thule were key features of Nazism. A similar theme is seen in the small modern cult of the World Church of the Creator.[75]

Other questions can be asked about the facts of human evolution and religious narratives. Is the rejection of the science validating the out-of-Africa theory observed in other faith traditions (Native American shamanism, Buddhism, Islam, Hinduism, Shinto)? The rejection of the land bridge hypothesis by many Native American traditionalists is well-known.[76] The GRID study referenced earlier did not collect data from self-identified members of Native American Nations or from persons who could be identified as belonging to the religious traditions of East Asia and the Indian subcontinent. In this regard, additional research is merited to determine the degree to which global faith communities respond to the scientific fact of human evolution and the out-of-Africa origins of the human race.

We can also ask whether it is possible for a sincere dialogue between science and religion to play a role in reducing racial ideology and the harm that it causes to our society. Does the realization that we are all "descendants of an ape" provide us with any agency over the most vexing of our societal problems, racism? If so, how do we get this message out? Finally, can we deploy this understanding to reduce societal racism, and if so, how? We know that increasing the public's scientific knowledge of human biological variation alone will not reduce societal racism.[77] However, efforts to increase this knowledge make it more difficult for racists to hide behind purported biological realities to support their agendas. For this reason alone, this enterprise is worth doing.

NOTES

1. Gallup, "Evolution, Creationism, Intelligent Design," accessed July 28, 2021, https://news.gallup.com/poll/21814/evolution-creationism-intelligent-design.aspx.
2. M. Berkman and E. Plutzer, *Evolution, Creationism, and the Battle to Control America's Classrooms* (New York: Cambridge University Press. 2010).
3. J. W. Martin, "Compatibility of Major U.S. Christian Denominations with Evolution," *Evolution: Education and Outreach* 3 (2010): 420–431, https://doi.org/10.1007/s12052-010-0221-5.
4. Berkman and Plutzer, *Evolution*, table 3.2, p. 73.
5. R. Pennock, *The Tower of Babel: The Evidence Against the New Creationism* (Boston: MIT Press, 2000).
6. Answers in Genesis, https://answersingenesis.org/racism/one-race/.
7. Crawford et al., "Loci Associated with Skin Pigmentation Identified in African Populations," *Science* 358, no. 6365 (2017): eaan8433. doi: 10.1126/science.aan8433; K. Adhikari et al., "A GWAS in Latin Americans Highlights the Convergent Evolution of Lighter Skin Pigmentation in Eurasia," *Nature Communications* 10, art. no. 358 (2019).
8. Adhikari et al., "A GWAS in Latin Americans."
9. S. Fan, M. E. Hansen, Y. Lo, and S. A. Tishkoff, "Going Global by Adapting Local: A Review of Recent Human Adaptation," *Science* 354, no. 6308 (2016): 54–59.
10. 1000 Genomes Project Consortium, "A Global Reference for Human Genetic Variation," *Nature* 526, no. 7571 (2015): 68–74, https://doi.org/10.1038/nature15393.
11. R. Moore, "The Revival of Creationism in the United States," *Journal of Biological Education* 35, no. 1 (2000): 17–21, https://doi.org/10.1080/00219266.2000.9655730.
12. S. Outram et al., "Genes, Race, and Causation: US Public Perspectives About Racial Difference," *Race and Social Problems* 10 (2018): 79–90, https://doi.org/10.1007/s12552-018-9223-7.

13. R. Jones, *White Too Long: The Legacy of White Supremacy in American Christianity* (New York: Simon & Schuster, 2020).
14. Brian M. Donovan, Rob Semmens, Phillip Keck, Elizabeth Brimhall, K. C. Busch, Monica Weindling, Alex Duncan, Molly Stuhlsatz, Zoë Buck Bracey, Mark Bloom, Susan Kowalski, Brae Salazar, "Toward a More Humane Genetics Education: Learning About the Social and Quantitative Complexities of Human Genetic Variation Research Could Reduce Racial Bias in Adolescent and Adult Populations," *Science Education* 103 (2019): 529–560.
15. J. L. Graves, "Why We Should Teach Our Students About Race," *The Natural Selection*: newsletter of the Biological Sciences Curriculum Study (Winter 2002), repr. in *Reports of the National Center for Science Education*, May/June 2002: 23–26.
16. A. Desmond and J. Moore, *Darwin: The Life of a Tormented Evolutionist* (New York: Norton, 1991), 442.
17. Desmond and Moore, *Darwin*, 442.
18. R. Knox, *The Races of Men: A Fragment* (Philadelphia: Lea & Blanchard, 1850).
19. L. Agassiz and A. A. Gould, *Principles of Zoology: Touching the Structure, Development, Distribution, and Natural Arrangement of the Races of Animals, Living and Extinct; with Numerous Illustrations* (Boston: Gould, Kendall, and Lincoln, 1848), ch. 13.
20. C. Darwin, *The Correspondence of Charles Darwin*, ed. F. H. Burkhardt (Cambridge: Cambridge University Press, 1990), 6:184, 189, 236.
21. J. S. Haller, "The Species Problem: Nineteenth Century Concepts of Racial Inferiority in the Origin of Man Controversy," *American Anthropologist* 72, no. 6 (1970): 1319–1329.
22. B. Braude, "The Sons of Noah and the Construction of Ethnic and Geographical Identities in the Medieval and Early Modern Periods," *William and Mary Quarterly* 54, no. 1 (1997)): 103–142.
23. J. Priest, *Bible Defense of Slavery and the Origin, Fortunes, and History of the Negro Race* (Glasgow, KY: Rev. W. S. Brown MD, 1852).
24. J. F. Blumenbach, *The Anthropological Treatises of John Friedrich Blumenbach* (London: Longman, Green, Longman, Roberts, and Green, 1865).
25. J. L. Graves, *The Emperor's New Clothes: Biological Theories of Race at the Millennium* (New Brunswick, NJ: Rutgers University Press, 2005), 129.
26. Graves, *The Emperor's New Clothes*, 25.
27. C. Caldwell, *Thoughts on the Original Unity of the Human Race* (New York: E. Bliss, 1830).
28. R. P. Forbes, "Slavery and the Evangelical Enlightenment," in *Religion and the Antebellum Debate Over Slavery*, ed. J. R. McKivigan and M. Snay (Athens: University of Georgia Press, 1998).
29. D. Malone, *The Public Life of Thomas Cooper* (New Haven, CT: Yale University Press, 1926), 285.
30. R. Horsman, *Josiah Nott of Mobile: Southerner, Physician, and Racial Theorist* (Baton Rouge: University of Louisiana Press, 1987), 100.
31. E. Lurie, "Louis Agassiz and the Races of Man," *Isis* 45, no. 3 (1954): 227–242.
32. J. Bachman, *The Doctrine of the Unity of the Human Race Examined in the Principles of Science* (Charleston, SC: C. Canning, 1850).
33. Lurie, "Louis Agassiz," 234.

34. G. B. Wood, *Biographical Memoir of Samuel George Morton, M.D.: Prepared by Appointment of the College of Physicians of Philadelphia, and Read Before That Body, November 3, 1852* (1852; repr. New York: Forgotten Books, 2017).
35. R. A. Smith, "Types of Mankind: Polygenism and Racism in the Nineteenth Century United States Scientific Community" (master's thesis, Pittsburgh State University, 2014), http://digitalcommons.pittstate.edu/etd/105.
36. C. Darwin, "To W. D. Fox, 4 September 1850," in *The Correspondence of Charles Darwin*, vol. 4, ed. F. Burkhardt and S. Smith (Cambridge: Cambridge University Press, 1989), DCP-LETT-1352.
37. B. Wallis, "Black Bodies, White Science," *American Art*, Summer 1995, 39–61.
38. C. Irmscher, *Louis Agassiz: Creator of American Science* (Boston: Houghton Mifflin Harcourt, 2013).
39. Irmscher, *Louis Agassiz*, Location 4198 Kindle Edition.
40. A. H. Dupree, "The First Darwinian Debate in America: Gray Versus Agassiz," *Daedalus* 88, no. 3 (1959): 560–569.
41. Dupree, "The First Darwinian Debate," 561.
42. Desmond and Moore 1991; A. Desmond and J. Moore, *Darwin's Sacred Cause: How a Hatred of Slavery Shaped Darwin's Views on Human Evolution*, (Boston: Houghton Mifflin Harcourt, 2009).
43. C. Darwin, *The Descent of Man and Selection in Relation to Sex* (Princeton, NJ: Princeton University Press, 1981), 215.
44. Graves, *The Emperor's New Clothes*, 64–73.
45. J. P. Jeffries, *The Natural History of the Human Races* (New York: E. O. Jenkins, 1869).
46. M. A. Miller, "The Neanderthal Demise: Was It Love, or Was It War (with Modern Humans)?," *Athena Review* 2, no. 4 (2001).
47. T. H. Huxley, "Further Remarks Upon the Human Remains of the Neanderthal," *Natural History Review* (n.s.) 4 (1864): 429–446.
48. K. P. Oakley and J. S. Weiner, "Piltdown Man," *American Scientist* 43, no. 4 (1955): 573–583, https://www.jstor.org/stable/27826700.
49. Darwin, *The Descent of Man*.
50. S. Dubow, "Human Origins, Race, Typology and the Other Raymond Dart," *Africana Studies* 55, no. 1 (1996): 1–30, https://doi.org/10.1080/00020189608707838.
51. F. Weidenreich, "Facts and Speculations Concerning the Origin of *Homo sapiens*," *American Anthropologist* 49, no. 2 (1947): 187–203.
52. B. Campbell, *Human Evolution: An Introduction to Man's Adaptations*, 4th ed. (New York: Aldine De Guyter, 1998); J. K. McKee, F. E. Poirer, and W. S. McGraw, *Understanding Human Evolution*, 5th ed. (Upper Saddle River, NJ: Pearson/Prentice Hall), 2005.
53. McKee, Poirer, and McGraw, *Understanding Human Evolution*.
54. Weidenreich, "Facts and Speculations," 202.
55. C. S. Coon, *The Origin of Races* (New York: Knopf, 1962).
56. J. P. Jackson, " 'In Ways Unacademical': The Reception of Carleton S. Coon's *The Origin of Races*," *Journal of the History of Biology* 34 (2001): 247–285.
57. M. H. Wolpoff, J. Hawks, and R. Caspari, "Multiregional, Not Multiple Origins," *American Journal of Physical Anthropology* 112, no. 1 (2000): 129–136.

58. Wolpoff, Hawks, and Caspari, "Multiregional, Not Multiple Origins," 130.
59. R. Cann, M. Stoneking, and A. Wilson, "Mitochondrial DNA and Human Evolution," *Nature* 325 (1987): 31–36.
60. See, for example, B. Harrub and B. Thompson, "The Demise of Mitochondrial Eve," © 2003 Apologetics Press, https://www.trueorigin.org/mitochondrialeve01.php#:~:text=As%20scientists%20have%20begun%20to,entire%20discussion%20a%20moot%20point.
61. F. J. Ayala, "The Myth of Mitochondrial Eve," *Science* 270, no. 5344 (1995): 1930–1936.
62. J. F. Kingman, "Origins of the Coalescent. 1974–1982," *Genetics* 156, no. 4 (2000): 1461–1463.
63. Ayala, "The Myth of Mitochondrial Eve."
64. M. C. Campbell et al., "The Peopling of the African Continent and the Diaspora Into the New World," *Current Opinion in Genetics and Development* 29 (2014): 120–132.
65. See, for example, K. L. Hunley and G. S. Cabana, "Beyond Serial Founder Effects: The Impact of Admixture and Localized Gene Flow on Patterns of Regional Genetic Diversity," *Human Biology* 88, no. 3 (2016): 219–231; and D. Shriner et al., "Ancient Human Migration After Out-of-Africa," *Scientific Reports* 6, art. no. 26565 (2016), https://doi.org/10.1038/srep26565.
66. D. Reich, *Who We Are and How We Got Here: Ancient DNA and the New Science of the Human Past* (New York: Pantheon, 2018).
67. M. O. Emerson and C. Smith, *Divided by Faith: Evangelical Religion and the Problem of Race in America* (New York: Oxford University Press, 2000).
68. R. A. Eve and F. B. Harrold, *The Creationist Movement in Modern America* (Boston: Twine, 1991).
69. A. Pavuk, "The American Association for the Advancement of Science Committee on Evolution and the Scopes Trial: Race, Eugenics and Public Science in the U.S.A.," *Historical Research* 91, no. 251 (2018): 137–160, https://doi.org/10.1111/1468-2281.12208.
70. C. Rosen, *Preaching Eugenics: Religious Leaders and the American Eugenics Movement* (New York: Oxford University Press, 2004).
71. American Association of Physical Anthropologists, Statement on Race and Racism, March 8, 2019, http://physanth.org/about/position-statements/aapa-statement-race-and-racism-2019/.
72. Episcopal Church of America, House of Bishops Pastoral Letter on Sin of Racism, March 1994, https://www.episcopalchurch.org/wp-content/uploads/sites/2/2020/07/awakening_pastoralletter.pdf.
73. T. Pileggi, "Embracing Racism, Rabbis at Pre-army Yeshiva Laud Hitler, Urge Enslaving Arabs," *Times of Israel*, April 30, 2019, https://www.timesofisrael.com/embracing-racism-rabbis-at-pre-army-yeshiva-laud-hitler-urge-enslaving-arabs/.
74. J. Kaplan, "Dreams and Realities in Cyberspace: White Aryan Resistance and the World Church of the Creator," *Patterns of Prejudice* 37, no. 2 (2003): 139–155, https://doi.org/10.1080/0031322032000084679.
75. G. Michael, "RAHOWA! A History of the World Church of the Creator," *Terrorism and Political Violence* 18, no. 4 (2006): 561–583, https://doi.org/10.1080/09546550600880633.
76. V. Deloria, *Native Americans and the Myth of Scientific Fact* (Golden, CO: Fulcrum, 1997).
77. J. L. Graves, "Human Biological Variation and the 'Normal,'" *American Journal of Human Biology* 33, no. 5 (2021): e23658, https://doi.org/10.1002/ajhb.23658.

PART IV

COHERENCE

AHMED RAGAB, TERENCE KEEL, AND MYRNA PEREZ SHELDON

If the subaltern can speak, can they be understood? The chapters in this part investigate the capacity and the need to be understood. bell hooks highlights how language creates barriers and expresses forms of distinction that emphasize marginalization along lines of race, gender, sexuality, and socioeconomic status.[1] In her analysis, comprehension is a function of discipline, whereby the marginalized are compelled to adhere to the manners of normative speech. In this way, translation itself is a vehicle of discipline. Translated, the subaltern speech is reproduced and reorganized to conform to the normative. Recalling hooks's analysis allows us to think about commensurability and legibility in relation not only to the debated magisteria of science and religion but also, more critically, to the contested spaces of colonial language and native or colonized meaning making.

This part takes on coherence itself as an object of analysis. Previous works on science and religion placed inherent value in commensurability and the capacity to talk across seeming divides—whether at the intellectual or philosophical level or in the spaces of social media or electoral politics. Yet this debate of commensurability assumes the naturalness of translatability and coherence—not as inherent qualities of a given discourse but as a quality in the eye of the beholder. The contributors in this part instead problematize coherence as a goal of translation. In translation, the colonized or marginalized speech is forced into the narratives of the colonizer not only through replacing words or seeking equivalence but also through forcing a structure of normative coherence, without which marginalized speech is seen to be incomprehensible. In this context, coherence becomes a moral category—a sign of modernity, civilization, and worth.

This part therefore explores the process through which coherence is produced, assumed, created, and transmitted. In reframing coherence as an object of analysis as opposed to a quality of speech making, it opens new spaces in the study of science and religion—ones that question debates around commensurability, dialogue, and magisteria as products and manifestations of the production of coherence. Our question therefore is not whether obeah or Native Hawai'ian beliefs about Mauna Kea are scientific or religious or how Native American "religion" can speak to Western "science." Rather, we ask how these categories come to coalesce, cohere, and make sense in their own spaces and in the process of their uptake (or distortion) by normative discourses.

Eli Nelson's chapter, "Kānaka Maoli Voyaging Technology and Geography Beyond Colonial Difference," focuses on discordance and the violence of categorical universals. It reflects on recent interest in animist theories of materiality within Western postmodern and posthumanist literatures. Partly in response to the existential threat of climate change, scholars in these fields have questioned traditional Western dichotomies between humans and nonhumans and between agential and nonagential actors. However, many Indigenous scholars argue that such dichotomies never existed in Indigenous worldviews and that this very lack prompted the colonial denigration of Indigenous practices. Critically, Nelson argues that "the apparent synergy" that now exists between Indigenous and postmodern thought belies their colonial relationship, in which "modernism and postmodernism, and their manifestations in science and religion, were built on the study and the making of the Native." He suggests that we must be wary of the flexibility of such intellectual colonialism—of the capacity of Western systems to always render themselves as the coherent neutral universal. Rather than submitting to a project of epistemological reconciliation, Nelson's chapter analyzes Kānaka wayfaring technology in order to bring forward its "productive incoherence." In doing so, he contends not only that Indigenous animism and Western objectivism are shaped by different cultural technologies but also, more profoundly, that each "literally changed the geography and space of those worlds."

Joanna Radin's chapter, "Speculation Is Not a Metaphor: More than Varieties of Cryobiological Experience," takes these themes into the heart of Western laboratory science. Radin asks how Indigenous theories of spirit, matter, and science force a reckoning with the speculation within Western science. She does this through a close reading of the career and work of the Catholic Jesuit scientist Basil Luyet, who was a leading figure in the early development of blood-freezing technology. She explores how Luyet's Catholicism, alongside his training in biology and physics, framed his speculative work on latency—the "ambiguous state between life and death." On the one hand, blood freezing was a Western colonial project wherein scientists collected and froze the blood of Indigenous peoples in order to preserve these so-called primitive cultures. On the other, it was a deeply Christian project that explored the limits and possibilities of bodily resurrection. Building with Indigenous thinkers, Radin's chapter breaks down intuitive boundaries around sacred and secular natural practices. She asks what changes in the history of cryobiology when we view Luyet's speculative materiality not as a metaphor but as a true feature of his Western science.

Katharine Gerbner's chapter, "Maroon Science: Knowledge, Secrecy, and Crime in Jamaica," examines the early history of the term *obeah* in the slave societies of the eighteenth-century Caribbean. Obeah refers to a set of spiritual and natural rituals utilized by enslaved healers and spirit workers during this period. Importantly, it was also the term used by colonial enslavers to denigrate and criminalize these practices. After 1760, Jamaican Maroons (descendants of runaway slaves) began to use the term *science* instead of *obeah* to name these practices, and their oral histories continue to emphasize science as crucial to their struggle for political autonomy within Jamaica. Gerbner questions how historians ought to narrate the relationship between obeah and science. Are they linguistic equivalents? Is there a meaningful difference between the two? Should one translate obeah as science? However, what Gerbner dwells on in the chapter is not how to discern what obeah really means but rather how to narrate colonial history through the lens of Maroon science in a manner that sets aside the intellectual mastery of archival documentation. In this, her analysis goes beyond even the methodological complexities of writing the history of slavery. Rather, she confronts the limitations of Western science and history in accounting for the realities and legacies of Jamaican Maroon society.

The final chapter in this part, "Obeah Simplified? Scientism, Magic, and the Problem of Universals" by J. Brent Crosson, also focuses on the relationship between obeah and science. In his anthropological study, Crosson observes that these words are used as synonyms by spirit workers in contemporary Caribbean communities. This is despite the history of colonial governments using obeah to define and criminalize subaltern religious practice as witchcraft or superstition. Akin to Gerbner, Crosson's purpose in the chapter is not to settle on a single translational relationship between obeah and science. Rather, he seeks to "defamiliarize some of the key 'universals' through which Atlantic modernity was forged: magic, science, and religion." To conclude that obeah is simply the Afro-Caribbean term for magic or religion is to flatten the word—and to disregard its ongoing use as an equivalent for science. Ultimately, Crosson's analysis reveals what is behind the desire for linguistic and epistemological translation: the need to make sense of the colonized culture in terms of the colonizer. But he argues that even this is a false promise, as "universals can never be universals in practice" because they are "premised on the geo-racial assumption that the West can be universal while the rest remain particular and in need of tutelage." If science is the universal and obeah is its partial shadow, then the world of Caribbean spirit workers is always and already illegitimate.

Throughout these chapters, we encounter the urgency of rethinking coherence as a quality often taken for granted, and we observe the work that analyzing coherence can do in understanding the spaces making science and religion as well as those between them.

NOTE

1. See, for example, bell hooks, "Language: Teaching New Worlds, New Words," *Estudos Feministas* 16, no. 3 (2008): 857, and *Talking Back: Thinking Feminist, Thinking Black* (Boston: South End Press, 1989).

CHAPTER 12

KĀNAKA MAOLI VOYAGING TECHNOLOGY AND GEOGRAPHY BEYOND COLONIAL DIFFERENCE

ELI NELSON

"WE ARE NOT ANTI-SCIENCE"

On July 17, 2019, the governor of Hawaii David Ige declared a state of emergency in the wake of the arrest of dozens of Kānaka Maoli (Native Hawaiian) activists, including thirty-three elders, for blocking the egress of construction vehicles primed to erect the Thirty Meter Telescope (TMT) atop Mauna Kea.[1] Non-Native scientists and financial backers in favor of building the massive telescope at the highest point in the Pacific cited its potential for groundbreaking observations of distant forming galaxies and the inner workings of far-away planetary atmospheres, as well as the potential for funding opportunities and youth science education on the island.[2] Kānaka Maoli activists countered that a construction project of this kind would contradict protocol for treading softly on Mauna Kea as sacred land and imperil its unique and diverse ecosystem. In a blog posted in 2015 when TMT was originally slated for construction, KAHEA (the Hawaiian-Environmental Alliance) representatives wrote:

> We use the word sacred to describe the mountain . . . a place where one treads carefully and reverently. . . . In the absence of that reverence, there is no meaning to our existence, and for Kānaka that meaningfulness is tied directly to our belief that we and the mountain share common ancestors. Even if not a single person had ascended that mountain in the last century to feel her quiet and yet powerful assurance, she would still have been waiting for us to remember.[3]

Mauna Kea, KAHEA argued, not only is sacred and deserving of reverence but also recognizes, waits, and welcomes; it is a genealogic relative in its own right. What kind of person would allow someone to cement over their relative for a better view?

The battle over Mauna Kea is now years long and still unfolding. It has been waged at road blocks and in courts, classrooms, and publications, and it is regularly referred to as a competition for the moral high ground. Reflecting on the first rounds of protests in 2015, the physicist Brian Koberlein conceded that science could not claim the moral high ground in the TMT case because scientists in the past had failed to "live up to the scientific ideal."[4] He lamented that this past moral failing would give Kānaka religious animists a rhetorical edge that could ultimately prevent him and other space enthusiasts from seeing TMT built in their lifetime. At the same time, KAHEA was quick to point out that the moral high ground had always already been assigned to science:

> TMT's claim to a *moral high ground* in the name of science has been made loudly and consistently to an audience trained to think of "science" as an undeniable force of innovation and an institution that has produced nothing but good for human beings. . . . Mauna Kea reminds us that there are other knowledges and understandings developed, honed and cherished by human beings which native peoples globally have been striving to recover after the long wave of European ideas and beliefs inundated our societies [italics mine].[5]

The TMT case does not pit the relative moral legitimacy of religious animists against scientists—they are *not*, as they say time and again, "anti-science."[6] Instead, the battle is political and epistemological—pitting Indigenous epistemologies, relations, and sovereignty against settler science, religion, and occupation.

Indigenous animism has historically been construed as prior to textual varieties of religion and, accordingly, as prior to science and civilization.[7] The cynical reading of TMT as a battle over a moral high ground evokes the core debate concerning animism in religious studies that seeks to identify how literally (and therefore how uncivilized) or metaphorically (or therefore how inauthentic) one ought to interpret modern Indigenous animism.[8] Recently, scholars in "new animism" have sought to correct older readings that locate animist worldviews in a less evolutionarily developed echelon of humanity, and many have addressed animism as an epistemological feature of Indigenous environmental knowledge and stewardship.⁹ This framing pits animism not against science per se but against objectivism, the belief in and concordant ways of knowing an inanimate world—a philosophical foundation that can be traced in European scientific and religious thought.¹⁰ Furthermore, new materialists have diagnosed this objectivist framework as having always been wrong and/or illusory to begin with, and in numerous disciplines, there have been concerted efforts to reassert the agency of nonhuman and non–traditionally sentient actors.¹¹ This mode of thought has gained in popularity as climate change continues to reveal the destructive nature of objectivism, and yet, as seen in the case of TMT, Indigenous animism is still framed as an inhibiting moral and religious check on science, not as a political ideology and epistemology in the same sphere. The cause of this impasse cannot be explained merely by a supposed disconnect among theory, public discourse, and practice, as evidenced by the extent to which Kānaka activist rhetoric, which intersects with the academy but is in no way sourced there, regularly communicates the same postmodern and posthuman arguments.¹²

This apparent synergy between Indigenous knowledges and postmodern theory belies multiple historical and categorical contingencies, perhaps the most significant being that these strands of thought are not isolated. Modernism and postmodernism, and their manifestations in science and religion, were built on the study and making of the Native, and this intimate dependence resulted in an inextricable trace of Indigeneity within these frameworks. In contrast, one of the most insidious strengths of colonialism has been its ability to implant its own logic as universal and common-sense, resulting in various forms of intellectual colonization and hybridity.¹³

Multiple Indigenous, postcolonial, and decolonial scholars have attempted to understand how knowledges meet and become enmeshed in this way, including most recently (and fruitfully) the use of the concept of braided knowledges in the works of Projit Mukharji and Robin Wall Kimmerer.¹⁴ As seen in the debate over TMT, the braid that has been composed of different strands of colonial and Indigenous knowledges and epistemologies produced on and with occupied Kānaka lands is categorically incoherent and contradictory. However, the experience of braiding and of being braided is historically traceable and can shed light on the intellectual, moral, and epistemological work that this incoherence does. The episteme that emerges from the experience of surviving weaving and objectification is what I call Native science—epistemologies that incorporate specific Indigenous knowledges, as well as the myriad tactics and vantages of being an object and a tool of colonial knowledge production. More so, Native science, as both a braid and a braiding, is radically incoherent in the traditional models deployed to understand and delimit science and religion.

This chapter explores the breadth of Kānaka epistemological animism and its productive incoherence by identifying the materials and process of braiding Kānaka geography, astronomy, and voyaging technology in Polynesia. I trace two key moments in the history of Kānaka animist technologies and their entanglement with colonial science and religion: first, Captain James Cook's expedition to the Pacific to track the transit of Venus in 1769 and, second, the 1974 construction of the *Hōkūle'a*, the first *wa'a kaulua* (double-hulled canoe) built in hundreds of years in Hawaii. Reading against the grain of Cook's journals, I argue that animism and objectivism not only profoundly shaped the technologies used to interact with either inert or animate worlds, like bark ships and canoes, but also literally changed the geography and space of those worlds in the first place, and vice versa. Then drawing on the accounts of the Polynesian Voyaging Society and published scientific papers, I analyze the *Hōkūle'a* as not only an animist technology but also a technology produced through Native science. Finally, I conclude by bringing these moments back to bear on Kānaka sovereignty, Mauna Kea, and new animism today by locating where (if at all) Kānaka epistemological animism sits in the decolonial framework of the geopolitics of knowledge and colonial difference.

THE VIEW FROM GREAT CANOE HARBOR

British-born Captain James Cook was no stranger to Indigenous technology when he first jotted down observations of the "Great Canoe Harbor" in Tahiti (Cook's designation) on April 14, 1769.[15] An agent of the Royal Society, he had earned his scientific reputation in North America years earlier where he charted the St. Lawrence River alongside Haudenosaunee canoes in the midst of the Seven Years War (1756–1763).[16] During expeditions to Pacific islands, including Tahiti, Aotearoa (New Zeland/Zealand), Hawaii, Tongo, Mo'orea, and Terra Australis (New Holland/Australia), among others, Cook and his crew would reference and sketch canoes hundreds of times.[17] *Wa'a kauluas*, in particular, were an oddity that looked nothing like Haudenosaunee models on the St. Lawrence, and Cook by turns despaired, scoffed, fumed, and marveled at their construction, meaning, and constant presence.

The most telling clue indicating Cook's assessment of these technologies was his records describing the actions he and his crew took against them. In October 1777, he described an encounter in Mo'orea. Upon being denied a goat by a man named Hamoah, who for his part denied having any knowledge of the goat in question, Cook ordered his men to burn all houses and canoes in the vicinity until the animal was delivered to him.[18] In the aftermath of this destruction, Omai, Cook's translator, told him to expect a "great many men" to retaliate. Instead, Cook wrote that a group of Mo'oreans approached him with plantain trees, "which they laid down at [his] feet and beg'd [he] would spare a Canoe that lay close by."[19] The following day, noting that "no account of the Goat had been received, so that all [he] had yet done had not had the desired effect," Cook sent a messenger to tell Hamoah that he "would not leave him a Canoe on the island and that [he] would continue destroying till it [the goat] came."[20] At the same time, Cook dispatched carpenters to break down yet more canoes to erect a new home for Omai. A goat was delivered to him by 7 p.m. that same day. Cook's preferred disciplinary measure was the destruction of canoes, especially after that fateful encounter.[21]

Cook was clearly acting on his observation that people valued their canoes dearly, perhaps even more so than their homes, but this only partially explains his fetish for burning them. Cook captained the HMS *Endeavor*, a bark-class sailing vessel, which at nearly ninety-eight feet long and with a hull depth of approximately eleven feet, was at least a third larger than the most substantial *wa'a kauluas* at the time. Still, these smaller vessels could carry dozens of people in addition to supplies and goods for trading, the latter of which Cook was in constant need.[22] As he maneuvered into island ports, he would describe with horror a recurring scene: hundreds or thousands of canoes surrounding his ship. In 1779, he wrote:

> At 11 AM anchored in the bay (which is called by the Natives Karakakooa) in 13 fathom water over a Sandy bottom and a quarter mile from the NE shore.... The ships very much Crouded [sic] with Indians and surrounded by a multitude [over 1,000] of Canoes. I have no where [sic] in this Sea seen such a number of people assembled at one place, besides those in the Canoes all the shore of the bay was covered with people and hundreds were swimming about the Ships like shoals of fish.[23]

Descriptions like these border on claustrophobic, and it is no coincidence that he likened Indigenous people to shoals of fish—both represented the (imagined and real) threats and obstructions of nature. He would develop theories regarding the intent of these smaller vessels, and he often feared theft, attack, or unwelcomed surveillance.[24] In fact, just prior to this scene, he reported showing off the power of the guns on his ship by firing over the crowd to prevent them from stealing his rudders.[25]

Canoes were simply nimbler, far easier to maneuver compared to Cook's practically brutish bark ship. At the same time, he noted variations and limitations, often with frustration, as he remained dependent on canoes for trade and short-distance travel. He recognized those canoes that were clearly designed to travel long distances on the open sea.[26] He noted with confusion the fact that Kānakas could not or would not set sail at high tide.[27] And he marveled at their ability to travel with "beasts" in such close quarters and with such proximity to the surface of the ocean.[28] As proximate technologies that were reliant on their medium, canoes were easy for Cook to judge as less powerful, but their resulting flexibility constituted both the basis of Cook's survival and a looming threat to him throughout his time in the Pacific.

The Royal Society first sent Cook to the region not to evaluate canoes but to observe and record the 1769 transit of Venus as part of a transnational European effort that sent colonizing scientists around the globe in order to record multiple perspectives that would enable the more precise calculation of longitude and the distance between the earth and sun.[29] Paralleling TMT rhetoric today, this European international project was considered such an inherent good for "all mankind" that the French military, despite their ongoing war with the British, declared that Captain Cook's ship was not to be interfered with on its travels.[30] And like TMT, Cook demanded space for his observations. He chose a spot in Tahiti to set up a viewing platform that was not to be disturbed, though he indicated in his journal that the "sandy shore" he chose seemed of no import to Tahitian onlookers.[31] However, obtaining reliable measurements proved to be more difficult than anticipated. Cook and his team observed a marked haze around Venus that prevented them from recording the precise times the planet crossed the sun.[32] Ultimately, to both his and the Royal Society's frustration, his expedition provided measurements that were incapable of eliminating longitudinal doubt.[33] This considerable disappointment aside, the auxiliary scientific mission behind Cook's first trip—to locate, explore, and lay claim to "Terra Nullius," as stated by the Royal Society—was an unmitigated success.[34] His next destinations after Tahiti were Aotearoa (New Zealand) and Terra Australus (Australia). Within a year, Great Britain had laid claim to swaths of both as settler colonies.

Cook's ship was first built as a collier (cargo) ship, initially christened the *Earl of Pembroke*. When it was recommissioned by the Royal Society for this scientific mission, it was renamed *Endeavor*.[35] To endeavor, to exert oneself to the utmost, evidently better captured the scientific spirit of Cook's mission to travel across the vast expanses of the Pacific to Terra Nullius in order to globalize the knowledge of the Royal Society. This vision of science as "proper and genuine exertion," in the words of the seminal British philosopher of science Francis Bacon, paired with the goal of a nonperspectival abstract truth (definitive longitude and planetary distance) to be known through the combined insight of multiple observations, was characteristic of an Anglo variety of colonial science.[36] This is part and parcel of what I call objectivist geography—an ideologically antirelational viewpoint paired with an epistemological valuing of exertion, of working against the chartable mediums and objects presumed incapable of cooperation and capable of resistance only in the form of obstruction.[37]

On the day Cook prepared for the transit of Venus, he made an innocuous comment in his journal: "Winds at East during the day, in the night a light breeze off the land, and as I apprehend it be usual here for the Trade wind to blow during great part of the Day from the Eastern board and to have it Calm or light breezes from the land that is Southerly during the night with fair weather, I shall only mention the wind and weather when they deviate from this rule."[38] This was strategic, of course. It saved him having to take detailed notes on the weather every day.[39] At the same time, it indicated the extent to which setting and perspective faded to the background when they were not obstructions, when their objecthood could be codified in their being subject to a rule. While rule-obeying weather was deemed unremarkable in Cook's journals, longitudinal and cardinal directions were a constant presence—the human, animate variable that could not be predicted and therefore was worthy of observation and record keeping.

After centuries of colonial dominance, the ubiquity of concepts like cardinal directions, latitude, and longitude serves to obscure their cultural specificity. In contrast, David A. Chang in *The World and All the Things Upon It*, which traces the history of Kānaka geography over the course of the eighteenth and nineteenth centuries, describes a perspectival geography:

> Kānaka knew about far-off places, reflected upon them, and valued connections of Hawai'i to them. This distinctive knowledge of the world was matched by a distinctive way of understanding space: a consistently perspectival way

of looking at the world. Kānaka understood geography from points of view. Kānaka did not describe the world as an abstract truth, but as a world seen from a perspective.[40]

Kānaka geography is incommensurate, contributing to a fundamentally different physical space, and Chang warns us to be wary of the "illusion of transparency" in translating Kānaka perspectival directions into English cardinal ones. He explains that Kānaka directions—*ka hikina* (the arriving), translated as "east"; *ka komohana* (the entering), translated as "west"; *ka ākau* (the right), translated as "north"; and *ka hem* (the left), translated as south"—do not *mean* east, west, north, or south.[41] They are perspectival, not cardinal, as evidenced by the fact that when Kānaka travel, they may employ different directional descriptions relative to cardinal ones based on where they are relative to the ocean and the sun.[42] Kānaka geography is composed of this perspectival way of understanding space and a valuing of the larger world not for its exotic potential but for its physical and conceptual connection to their position. This is a Kānaka variety of what I call animist geography—animist in its conception of consequential, shifting, and cooperative terrain and specifically Kānaka in its genealogy, legacy of connection through voyaging, and (literal) fluidity.

Geography and navigation are inexorably tied to forms of transportation technologies, and vice versa.[43] Cook's bark ship—with its huge stature, inflexibility, anchor, and methods of traveling against the tide—was an objectivist technology that sailed according to directions that never changed through a world populated by inanimate features that merited detail only if and when they needed to be overcome. Kānaka canoes, in contrast—with their flexibility, size that scaled according to need, proximity to the elements, and inability to travel against the tide—were animist technologies. Canoes reflected an epistemological valuing of flexibility over exertion and reflected a worldview wherein what was worth detailing was based on proximity and repetition rather than potential for abstraction. Canoes operate in a vastly different geographical space—an animate and shifting world, where minute variation and limitation not only mattered but also were not problematized. In fact, to the contrary, they were celebrated and intimately known.

Cook set the violent tone for how colonizers would understand and think about canoes for centuries to come, and this, in addition to the ensuing genocide of Kānakas over the course of the next centuries, resulted in the decline of animist canoe building, wayfinding, and geography.[44] But animism and pacifism are not equivalent categories, and objectivism does not eliminate vulnerability. As evidence, take Cook's demise in Hawaii. In 1779, after he kidnapped the Hawaii (the big island) *ali ʻi nui* (grand chief) *Kalaniʻōpuʻu* in retaliation for yet another suspected instance of theft, Kānakas killed the captain. He was stabbed to death on the beach while attempting to make it back to his ship, anchored offshore.[45] His translator's earlier warning of retaliation proved accurate, if much delayed, and his ability to sail against the tide proved useless, as his ship was anchored, stationary and looming, out of reach.

STAR OF GLADNESS

There are two interconnected origin stories for the *Hōkūleʻa*, meaning "Star of Gladness," the first *waʻa kaulua* built and sailed in over six hundred years.[46] One focuses on its important position as a scientific object, a laboratory even, that allowed non-Kānaka anthropologists and Kānakas to collaboratively challenge dominant racist and colonial scientific narratives regarding Indigenous seafaring and canoes. The other tells of *Hōkūleʻa*'s unifying role in the Hawaiian renaissance of the 1970s, when traditions and diasporas were welcomed home again via voyaging, among other practices, in a political resurgence movement that is ongoing. The ways in which the TMT debate plays out today is the legacy of both of these. Kānaka animism as it manifested in the *Hōkūleʻa* drew on, challenged, and repurposed colonial scientific objectivism. As such, it is a Native technology just as it is a Kānaka one, and while these are interconnected, they have different theoretical and political consequences.

In settler science publications, the *Hōkūleʻa* has been represented as the achievement of one forward-looking settler anthropologist, Ben Finney.[47] As a master's student in anthropology at the University of Hawaii in the early 1960s, he worked on the history of Polynesian surfing. In this environment, he regularly encountered and was baffled by the accidental settlement hypothesis, which, citing the challenge of westward-blowing trade winds and windless doldrums, denied the capacity of Indigenous people and *waʻa kauluas* to purposefully travel tens of thousands of miles across the Pacific Ocean from Asia. At the time, this position was defended most famously by Thor Heyerdahl, who insisted Indigenous Pacific Islanders accidentally floated on rafts across the Pacific from

South America. His hypothesis directly contradicted what Indigenous Pacific Islanders have always asserted—that they migrated from Asia thousands of years before Europeans reached the area—and it even contradicted the majority of non-Native scientific findings that verified Asian origins, if for reasons other than valuing Indigenous testimony.[48] Still, when Heyerdahl floated on a raft across the Pacific from South America to the Tuamoto Islands in 1948 to prove his theory, the (*spectacularly* stupid and dangerous) stunt increased the popularity of accidental drift hypotheses in the academy and beyond.[49] Finney objected to the hypothesis but made no headway in theoretical arguments. Heyerdahl's work had been tested and verified, but Finney had no access to *wa'a kauluas* to try the same. All Heyerdahl had to do was prove that migration could be achieved (or, more accurately, could be survived) accidentally. Finney, to counter, had to prove both method and intent, two characteristics rarely attributed to Indigenous people in settler sciences.[50]

As a doctoral student at Harvard University working on Pacific migration a couple years later, Finney concluded that only a demonstration of Polynesian seafaring could positively evidence his theories of purposeful eastward migration. He started out with small canoes, built with funding from the National Science Foundation, that he took to Hawaii to test whether or not paddling small vessels over tens of thousands of miles was feasible.[51] It did not take him long to conclude that paddling would have been completely out of the question due to the enormous amount of transported food (i.e., fuel) it would have required. Finney tried again in 1971, this time connecting with a Kānaka collaborator, Herb Kane, on the construction of a larger, double-hulled sailing canoe. A diasporic Kānaka who had been conducting his own research into canoes and wayfinding, Kane was deeply curious about the scientific conversations to which Finney was a party. However, Kane's highest priority was to heal and unite Kānakas in a future-oriented revitalization and sovereignty project. Kane was, first and foremost, a mentor and youth advocate. Nainoa Thompson, who would go on to serve as navigator on the *Hōkūle'a*, described his introduction to Kane and the world of canoes as self- and life-altering:

> Growing up in Hawai'i in the 1960s, I found my Hawaiian culture ebbing away. It was a confusing time for me and I felt lost between worlds that seemed in conflict. All that changes one night when Herb Kane introduced me to the stars and explained how my ancestors had used them to find their

way across a vast ocean to settle all of Polynesia. At that moment, my vision of my ancestry became timeless and alive in those stars.... It has been a process of finding ourselves.⁵²

For Kane, Thompson, and other Indigenous wayfinders, it would be a similarly expansive and global, even extraplanetary, project—but one that was fundamentally tied to Kanaka perspective and identity. This dream and project is the second origin of the *Hōkūle'a* and the more far-reaching one.

Kane and Finney committed to building a large double-hulled canoe that they would sail from Hawaii to Tahiti and back.⁵³ In 1971, they founded the Polynesian Voyaging Society, a nonprofit corporation, which secured $100,000 worth of grant money and donations for the project, many based on applications that highlighted the project's potential in "experimental archeology," then an emergent field.⁵⁴ They determined the design of the canoe based on anthropological and early colonial drawings and accounts, guided by the age distribution theory to determine the oldest and most widespread models.⁵⁵ Once they had settled on a *pahi* canoe with rounded sides and a v-shaped bottom, they determined that despite their initial desire to use the same construction tools and materials Kānakas did hundreds of years ago, the least cost- and time-prohibitive option was to use plywood coated in plexiglass.⁵⁶ Decades later Nainoa Thompson characterized this as a traditionalist choice: if Kānaka ancestors had had the option to use Plexiglas, he argued, they would have. There is nothing traditional about rejecting useful things in the world, especially not for Kānakas.⁵⁷ They launched the *Hōkūle'a* on her first voyage between Hawaii and Tahiti in 1976, and she has subsequently traversed the globe many times over.

In 1992, the Polynesian Voyaging Society and the University of Hawai'i organized a conversation for university students and the *Hōkūle'a* crew with the crew of the space shuttle *Columbia*, then in orbit. A student asked Lacy Veach, the pilot of the space shuttle who was from Hawaii, about the differences and similarities between canoes and space travel. Veach responded, "Both are voyages of exploration. *Hōkūle'a* is in the past. *Columbia* is in the future."⁵⁸ How the *Hōkūle'a* was in the past despite its immediate presence (half the conversation took place literally on the canoe) was not discussed.⁵⁹ What this interaction speaks to is how difficult it is to locate and conceive of the *Hōkūle'a* outside other scientific temporalities and imaginations. Nearly every element of the *Hōkūle'a*'s planning and construction was shaped by the extent to which both Kānakas and their technologies had been rendered as scientific objects for centuries: the Polynesian Voyaging Society was funded largely through science grants, the *Hōkūleʻa* was designed according to anthropological models and sketches, and so on. Even the personal and political motivations expressed by Kane and others were rooted in the trauma of loss at the hands of colonial scientific progress.

At the same time, the *Hōkūleʻa* was a profoundly non–settler scientific endeavor. Kane named the *Hōkūleʻa* through a process of dreaming and spiritual consultation.[60] The *Hōkūleʻa* was known by its crew as the "spaceship of the ancestors" due to its ability to collapse both time and space and make them manifest. Its success ultimately was judged not by Finney's output but by its ability to strengthen Kānaka relations and to then connect the Kānakas with ancestors and future generations, with the planet, and with each other.[61] Kane declared that the *Hōkūleʻa*'s first trip would be a success only if it inspired people to come back to their communities and traditions.[62] When the *Hōkūleʻa* made it to Tahiti in 1975, that inspiration and enthusiasm were palpable. Over half the island's residents came to greet the *Hōkūleʻa*.[63] This was a classic dynamic for what I call Native science—a mode of epistemic production that is so structured by being a scientific object in an occupied state while simultaneously making space for and cultivating elements of Indigenous knowledge, relations, and history.[64] Kānaka animism has been profoundly impacted—conceptually, discursively, and structurally—by the experience and vantage of its producers as themselves being the objects of an objectivist knowledge regime for centuries. Indeed, for Kānakas, Native science in this sense has served only to elevate animism as a core epistemic value, as it has reinforced their lateral relations to other objectified beings.[65]

The categories of science and religion, like modernity, are built on the conceptual and geographic work of creating Indigenous and other colonized people as objects of and archives for colonial power. Kānaka animism predates Kānakas being reimagined by colonial powers as objects, *as Natives*. The *Hōkūleʻa* is both a Kānaka animist technology and a Native one, and while these are not synonymous, they are interconnected and mutually reinforcing. As a Native technology, the *Hōkūleʻa* draws on its position as an object to create guarded space where Kānaka perspectives and relations can persist in the face of genocide. This dynamic in Native and settler/colonial epistemologies is true of the economic and political components of colonialism and Indigenous nationhood, too. In *Mohawk Interruptus*, Audra Simpson finds that Indigenous sovereignty in a settler colonial state is a nested sovereignty—strangulated but sovereignty nonetheless.

Sovereignty is a Western concept that is reliant on notions of absolute power, and yet nested sovereignty persists and refuses to be otherwise. In claiming sovereign status, Indigenous people not only assert political independence from a settler state but also challenge the very notion of political independence in the first place.[66]

Claiming science or religion for Kānakas and other Indigenous peoples simultaneously reorients the power of these terms and undoes them, as these terms were never meant to be wielded by the objects whose absence brought them into being. Critically claiming sovereignty, science, or religion, among other key categories, is not only a strategy. The choice constitutes the shape of the space carved for Indigenous modern governance, epistemology, and ontology. These are amplified and resonate and learn new meaning through the experience of strangulation, survival, refusal, and resurgence. These terms are chosen according to and emerge from historical relationships and political positionalities.

GEOGRAPHY BEYOND COLONIAL DIFFERENCE

In conclusion, I want to return to Mauna Kea and locate it in the global geopolitics of knowledge. More importantly, I want to return to the Kānakas' animism, their relation to Mauna Kea, and the miraculous refusal of each to abandon the other, the refusal to reorient according to other ontological, epistemological, and geographic axes. Native science and technology, like the *Hōkūle'a*, resemble what the decolonial theorist Walter Mignolo calls border thinking, or the epistemologies of the subaltern produced within diverse geopolitics of knowledge and different vantages of irreducible colonial difference that engender a sort of epistemic double-consciousness for colonized people.[67] For Mignolo, the decolonization of knowledge requires both an attentiveness to border thinking and a critique of Western civilization from sources geopolitically outside of it. Postcolonial and decolonial frameworks, like colonial ones, are inherently spatial. Even the term *border thinking* implies a space through which knowledges move. However, if we are to take geopolitics seriously, as I have tried to in this chapter, then we confront a seemingly impenetrable wall of incommensurability. How can one seriously hold both border thinking and geopolitical location at once, when taking subaltern epistemology seriously could completely transform the literal geographic lay of the land?

In exploring moments in the history of how animism has itself animated Kānaka geography and travel technology, I find that Hawaii is impossible to locate on the axis of colonial power and difference. Kānakas' perspectival geography and their sense of the world as proximate, even over spans of otherwise unthinkable time and distance, are not merely details of a unique subaltern epistemology that critiques colonial knowledge/power and its posts and newisms. Their perspectival geography and its roots in animism are an incommensurate state of Indigenous exception.[68] Both Anglo and Portuguese colonizers attempted and failed to locate Kānakas historically and geographically, and centuries later colonizers are still attempting to locate them.[69] In the meantime, the *Hōkūleʻa* has traversed the globe, brought ancestors to life, and animated a renewed commitment to perspectival global thinking for Kānakas. A center and a periphery are unthinkable in Kānaka geography, and that geography is securely made manifest in Kānaka Native science and technology today. In valuing flexibility and mobility, in valuing a world always emanating outward toward a horizon they could explore but not meet, Kānakas do not have a double epistemic consciousness. They have an animist epistemology that is bolstered, if battered, by colonial objectivism and cardinality.

Indigeneity in general is difficult to locate in decolonial theory, though Mignolo and others have been careful not to exclude Indigenous scholars from their analysis.[70] Even the concept of subaltern is a complicated and not fully useful one in critical Indigenous theory.[71] This is largely due to critical Indigenous theory's profound disinterest in examining colonial difference and in being known.[72] Indigenous is a diverse category, referring to people in settler, colonial, and postcolonial contexts, all with radically different histories and cultures, but perhaps the most constant unifying characteristic is this Indigenous stubbornness and exception. Audra Simpson, Leanne Betasamosake Simpson, and Gerald Viznor have all developed words for this stubbornness—refusal (A. Simpson), resurgence (L. Simpson), and survivance (Viznor).[73] The commonality among all three is a radical rejection of colonial recognition. All revolve around what it means to survive genocide—to be made other in the process of being appropriated and erased and, most importantly, *to still be here.*

Leanne Betasamosake Simpson, in particular, frames contemporary indigeneity as a miracle that collapses linear time between generations and reinforces relational space and responsibility. That genealogical subjectivity born of this miracle is where I will conclude, as it brings us back to Mauna Kea and its genealogical relatedness to Kānakas. Kānakas note that Mauna Kea would be waiting for them to remember it, even if they did not visit for generations. This is the stubbornness of animist epistemology. Mountains are patient. In remembering, Kānakas tap into the temporalities of that patience, as well as its geography. This animism is not antiscience. It is neither a premodern religious belief nor the cure for postmodernist regret. It is neither a moral high ground nor a decolonial critique, though it does insist on radical Indigenous sovereignty. It is the stubbornness of flexibility: like the traditionalist Plexiglas that lines the *Hōkūleʻa*, Kānakas will bend, but they will not break.

NOTES

1. Hawaii News Now, "Read the Governor's Emergency Proclamation Following TMT Arrests."
2. Overbye, "Under Hawaii's Starriest Skies, a Fight Over Sacred Ground."
3. Osorio, Muneoka, and Fujikane, "A Sacred Mountain, Scarred by Ambition."
4. Koberlein, "Ignorant Savages."
5. Osorio, Muneoka, and Fujikane, "A Sacred Mountain, Scarred by Ambition."
6. Goodman and Case, "'We Are Not Anti-Science.'"
7. See Tylor, *Primitive Culture*.
8. See Robinson, "Animism as a World Hypothesis"; Harvey, "Things Act"; and Willerslev, "Taking Animism Seriously, but Perhaps Not Too Seriously?"
9. See Bird-David, "'Animism' Revisited"; Burley, "'A Language in Which to Think of the World'"; Eldridge, "Is Animism Alive and Well?"; and Wilkinson, "Is There Such a Thing as Animism?"
10. *Objectivism* is not necessarily the same as *objective*, whether typically used as an adjective meaning "impartial or nonsubjective" or more specifically used in nineteenth- and twentieth-century scientific discourse as a mode of representation—though both of these are related and make sense only in an objectivist framework. See Daston and Galison, *Objectivity*. Much like in animist worldviews, there is room for debate over whether objectivism ought to be understood as literal or metaphorical, and there is space for understanding it as incomplete, as many scholars in science and religion, especially those concerned with questions of Western and/or modern (dis)enchantment, have done. See Josephson-Storm, *The Myth of Disenchantment*.
11. See Latour, *We Have Never Been Modern*; Tsing, *The Mushroom at the End of the World*; Cruikshank, *Do Glaciers Listen?*; and Haraway, *Staying with the Trouble*. An exception to this body of literature that bears more in common with Indigenous critiques than postmodernist ones is Mel Chen's work. In *Animacies*, they conduct a beautifully incisive critique and deconstruction of the racial, queer, and crip work that notions and embodiments of animacy entail.

12. For a more in-depth discussion of the problem of the false "newness" of new materialism, see TallBear, "Beyond the Life/Not Life Binary."
13. For more on the colonial use of *common sense*, see Rifkin, *Settler Common Sense*. For more on hybridity, see Bhabha, *The Location of Culture*.
14. Mukharji, *Doctoring Traditions*; Kimmerer, *Braiding Sweetgrass*.
15. Cook and Edwards, *The Journals of Captain Cook*, 41.
16. Indeed, this connection—which Cook makes in his journals in comparing "American" canoes to those found throughout the Pacific, especially as he rounded back on the west coast of Turtle Island—was my first introduction to the history of Indigenous Pacific canoes as I researched those of my own (Haudenosaunee/Kanien'kehá:ka) context. See Lockett, *Captain James Cook in Atlantic Canada*.
17. See Cook and Edwards, *The Journals of Captain Cook*.
18. He noted five to eight homes and three "war canoes" being burned first, followed by an additional five war canoes as his men made their way back to their ship. Cook and Edwards, *The Journals of Captain Cook*, 517.
19. Cook and Edwards, *The Journals of Captain Cook*, 517.
20. Cook and Edwards, *The Journals of Captain Cook*, 518.
21. For more examples, see Cook and Edwards, *The Journals of Captain Cook*, 48, 49–50, 61, 75, 82–83, 94, 110, 273–277, 339–342, 357, 371, 466, 494, 539.
22. This estimate is based on the size of the *Hōkūle'a*, modeled off the *wa'a kauluas* Cook wrote about and sketched. Of course, the *Endeavor* fully towered over smaller single-hull canoes. See Winfield, *British Warships in the Age of Sail 1714–1792*, 354; Polynesian Voyaging Society, "Hōkūle'a."
23. Cook and Edwards, *The Journals of Captain Cook*, 604.
24. Cook and Edwards, *The Journals of Captain Cook*, 41, 61, 75, 82, 82–83, 94, 301, 339–342, 346–348, 551–353, 605.
25. This assertion of his technological dominance fits with Cook's assessment of swimming people compared to animals. Again, the narrative of Indigenous people being aligned with natural obstructions is a long-standing one.
26. Cook remarked in September 1777 that Omai received a "very fine double Sailing Canoe completely equipped, Man'd and fit for the Sea." The next day Cook was asked to transport another canoe as a gift to the "Earee rahie no Pretane," whom Cook identified as royalty. He dismayed that this canoe, which he mistook at first for a "vessel of war" due to its intricate carving and overall craftsmanship, was not intended for him, indicating that Cook was not the most highly regarded person in the vicinity. Cook and Edwards, *The Journals of Captain Cook*, 511.
27. Cook and Edwards, *The Journals of Captain Cook*, 535.
28. Cook and Edwards, *The Journals of Captain Cook*, 497–499.
29. See Andrew S. Cook, "James Cook and the Royal Society," in Williams, *Captain Cook*.
30. Robin Inglis, "Successors and Rivals to Cook" in Williams, *Captain Cook*, 165.
31. Cook and Edwards, *The Journals of Captain Cook*, 41.
32. This would later be called the *black drop effect*, believed by astronomers to be caused by interference from earth's atmosphere. It seems that even combining perspectives could not eliminate the fundamentally local character of a global perspective. See Pasachoff, Schneider, and Golub, "The Black-Drop Effect Explained."

33. Cook, "James Cook and the Royal Society."
34. *Terra nullius*, the notion of empty land, is central to Anglo settler ideology, especially in the Pacific, and was central to Cook's charge. See Banner, *Possessing the Pacific*.
35. See Hosty and Hundly, *Preliminary Report on the Australian National Maritime Museum's Participation in the Rhode Island Marine Archaeology Project's Search for HMB Endeavour*.
36. Bacon, *Novum Organum*, CXXX.
37. Graham Harvey has adopted a similar use of *objectivism*. In his telling, when caught between objectivism and subjectivism, animism and its relationality emerge as the preferable escape from a destructive modernity and an unanchored postmodernity. See Harvey, *The Handbook of Contemporary Animism*, 249.
38. Cook and Edwards, *The Journals of Captain Cook*, 42.
39. It is telling that these very detailed observations of minute changes at the same place over the course of a very long time are exactly what the Hawaiian epistemologist Manulani Meyer identifies as "gross knowledge" and a critical component of Hawaiian epistemology. See Meyer, "Our Own Liberation."
40. Chang, *World and All the Things Upon It*, 20.
41. Chang, *World and All the Things Upon It*, 21.
42. Chang, *World and All the Things Upon It*, 21.
43. See Rankin, "The Geography of Radionavigation and the Politics of Intangible Artifacts"; Louter, *Windshield Wilderness*; and White, *Railroaded*.
44. Even at the time Cook was writing, larger open-ocean double-hulled canoes were rare, as Kānakas had not voyaged in search of other unvisited islands since the 1200s; see Chang, *World and All the Things Upon It*.
45. Cook and Edwards, *The Journals of Captain Cook*, 610.
46. Polynesian Voyaging Society, "The Story of Hōkūleʻa."
47. Finney, *From Sea to Space*; Finney, *Hokuleʻa*; "Ancient Navigators' Methods Confirmed."
48. Haddon and Hornell, *Canoes of Oceania*; Hiroa and Buck, "The Disappearance of Canoes in Polynesia."
49. During Finney's time as a graduate student, the theory was also making the rounds due to Andrew Sharp's work in the 1950s, which theorized that Indigenous people did migrate from Asia but that they must have done so accidentally in several short travels where they were blown off by storms or diverted by other obstructions. Though Heyerdahl was more fantastical in his claims, Sharp's were far more consequential in the academy. See Heyerdahl, *Kon-Tiki*; Heyerdahl, *Early Man and the Ocean*; Finney, "Contrasting Visions of Polynesian Voyaging Canoes"; and Sharp, *Ancient Voyagers in the Pacific*.
50. This challenge is not unique to questions of Indigenous origins in the South Pacific. Colonial (especially settler) sciences, from ethnology and archeology to DNA science, have been rooted in concerted efforts to determine Indigenous origins and migration patterns through history, despite nearly unanimous insistence from Indigenous peoples across the globe that their origins are already well-known in their own traditions and are not in any need of scientific investigation. Another prominent example of this can be found in the long and circuitous history of the Bering land bridge migration theory, which posited that Indigenous peoples in North America migrated from Asia across a land bridge approximately twenty thousand years ago. Despite Indigenous objections and scientific counterevidence, the theory's popularity, both in the academy and outside,

has remained strong. In both instances, these scientific investigations serve first and foremost to undermine Indigenous land rights and political legitimacy, as they undergird settler conceptions of history, ontology, and rightful political hegemony. For more on recent settler scientific work on (and refusals to abandon) the Bering land bridge migration theory, see Pedersen et al., "Postglacial Viability and Colonization in North America's Ice-Free Corridor"; and MacEachern, "The Bering Land Bridge Theory." For more on Indigenous critiques, see Deloria, *Red Earth, White Lies*.

51. Finney, "Tracking Polynesian Seafarers."
52. Lowe, *Hawaiki Rising*, ix.
53. Lowe, *Hawaiki Rising*, 36.
54. Polynesian Voyaging Society, *PVS Newsletter*, vol. 1, no. 1, April 1974, http://archive.hokulea.com/index/newsletters/1974_04.html.
55. Kane, "In Search of the Ancient Polynesian Voyaging Canoe."
56. Kane, "In Search of the Ancient Polynesian Voyaging Canoe."
57. Nainoa Thompson, "The Land and the Waters Are Speaking" (lecture at Harvard Divinity School, April 4, 2019).
58. Hawai'i Space Grant Consortium, "Astronaut Lacy Veach Day of Discovery."
59. It is unlikely that Veach intended to offend or even to minimize the importance of the *Hōkūle'a*. Many years later, in a talk given at the Harvard Divinity School, Naoni Thompson spoke glowingly of Veach's support of the Polynesian Voyaging Society. See Thompson, "The Land and the Waters Are Speaking."
60. Kane, "In Search of the Ancient Polynesian Voyaging Canoe."
61. Lowe, *Hawaiki Rising*, 61–79.
62. Lowe, *Hawaiki Rising*, ix.
63. Polynesian Voyaging Society, "The Story of Hōkūle'a."
64. See Nelson, "Making Native Science."
65. Religion had a similar but distinct edifying effect, prompting many Indigenous peoples to seek political rights through the language of religious rights. For more on Indigenous claims to religion in the context of legal rights in the United States, see Wenger, *We Have a Religion*.
66. Simpson, *Mohawk Interruptus*.
67. Mignolo, "The Geopolitics of Knowledge and the Colonial Difference."
68. Jodi Byrd describes indigeneity as a state of exception very different from, though not necessarily in opposition to, that of Giorgio Agamben. Indigenous theoretical exception refers to how Indigenous peoples, as sites of transit, remain subjects, even in subjectless critique. They are conceptually irreducible. See Byrd, "*Loving* Unbecoming." In this case, I refer to a state of exception also to indicate the ways in which Indigenous peoples do not map onto Mignolo's and other decolonial critiques of the geographical and historical loci of postcolonial theory. See Bhambra, "Postcolonial and Decolonial Dialogues." Perhaps even more importantly, these critiques do not align with the ways in which decolonization is used in Indigenous theory, which in many aspects harken back to earlier postcolonial thought, like that of Frantz Fanon. Decolonial theory in critical Indigenous studies highlights the literal ongoing occupation of Indigenous nations and the extent to which decolonization means nothing without the unequivocal dissolution of colonial and settler colonial powers. See Tuck and Yang, "Decolonization Is Not a Metaphor."

69. For Portuguese accounts of Polynesian seafaring and canoes, see Queirós, *Terra Australis Incognita*. For a more general account of colonial attempts to possess, know, and therefore locate the Indigenous Pacific, see Arvin, *Possessing Polynesians*.
70. It is important to note that Kānakas went through a process of affinity and solidarity with Indigenous peoples in other parts of the U.S. settler colonial empire—their identity as Indigenous is through this specific settler colonial lens (see Chang, "We Will Be Comparable to the Indian Peoples"). As for Mignolo's engagement with Indigenous people, his main Indigenous interlocutor is the prolific Vine Deloria Jr., who was at the heart of what would become the Native science movement on Turtle Island and in the Pacific, and *Hōkūle'a* is a prime example of said movement. Mignolo describes Deloria predominantly as a legal scholar, though later in his career his primary interests were in science and religion (see Deloria, *Red Earth, White Lies*).
71. The object of critical Indigenous studies is to analyze, historicize, and interpret the world through Indigenous lenses. It is a field that by its premise situates Indigenous people as present and Indigenous knowledges not as necessarily knowable or not but as knowing. While there are important conversations about the extent to which Indigenous knowledge could ever be produced or *heard* in the settler academy, these institutional critiques take rather different routes than subaltern studies like Spivak and Morris, *Can the Subaltern Speak?*, and these largely form the basis of antitheory positions altogether in Indian country.
72. This is not to say that Indigenous theory is not in conversation with postcolonial and decolonial theories, though it has not always been fruitful. Instead, Indigenous theory, as it has been expressed in the academy most often in contemporary Anglo settler states, has focused on engaging colonial difference through theorizing settler colonialism. Settler colonialism draws on colonialism, but its methods and discourses are not the same. Therefore, Indigenous theorists have been less attentive to colonial difference and historicity. Perhaps the most important distinction between the two is a focus on the establishment of the center and periphery in postcolonial theory compared to a greater focus on the dangers of assimilation in Indigenous theory. For more on settler colonial studies in Indigenous theory, see Kauanui, "'A Structure, Not an Event'"; Rifkin, *Settler Common Sense*; and Vimalassery, Pegues, and Goldstein, "Introduction."
73. Simpson, *Mohawk Interruptus*; Simpson, *As We Have Always Done Indigenous Freedom Through Radical Resistance*; Vizenor, *Manifest Manners*.

WORKS CITED

"Ancient Navigators' Methods Confirmed." *Washington Post*, June 9, 1976.
Arvin, Maile Renee. *Possessing Polynesians: The Science of Settler Colonial Whiteness in Hawai'i and Oceania*. Durham, NC: Duke University Press, 2019.
Bacon, Francis. *Novum Organum*. Oxford: Clarendon Press, 1878.
Banner, Stuart. *Possessing the Pacific: Land, Settlers, and Indigenous People from Australia to Alaska*. Cambridge, MA: Harvard University Press, 2007.
Bhabha, Homi K. *The Location of Culture*. London: Routledge, 1994.
Bhambra, Gurminder K. "Postcolonial and Decolonial Dialogues." *Postcolonial Studies* 17, no. 2 (April 3, 2014): 115–121. https://doi.org/10.1080/13688790.2014.966414.

Bird-David, Nurit. "'Animism' Revisited: Personhood, Environment, and Relational Epistemology." *Current Anthropology* 40, no. S1 (February 1, 1999): S67–S91. https://doi.org/10.1086/200061.

Burley, Mikel. "'A Language in Which to Think of the World'—Animism, Indigenous Traditions, and the Deprovincialization of Philosophy of Religion, Part 1." *Religious Theory* (blog), October 29, 2018. http://jcrt.org/religioustheory/2018/10/29/a-language-in-which-to-think-of-the-world-animism-indigenous-traditions-and-the-deprovincialization-of-philosophy-of-religion-part-1-mikel-burley/.

Byrd, Jodi A. "*Loving* Unbecoming: The Queer Politics of the Transitive Native." In *Critically Sovereign*, ed. Joanne Barker, 207–227. Durham, NC: Duke University Press, 2017. https://doi.org/10.1215/9780822373162-007.

Chang, David A. "'We Will Be Comparable to the Indian Peoples': Recognizing Likeness Between Native Hawaiians and American Indians, 1834–1923." *American Quarterly* 67, no. 3 (September 21, 2015): 859–886. https://doi.org/10.1353/aq.2015.0045.

———. *The World and All the Things Upon It: Native Hawaiian Geographies of Exploration*. Minneapolis: University of Minnesota Press, 2016.

Chen, Mel Y. *Animacies: Biopolitics, Racial Mattering, and Queer Affect*. Durham, NC: Duke University Press, 2012.

Cook, James R., and Philip Edwards. *The Journals of Captain Cook*. Abridged ed. London: Penguin, 2000.

Cruikshank, Julie. *Do Glaciers Listen? Local Knowledge, Colonial Encounters, and Social Imagination*. Vancouver: University of British Columbia Press, 2005.

Daston, Lorraine, and Peter Galison. *Objectivity*. New York: Cambridge, Mass.: Zone Books ; Distributed by the MIT Press, 2007.

Deloria, Vine. *Red Earth, White Lies: Native Americans and the Myth of Scientific Fact*. Wheat Ridge, CO: Fulcrum, 1995.

Eldridge, Richard. "Is Animism Alive and Well?" In *Can Religion Be Explained Away?*, ed. D. Z. Phillips, 3–25. London: Palgrave Macmillan, 1996. https://doi.org/10.1007/978-1-349-24858-2_1.

Finney, Ben. "Contrasting Visions of Polynesian Voyaging Canoes: Comment on Atholl Anderson's 'Traditionalism, Interaction and Long-Distance Seafaring in Polynesia.'" *Journal of Island and Coastal Archaeology* 3, no. 2 (October 27, 2008): 257–259. https://doi.org/10.1080/15564890802340067.

———. *From Sea to Space*. Palmerston North, New Zealand: Massey University, 1989; Honolulu: University Press of Hawaii, 1992.

———. "Tracking Polynesian Seafarers." *Science* 317, no. 5846 (2007): 1873–1874.

———. *Hokule'a: The Way to Tahiti*. New York: Dodd, Mead, 1979.

Goodman, Amy, and Pua Case. "'We Are Not Anti-Science': Why Indigenous Protectors Oppose the Thirty Meter Telescope at Mauna Kea." *Democracy Now!*, July 22, 2019. https://www.democracynow.org/2019/7/22/why_indigenous_protectors_oppose_the_thirty.

Haddon, Alfred C., and James Hornell. *Canoes of Oceania*. Honolulu: Bishop Museum Press, 1936.

Haraway, Donna Jeanne. *Staying with the Trouble: Making Kin in the Chthulucene*. Durham, NC: Duke University Press, 2016.

Harvey, G. "Things Act: Casual Indigenous Statements About the Performance of Object-Persons." In *Vernacular Religion in Everyday Life: Expressions of Belief*, ed. Marion Bowman and Ulo Valk, 194–210. London: Routledge, 2012.

Harvey, Graham. *The Handbook of Contemporary Animism*. London: Routledge, 2014.
Hawaii News Now. "Read the Governor's Emergency Proclamation Following TMT Arrests," July 18, 2019. https://www.hawaiinewsnow.com/2019/07/18/read-governors-emergency-proclamation-following-tmt-protest-arrests/.
Hawai'i Space Grant Consortium. "Astronaut Lacy Veach Day of Discovery." Accessed May 4, 2018. http://www.spacegrant.hawaii.edu/old/Day-of-discovery/
Heyerdahl, Thor. *Early Man and the Ocean: A Search for the Beginnings of Navigation and Seaborne Civilizations*. Garden City, NY: Doubleday, 1979.
———. *Kon-Tiki: Across the Pacific by Raft*. Chicago: Rand McNally, 1950.
Hiroa, Te Rangi, and Peter H. Buck. "The Disappearance of Canoes in Polynesia." *Journal of the Polynesian Society* 51, no. 3 (September 1942): 191–199.
Hosty, Kieren, and Paul Hundly. *Preliminary Report on the Australian National Maritime Museum's Participation in the Rhode Island Marine Archaeology Project's Search for HMB Endeavour*. Sydney: Australian National Maritime Museum, 2003.
Josephson-Storm, Jason A. *The Myth of Disenchantment: Magic, Modernity, and the Birth of the Human Sciences*. Chicago: University of Chicago Press, 2017.
Kane, Herb. "In Search of the Ancient Polynesian Voyaging Canoe." Polynesian Voyaging Society. Accessed July 31, 2019. http://archive.hokulea.com/designing.html.
Kauanui, J. Kehaulani. " 'A Structure, Not an Event': Settler Colonialism and Enduring Indigeneity." *Lateral* 5 (May 1, 2016). https://doi.org/10.25158/L5.1.7.
Kimmerer, Robin Wall. *Braiding Sweetgrass: Indigenous Wisdom, Scientific Knowledge and the Teachings of Plants*. Minneapolis, MN: Milkweed, 2013.
Koberlein, Brian. "Ignorant Savages." *One Universe at a Time* (blog), December 6, 2015. https://archive.briankoberlein.com/2015/12/06/ignorant-savages/index.html.
Latour, Bruno. *We Have Never Been Modern*. Cambridge, MA: Harvard University Press, 1993.
Lockett, Jerry. *Captain James Cook in Atlantic Canada: The Adventurer and Map Maker's Formative Years*. Halifax, Nova Scotia: Formac, 2010.
Louter, David. *Windshield Wilderness: Cars, Roads, and Nature in Washington's National Parks*. Seattle: University of Washington Press, 2006.
Low, Sam. *Hawaiki Rising: Hōkūle'a, Nainoa Thompson, and the Hawaiian Renaissance*. Waipahu, HI: Island Heritage, 2013.
MacEachern, Alan. "The Bering Land Bridge Theory: Not Dead Yet." Active History, September 6, 2016. http://activehistory.ca/2016/09/the-bering-land-bridge-theory-not-dead-yet/.
Meyer, Manulani Aluli. "Our Own Liberation: Reflections on Hawaiian Epistemology." *Contemporary Pacific* 13, no. 1 (January 1, 2001): 124–148. https://doi.org/10.1353/cp.2001.0024.
Mignolo, Walter. "The Geopolitics of Knowledge and the Colonial Difference." *South Atlantic Quarterly* 101, no. 1 (January 1, 2002): 57–96.
Mukharji, Projit Bihari. *Doctoring Traditions: Ayurveda, Small Technologies, and Braided Sciences*. Chicago: University of Chicago Press, 2016.
Nelson, Eli. "Making Native Science: Indigenous Epistemologies and Settler Sciences and the United States Empire." Cambridge, MA: Harvard University Press, 2018.
Osorio, Jonathan, Shelley Muneoka, and Candace Fujikane. "A Sacred Mountain, Scarred by Ambition." KAHEA, April 19, 2015. http://kahea.org/press-room/press-clips/a-sacred-mountain-scarred-by-ambition.

Overbye, Dennis. "Under Hawaii's Starriest Skies, a Fight Over Sacred Ground." *New York Times*, October 3, 2016. https://www.nytimes.com/2016/10/04/science/hawaii-thirty-meter-telescope-mauna-kea.html.

Pasachoff, Jay M., Glenn Schneider, and Leon Golub. "The Black-Drop Effect Explained." In *Proceedings of the International Astronomical Union*, 242–253. Cambridge: Cambridge University Press, 2004.

Pedersen, Mikkel W., Anthony Ruter, Charles Schweger, Harvey Friebe, Richard A. Staff, Kristian K. Kjeldsen, Marie L. Z. Mendoza et al. "Postglacial Viability and Colonization in North America's Ice-Free Corridor." *Nature* 537, no. 7618 (September 1, 2016): 45–49. https://doi.org/10.1038/nature19085.

Polynesian Voyaging Society. "Hōkūle'a." Accessed July 24, 2019. http://www.hokulea.com/vessels/hokulea/.

———. *PVS Newsletter* 1, no. 1, April 1974, http://archive.hokulea.com/index/newsletters/1974_04.html.

———. "The Story of Hōkūle'a." Accessed July 30, 2019. http://www.hokulea.com/voyages/our-story/.

Queirós, Pedro Fernandes de. *Terra Australis Incognita, or A New Southerne Discouerie, Containing a Fifth Part of the World. Lately Found Out by Ferdinand de Quir, a Spanish Captaine. Neuer Before Published*. Trans. W. B. London: Iohn Hodgetts, 1617.

Rankin, William. "The Geography of Radionavigation and the Politics of Intangible Artifacts." *Technology and Culture* 55, no. 3 (2014): 622–674.

Rifkin, Mark. *Settler Common Sense: Queerness and Everyday Colonialism in the American Renaissance*. Minneapolis: University of Minnesota Press, 2014.

Robinson, Elmo A. "Animism as a World Hypothesis." *Philosophical Review* 58, no. 1 (1949): 53–63. https://doi.org/10.2307/2181835.

Sharp, A. *Ancient Voyagers in the Pacific*. Harmondsworth, UK: Penguin, 1956.

Simpson, Audra. *Mohawk Interruptus: Political Life Across the Borders of Settler States*. Durham, NC: Duke University Press, 2014.

Simpson, Leanne. *As We Have Always Done Indigenous Freedom Through Radical Resistance*. Minneapolis: University of Minnesota Press, 2017.

Spivak, Gayatri Chakravorty, and Rosalind C. Morris. *Can the Subaltern Speak? Reflections on the History of an Idea*. New York: Columbia University Press, 2010.

TallBear, Kimberly. "Beyond the Life/Not Life Binary: A Feminist-Indigenous Reading of Cryopreservation, Interspecies Thinking and New Materialism." In *Cryopolitics: Frozen Life in a Melting World*, ed. Joanna Radin and Emma Kowal. Cambridge, MA: MIT Press, 2017.

Tsing, Anna Lowenhaupt. *The Mushroom at the End of the World: On the Possibility of Life in Capitalist Ruins*. Princeton, NJ: Princeton University Press, 2015.

Tuck, Eve, and K. Wayne Yang. "Decolonization Is Not a Metaphor." *Decolonization: Indigeneity, Education, and Society* 1, no. 1 (2012). https://jps.library.utoronto.ca/index.php/des/article/view/18630/15554.

Twain, Mark, Samuel Langhorne Clemens, and A. Grove Day. *Mark Twain in Hawaii: Roughing It in the Sandwich Islands, Hawaii in the 1860's*. Honolulu: Mutual, 1990.

Tylor, Edward Burnett. *Primitive Culture: Researches Into the Development of Mythology, Philosophy, Religion, Art, and Custom*. London: J. Murray, 1871.

Vimalassery, Manu, Juliana Hu Pegues, and Alyosha Goldstein. "Introduction: On Colonial Unknowing." *Theory and Event* 19, no. 4 (October 12, 2016). https://muse.jhu.edu/article/633283.

Vizenor, Gerald Robert. *Manifest Manners: Postindian Warriors of Survivance*. Hanover, NH: University Press of New England, 1994.

Wenger, Tisa. *We Have a Religion: The 1920s Pueblo Indian Dance Controversy and American Religious Freedom*. Chapel Hill: University of North Carolina Press, 2009.

White, Richard. *Railroaded: The Transcontinentals and the Making of Modern America*. New York: Norton, 2011.

Wilkinson, Darryl. "Is There Such a Thing as Animism?" *Journal of the American Academy of Religion* 85, no. 2 (June 1, 2017): 289–311. https://doi.org/10.1093/jaarel/lfw064.

Willerslev, Rane. "Taking Animism Seriously, but Perhaps Not Too Seriously?" *Religion and Society* 4, no. 1 (2013): 41–57.

Williams, Glyndwr. *Captain Cook: Explorations and Reassessments*. Woodbridge, UK: Boydell and Brewer, 2004. https://www.jstor.org/stable/10.7722/j.ctt163tb8h.

Winfield, Rif. *British Warships in the Age of Sail 1714–1792: Design, Construction, Careers and Fates*. Barnsley, UK: Seaforth, 2007.

CHAPTER 13

SPECULATION IS NOT A METAPHOR

More than Varieties of Cryobiological Experience

JOANNA RADIN

How does the way life is materialized matter? Is it possible for scholarship that is situated in the Western academy to fully take on a radically different understanding of life? If so, what would this look like? And if not, what might the history of efforts to freeze life reveal about the limits of such a project?

These questions first became meaningful to me as I reckoned with Cold War era efforts to achieve control over human bodies such that they could become perpetual resources for an emerging biomedical infrastructure. This was the subject of my 2017 book, *Life on Ice: A History of New Uses for Cold Blood*.[1] The ability of physicists to produce unprecedented amounts of concentrated heat by splitting atoms created anxieties about the effects of radiation on bodies and environments. One group of scientists, human biologists, responded by endeavoring to collect and freeze blood from Indigenous peoples around the world, whom they imagined as both closer to nature and imperiled by the forces of modernity. These white scientists, whose labs were based in the United States, the United Kingdom, and Australia, believed that molecular keys to their own salvation would be found in blood they "salvaged" from those they feared were disappearing. In doing so, they reinvented Rousseau's noble savage for the nuclear age.[2]

In that book, I focused on how freezing blood became the "right tool for the job" of creating an ark for a very specific kind of supposedly secular human future.[3] I was also effectively tracking projects that were devoted to enrolling Indigenous bodies into the epistemic regimes of a Christian-inflected, neocolonial science. In such a science, Indigenous cosmologies and philosophies of living relation—if they were even considered at all—were not considered as valuable as Indigenous body parts. In the years since I published that book, my interest in the role of religious projects of so-called scientific salvage has brought me to a more fundamental question: What might become visible by narrating the use of Indigenous bodies in biomedical research as part of a longer history of Christian mission and violent conversion? And as an approach to a kind of situated history, how might I, a white historian of science descended from Jewish settlers in the United States, reckon with dimensions of colonialism, existence, and vitality that exceed and subvert the molecular forms that have dominated biology since the Cold War?

Here, I consider these questions and the ones with which I began this chapter through an examination of the scientist and Catholic priest Basile Luyet (1898–1974). He has come to be regarded as a father of cryobiology, having made possible the long-term low-temperature preservation of blood and other tissues through both his theoretical work and his institution building. Beginning in the 1930s, he sought to merge physics and biology to probe what he understood to be the boundary between life and not-life. In doing so, he argued for the importance of an experimental approach to supplant the "unverifiable speculation" that had previously suffused and, in his view, confused studies of the nature of matter.[4] He endeavored to articulate a less dichotomous view of life by retheorizing the cell as a "symbiotic colony," in which vitality emerged from an ensemble of relations, and by promoting the concept of latency, which he articulated as an ambiguous state between life and death.

Latency was a term Luyet borrowed from one of his teachers, Alexis Carrel, who had won the Nobel Prize for his work with tissue culture.[5] (Carrel, raised as a devout Catholic, had left his native France for the United States when his reports of mystical experiences tarnished his credibility with the French scientific elite.)[6] As a commentator from the late 1950s observed of the phenomenon of latency, "Whether life under certain conditions may be a discontinuous process [was one of the oldest questions] reflected in almost all religions, in some legends and even fairy stories."[7]

The feminist philosopher of science and biologist Donna Haraway has observed, "The discourses of genetics and information sciences are especially replete with instances of barely secularized Christian figurative realism at work."[8] I focus on how the intellectual frameworks available to Luyet, which circulated through Euro-American philosophical communities, of which the twentieth-century Catholic Church was but one, made certain kinds of thinking and practices possible while effacing others.[9] I argue that Luyet's privileging of experiment over speculation in early twentieth-century biophysics—the precursor to the informatic and genetic paradigm that swamped other approaches to thinking about matter—perpetuated a fixation with Christian (and particularly Catholic) imaginaries, such as that of bodily resurrection and transubstantiation.[10] I also demonstrate the unfulfilled, or even latent, potential within his retheorizations of life to break with such enduring narratives.

The significance of Luyet's efforts to rationalize existence, to pinpoint indeterminacies in transformations of matter, became amplified as applications of his work to induce states of suspended animation through freezing were taken up by various other groups, from cattle breeders seeking to standardize herds through assisted reproductive therapy to human biologists interested in salvaging Indigenous blood to create arks of biological variation. Such varied applications made assumptions about the limits of life matter differently for different communities, human and otherwise. That Luyet's experimental philosophy yielded practical techniques that transcended the circumstances of their origins allows me to trace the fecund intersections of self-identified religious and secular *institutions* to better understand the shared worldview they have helped produce—one in which body parts can be made to speak in languages foreign and even violent toward the persons from whom they have been extracted.

Indigenous peoples' blood has been made by scientists to tell stories about origins, to confess forms of relation that undermine the very cosmologies that connect Indigenous communities to the land and each other.[11] The question then becomes not whether Luyet's biology was any more Christian because he was a Catholic priest but to identify specific ways in which Western science's attempts to take seriously the animacy of materiality have nonetheless expressed themselves as extensions of colonial power. In this sense, Luyet's personal religious beliefs, which remain obscure, are perhaps less relevant than values shared by both Western science and Christianity.

The stakes of my approach become apparent when Luyet's theories about and manipulations of matter are compared with Indigenous and decolonial scholars' provincialization of Western scientific modes of mattering. The Indigenous science and technology studies scholar Kim TallBear has recently asked what it would mean to refuse the entire ontological project of "life" as it has been posed by both Judeo-Christian religions and the secular modes adopted by science. Inspired by the queer theorist Mel Chen, TallBear reinterprets a life/not-life binary as an animacy hierarchy that refers to the relative greater and lesser degree of an entity's sentience, aliveness, (self-)awareness, and agency.[12] She rejects such a hierarchy in favor of a view of the interrelatedness of all things. "Indigenous standpoints," she writes, "accord greater animacy to nonhumans, including nonorganisms, such as stones and places, which help form (indigenous) peoples as humans constituted in much more complex ways than in simply human biological terms."[13]

Of equal significance is the fact that TallBear's approach seeks to disrupt another binary that has served Western science, that of the secular versus the religious. She retains the category of religion within what she calls an "indigenous metaphysic" both because she seeks to affirm a broader array of relations than can be accommodated by secular logics *and* because doing so holds open space to disrupt the assumption that Western science is uninflected by religion. "The advantage of indigenous analytical frameworks that are not secular," she writes, "is that they are more likely to have kept sight of the profound influence in the world of beings categorized by Western thinkers (both the church and science) in hierarchical ways as animate, or less animate."[14] Her goal is not to abandon science but, in emphasizing its historical collusion with Christianity and colonialism, to demonstrate that the relations of matter that emerge from Indigenous worldviews should be privileged to restore agency to those humans and nonhumans who have too often been regarded as research materials rather than reliable knowers.

For these reasons, I want to return to Luyet's experimental philosophy as a means of reckoning with the colonial logics that aligned the Catholic Church with a so-called secular science of life. Doing so not only helps to provincialize biomedical worldviews that are parasitic on Indigenous bodies without consideration of Indigenous thought but also may open new avenues for historicizing the pro-life positions that brought the Church into politics in the late twentieth century after Vatican II.[15] Moreover, it demonstrates that speculation is not an unmarked category of unreality or liberation, but can be a powerful force in maintaining the status quo—often especially in circumstances where it is explicitly devalued. Speculation, I argue, following the work of Indigenous scholars of colonization, is not a metaphor.[16]

In what follows, I describe Luyet's experimental approach to discerning the relationship between life and death. Though I provide biographical information to situate Luyet, it is less important for me to pinpoint exactly what was religious about his ideas than to trace the boundaries of the thinkable that existed for him as a scientist who was not only religiously affiliated but also a member of the clergy. As a priest, he was authorized to preside over sacrifice to God; in the Catholic Church, this means the Eucharist, the transformation of matter into something more than itself. Examining how Luyet negotiated the overlapping boundaries presented to him by his science *and* his Catholicism contributes to a rescaling of Western science in ways that better account for its reliance on speculation, verifiable and otherwise. This enables a reckoning with the multivalent work that speculation does in ways that are less Catholic than catholic—in the sense of varied and all-embracing.[17]

AN EXPERIMENTAL LIFE

Born in Saviese, Switzerland, in 1897, Luyet demonstrated early on the qualities that would characterize him as an inveterate experimenter and institution builder. In his youth, he was an amateur ethnologist who established the journal *Cahiers de valasains de folklore*, dedicated to salvaging the traditions and culture of his beloved alpine homeland.

At the age of twenty-four, Luyet joined the Catholic order of St. Francis de Sales, likely because it provided him with the ability to pursue a life of the mind rather than because he was responding to any deeper spiritual calling.[18] Soon after, he enrolled at the University of Geneva, where he was its first Catholic priest to be granted a doctorate. Technically, he was granted two doctorates: one in natural sciences and the other in physics. This work, which involved applying theories of material structure to living beings, earned him Yale's Seesel Biology Prize in 1928. The prize supported him for a year of postgraduate study at Yale in 1929, where he became interested in the biological definition of life.

Luyet's path into this morass was to start by asking what biological life was *not*.[19] He reasoned that since death was the destruction of life, he would approach the question from the vantage point of the end or near-end of life. At Yale, he worked in the laboratory of Ross Granville Harrison, famous for being the first to grow cells outside of the body from which they were generated—a precursor to the stem cell technologies that have become a twenty-first-century flashpoint of religious opposition to biological research.[20] These interests soon brought Luyet to the Rockefeller Institute in New York City, where, as a visiting fellow, he became acquainted with Alexis Carrel.

Carrel's early experiments with such precarious forms of life—which he referred to as "latent"—involved freezing, thawing, and transplanting pieces of dog artery. In a 1910 article, he explained that "a tissue is in latent life when its metabolism becomes so slight that it cannot be detected, and also when its metabolism is completely suspended. Latent life means, therefore, two different conditions: unmanifested actual life and potential life."[21] His investigations into latent life built on a longer tradition of research into suspended animation by Catholic experimentalists such as the eighteenth-century priest Lazzaro Spallanzani, who discredited the idea of spontaneous generation while also making practical advances in food preservation.[22]

Luyet carried this tradition forth as he popularized the idea of latent life in his writings about cryobiology, beginning in the 1930s when he first became interested in using new technologies of rapid freezing—a process he referred to as vitrification (from the Latin for glass). Around the same time, he accepted a faculty position at the Jesuit St. Louis University in Missouri, where he would teach for three decades. Upon his retirement, he was hailed by its leadership as biological savant and genius, "symbolic of true University life. He has sought truth, taught truth, and conserved truth."[23]

In Luyet's early years at St. Louis University, he founded *Biodynamica*, described in its front matter as a "scientific journal for the elaboration and the experimental study of working hypotheses on the nature of life." His aim was to import the physical sciences' practice of theoretical speculation backed up by experimental verification into biology, which was then still largely beholden to the observational techniques of natural history.[24] In his introduction to *Biodynamica*, he explained his approach:

> During the last forty years the physical sciences have progressed to an outstanding degree, some of the most fundamental problems on the structure of matter have been solved. The study of a number of these problems started with mathematical speculation and ended in an experimental verification. Since this method—which is, in the last analysis, a method of "working hypotheses"—has been so successful in the hands of physicists, we plan to use it in biology; but instead of the mathematical type of speculation which is rather foreign to the habits of biologists, we will resort, in general, to a more descriptive kind of theorization. . . . *Unverifiable speculations are excluded.*[25]

The rejection of *unverifiable* speculations was not directed toward his Catholic brethren but rather toward his scientific peers. Luyet's embrace of experiment represents an inflection point in the emergence of a new episteme for the study of life, one that was struggling against assumptions that had guided the practice of natural history, itself profoundly shaped by biblical narratives.[26] In other words, Luyet sought to redefine and delimit speculative practices to those that could yield experimentally verifiable results. He was effectively renegotiating the terms under which life could be understood in ways that constricted what could count as life.

The stakes of this reorientation become clear in the journal's very first article, published in October 1934. It is worth discussing in detail, as it demonstrates how Luyet understood the challenges facing his emergent experimental approach. Let me begin with his statement of the working hypothesis that he set for his research program:

> that the processes that we call life takes place in some formed components of the cell, such as the genes or chromosomes . . . and that the fundamental . . . mass of the cell is not living matter. The living units would then constitute only a relatively small fraction of the entire cell. . . . The different kinds of formed constituents of the cell might possess different degrees of life.[27]

This was a significant departure from the scientific orthodoxy of cell theory, which for over one hundred years had held that the cell itself was the necessary precondition for life. Rather than seeing the cell as an organic whole from which life emanated, Luyet recast it as "a *symbiotic colony* of biological units which possess life in different degrees."[28] Here, he was referring not to the thermal properties of life but rather to its widely varying constituents. He was arguing for a theory of life in which vitality emerged from the coordination of differently animate forms. His symbiotic approach, which sought to reimagine the cell as not a fundamental unit of life but an ensemble of relations, was radical.[29] And he knew it.

Promoting acceptance for this view in the scientific community was going to require Luyet to do more than conduct experiments and hope their results spoke for themselves. He was going to need to build a new set of institutional structures, a new science of biophysics with its own journal. And in order to attract an audience, he was also going to have to provoke his peers in a journal that they already read. In making his case against the cell theory, which he eventually published in the eminent journal *Science*, he found it necessary to begin by commenting on the very practice of theory making.[30]

I quote from Luyet's article in *Science* at length, since his choice of words is evocative. Notably, he began his appeal to his fellow scientists by invoking their piety, observing that "since faith in theories has done much harm to human thought, the dangerous aspect of any theory, namely, to inspire belief, will be emphasized in the following discussion."[31] As he prepared to depose the cell as the center of life, he waxed philosophical about theory:

> A theory is often considered acceptable . . . if it allows one to foresee unknown facts. It is clear, however, that, in the last analysis, we require more than that from a theory; we want it to represent the truth. . . . When it becomes evident that a theory does not represent the truth, even if it has been useful and is still useful in the discovery of new facts, we abandon it and try a new theory which might have more chance of being the true one. In general, a theory is useful in proportion to its nearness to the truth, but there are examples of theories which have been useful for centuries and finally had to be abandoned as inadequate to explain newly discovered facts. . . . In the last analysis, then, the decision as to the acceptance or maintenance of a theory depends only on the answer to the question: Has this theory a chance of being true or not?[32]

It was only then that Luyet made his move from the general to the particular to argue that

> the more we learn about life, the more the cell theory loses its chances of being true. The discovery, or the more complete observation of a number of facts during the 100 years which have elapsed since the formulation of the cell theory, as well as a more synthetic comprehension of these facts, make it now highly probable that the cell is not the necessary structural unit of any living matter.[33]

Life, for Luyet, could never be "itself" because it was fundamentally an ensemble of relations, a "symbiotic colony" bounded by the cell but not coextensive with it. By configuring life as a product of relational entities possessed of vitality in varying degrees, he was straining against mainstream views that emphasized autonomy in biological and social relations. And yet he did not feel pressed to offer any reflections on the political ramifications of the logics of colonialism, let alone symbiosis, for his science.[34]

He did, however, think deeply about the existential dimensions of his inquiries. An undated tall, brown notebook (likely from the early 1950s) contains the final handwritten paragraph of a lecture with the title "Human Destiny," outlined in the middle of what seem to be notes for a course covering electricity, heat, and space:

> All our laws are about perceptions, not about objective phenomena. We don't know how psychological phenomena are related to vital, vital to molecular, molecular to atomic, atomic to electrons. The properties of electrons do not explain those of molecules. We analyze but this is of no use to what we want to know: man. The scale of observation creates the phenomena, and the phenomena is human man-made. The objective world is one, not scale-conditioned, man's brain splits it in compartments and scales.[35]

This observation is resonant with the ideas Luyet expressed in his essay in *Science*, in which he sought to fracture the cell theory. Stated here, in explicitly phenomenological terms, his perspective unfolds in a language aligned with idealism/realism debates, which reverberated in European Catholic intellectual worlds. The historian of philosophy Edward Baring has recently shown that not only did Catholics "introduce phenomenology to many better known thinkers . . . they served as important interlocutors; and they provided a receptive audience for phenomenological and existential texts."[36] Luyet, whose notebooks record his readings of existentialist philosophers like Jean-Paul Sartre—who, Baring argues, got his ideas about the importance of "situation" from the Catholic thinker Gabriel Marcel—seems to have been both audience and interlocutor, aligning his epistemology of biology with phenomenological ideas and language.[37] Could Luyet have been attempting to work out in biophysics what other Catholic thinkers were trying to do in the terrain of phenomenology?[38] Here is something exciting then: way in which new concerns about the nature of matter and existence were being metabolized through philosophy and biology via Catholic institutions and scaled up into the world through practical applications of Luyet's epistemology.

THE USES OF LATENT LIFE

In 1948, Luyet received an invitation to participate in a Harvard conference on blood preservation. The meeting, organized for the U.S. military, the National Research Council, and the Red Cross, aimed to improve their ability to provide whole blood to transfuse soldiers injured in far-flung theaters of war—and perhaps eventually civilians at home. Luyet was recruited to apply his ideas about rapid freezing to the transformation of human blood into a biomedical resource that could be stockpiled indefinitely, available whenever and however it might be needed. The process that emerged was called fractionation, for the ways it broke blood into a series of constituents that could subsequently be amplified and transfused. Not only did Luyet's ideas break open the cell, but also they demonstrated that blood—long regarded as a sacred fluid, the "juice" of life—could be mined by potentiating different components of the symbiotic colonies of which it was comprised.[39]

By the early 1950s, Luyet's work began to capture the attention of nonscientists. He was profiled in *Time* magazine in 1952. The article, titled "Deep Freeze," observed that Luyet's "modest report of his work furthered some extravagant speculation," with the author suggesting that "men might be put into deep-freeze and revived thousands of years later. At the very least, spermatozoa from exceptional males could be saved to fertilize females of the future."[40]

This prediction would soon prove true when Luyet was recruited by Rockefeller Prentice, the owner of American Breeders' Service (ABS)—a cattle production business founded in Madison, Wisconsin. (In 1949, Prentice had purchased eighty acres on the outskirts of the city to support the newly created American Foundation for the Study of Genetics, a not-for-profit corporation devoted to the science of breeding. The industrial farm soon became a site for the commingling of cryobiological and eugenic thought and practice.) Luyet accepted an invitation to work at ABS while on sabbatical from St. Louis University. Lured by the promise of facilities, financial support, and the freedom to pursue his research without any other commitments, in 1956 Luyet decamped to Madison. He instructed that his laboratory facilities there be built in dimensions all using the golden ratio, in keeping with his belief that it was the structure or form of matter that produced vital phenomena, which he likened to an architecture.

But how did Luyet understand the moral implications of his retheorization and instrumentalization of life? One clue exists in a piece of ephemera, also from 1956, found in his archives. It is a program commemorating the 138th year of St. Louis University, which featured a special statement on "Social Responsibility in an Age of Technology":

> In a universe which achieves its purpose through men, scientists have with astonishing success brought the material world under the control of man's intelligence, making his purposeful intervention in the process of nature more spectacularly direct and effective. Their accomplishments have brought about a new social dimension, for their leadership is sought not only for the provision of means, but also for the determination of the proper ends to which they shall be put. The scientists, as men and as members of society, must share the total responsibility for their work, and society must recognize the proper place of the scientists. To the advancement of these objectives Saint Louis University is dedicated.[41]

From the vantage point of the twenty-first century, it seems obvious that Luyet must have recognized that his approach to biology was poised to raise profound existential and ethical questions. However, in the late 1950s, so many of the ideas about the concerns appropriate to science were registered in terms of physics' relationship to the bomb. They were focused on thermal extremes of heat, not cold. The philosopher Hannah Arendt recognized this when, in *The Human Condition*, published in 1958, she wrote with concern that scientists were using frozen germplasm to manipulate the possibilities for human life without adequately reckoning with their responsibilities to society.[42]

Luyet seemed to interpret his own responsibility in a different register. By the mid-1960s, as the ability to freeze life was put to an increasingly wide array of uses, from agriculture to biomedicine to conservation, he once again sought to create a space to share information about the refinement of technique. The question of form and its relation to function was the basis for what amounted to a cosmology of cold, a belief that the study of life at low temperature could be scaled up to an all-encompassing view about the order of nature. He expressed the contours of this cosmology in the first issue of the journal *Cryobiology*, published in 1964. It comes closest to articulating what might be understood as Luyet's own personal creed about the nature of life, a phenomenon so varied and distributed that it appeared to warrant a new origin story.

For Luyet, the effort to "share the total responsibility" seemed to require nothing less than a new theory of society organized around cold:

> The universe may be looked upon as consisting of two kinds of bodies, some intensely hot and incandescent (the stars), others cooler and nonluminous (the planets, their satellites, and miscellaneous minor bodies).... It would be of great importance for cryobiologists to know the history of the evolution of hot and cold bodies in the immensely large universe, during the immensely long time of its existence, and thus to fit the present stage of our sun and our life-carrying earth into the great evolutionary scheme in God's creation. But our knowledge on that matter is very limited; we can hardly do more than suggest as a possibility that the substance of the universe undergoes rhythmic fluctuations of temperature and that the solar system and the earth are now in a cooling phase of that cycle.[43]

There is much to remark on in this passage, and, certainly, that could be the basis for an entirely different paper. What I want to call attention to is that Luyet saw no incommensurability with his schema—one in which life is latent and forged from a symbiotic colony of vital entities—and the existence of a Christian God. Indeed, this is the only place I have found in any of his published writings where he invokes God by name. He does not endeavor to disprove the existence of a divine creator, but he urges that cryobiologists work only with the experimental data they possess about the nature of the universe and its origins.[44] This is consistent with the rejection of unverifiable speculations he called for in the first issue of *Biodynamica*.

How could Luyet at once take God's existence as given and yet dismiss that which could not be verified experimentally? Perhaps he did not take God's existence as given. The conditional phrasing of this passage is revealing: "It would be of great importance" for cryobiologists to know where we fit in the history of creation. But they do not: "our knowledge on that matter is very limited." In the absence of experimentally verifiable knowledge about Creation, Luyet turns not to a strictly thermodynamic vision of the universe, in which entropy leads to heat death, but to a consideration of flux, in which cooling is not an endpoint but a phase in a potentially eternal cycle.

Moreover, Luyet's recourse to the "great evolutionary scheme" revealed itself to be more than an innocent theory of change.[45] Having situated the earth in the cosmos, he turned his attention to the humans who came to inhabit it. His social theory tracked to midcentury adaptationist theories of evolution via natural selection. Yet he made a fascinating narrative inversion. Rather than telling a Promethean story about what fire enabled humans to do, he described how their relationship to cold shaped who they became.

> Men learned to fight cold by preserving and making fire. . . . The ability to keep fire gave rise to a sort of social institution, the gathering around the fire, which is designated by anthropologists as "community fire." The repetition of such gatherings naturally led to the transmission of knowledge about the fire from the better-informed members of the community to the others and from the older to the younger generation. The Community Fire institution this became a school. In a sense, we may say that cold started a school system. . . . In addition to stimulating man's intelligence and inaugurating a school system, cold exerted a long-term influence on both men and animals, either causing temporary changes in biological functions (acclimatization) or leading to a selection of more permanent evolutionary changes (adaptation).[46]

There is a hierarchy in this discussion of acclimatization and adaptation that was echoed in other research of the time. Such research focused on how Indigenous groups had supposedly adapted to life in colder climes more effectively than the scientists who studied them.[47] While Luyet did not make any explicit statements about racialization, his narrative indicated that the wisdom emerging from the community fire flowed to and from ancient Greece, claimed as the root of Western secular thought:

> Whereas other civilizations had placed emphasis on ways of life, the Greek thinkers pointed out the intelligibility of the world, a concept which provided men with new ways of enjoying life. Using the direct approach, that is, looking mysteries right in the face and asking them bluntly what was their place in the world, the Greek thinkers raised the question: What is cold? But then, when they tried to answer this question (and others), they indulged in

generalities rather than undertaking thorough and extensive studies of the factual evidence. They thus came to the conclusion that cold is one of the four qualities of the bodies, the other three being: warm, dry and wet, and that the category quality to which the four belong is just one of the ten categories into which all entities can be classified.[48]

Luyet's narrative led the reader from the Greeks into the humoral theories of the Middle Ages, the time when the Catholic Church assumed unprecedented authority. And from there, his narrative swiftly arrived at the Enlightenment of the eighteenth century, during which the science of cryobiology emerged in northern Europe, reaching its apotheosis with the development of practical applications—namely, mechanical methods of refrigeration. Arriving in the near past, Luyet celebrated "the intensification and organization of the research and the accumulation of data on the multiple aspects of preservation at low temperatures."[49] In a final, remarkable, passage he reflects: "To tell the preceding story in a few words, cryobiology, after having been in a state of latent life, as a seed, for some hundred thousand years, has undergone its embryonic development and growth to maturity during the last few hundred years and is now in full bloom."[50]

Latent life is revealed to have a telos, which is to bring about its own possibility as a phenomenon to be both catalogued and put to use in a Eurocentric mid-twentieth-century present. That Luyet's discussion of human encounters with cold should culminate in the bloom of a plant is, perhaps, a gesture to his understanding of the proper role of the scientist as not merely an idealist but also a materialist, a knower and a doer. Rather than see the two in tension, a contradiction in one's calling, this cosmology represented an effort at accommodation. I do find it fascinating and not a little disturbing that in his prodigious efforts to chart the bounds of the world of knowing and making in which he found himself, Luyet eschewed concerns about the moral problems presented by a cryobiological future, as expressed by Arendt, and instead provided an account of the past that was perhaps just as speculative. It was this final story that Luyet rehearsed not only in the pages of his journal but also in speeches for audiences around the country, including at the National Institutes of Health—one of the most significant funders of biophysics research and the site, today, of the largest collection of frozen blood in the United States.[51]

CONCLUSION: VARIETIES OF CRYOBIOLOGICAL EXPERIENCE?

Though the salvage biologists I wrote about in *Life on Ice* were unlikely to have read anything that Luyet wrote, they represented a particular instantiation of efforts to use freezers to extract potential from life in the latent state. His work with cattle breeders and blood fractionators contributed directly to the material conditions of possibility that enabled the stockpiling of Indigenous blood samples. Their freezers came to function like reliquaries in the way they sacralized the Indigenous body as a panacea for modernity.[52]

What I found especially fascinating was that the same blood was mined for different life forms in a relatively short period of decades. In ways that evoked Luyet's argument against the cell theory, blood itself came to be seen by late-twentieth century scientists as a kind of symbiotic colony in which an array of extrahuman life forms—from viruses to DNA—coexisted. Indigenous blood was initially valued for its ability to reveal a small number of protein variants, known in the 1960s as polymorphisms (literally, varied forms). By the 1990s, it was being revalued for the human DNA contained within. By the end of the decade, DNA from microbes like malaria was also detected, allowing scientists to recast these frozen droplets of blood as microcosms reflective of a larger, but no longer existent, macrocosm.

Luyet's concept of the symbiotic colony reflected his effort to displace the cell as the fundamental unit of life in favor of a model in which life emerged from an ensemble of relations between vital forms. Such a view allowed for the persistence of certain aspects of the Catholic idea of transubstantiation—the idea that matter, under particular circumstances, could become more than itself. However, his theory of the colony did not come with a concomitant theory of power, leaving it to absorb the excesses of meaning and animation that might have previously been attributed to spirit. Moreover, his emphasis on experiment functioned to naturalize a particular set of commitments about what it meant to make claims about how matter is materialized. In his devaluation of what he dismissed as forms of unverifiable speculation, Luyet participated in a restriction of possibilities for more deeply imagining the ways that matter is more than matter and for being accountable for the consequences of regarding it as an experimental substrate. To engage in some speculation of my own, what would it have meant for Luyet to deploy the language of the symbiotic colony in a way that was also attentive to the historical power relations inherent to colonialism?

I have called this conclusion "Varieties of Cryobiological Experience" in an explicit reference to the philosopher of religion William James.[53] To view the history of Luyet's ideas only from his own perspective is to miss the ways in which their uses have generated profoundly different experiences. For instance, after the turn of the millennium, when a globalizing biotech industry scrambled to gain access to these supposedly precious portals to the past, Indigenous peoples—who had not disappeared as Cold War scientists had predicted—began to demand their return. In some instances, Indigenous peoples described scientists' efforts to collect and freeze their blood as vampirism, an extremely parasitic form of symbiosis.[54] Indeed, Indigenous peoples who have more recently sought to have their ancestors' blood released from the freezer are also, in effect, demanding to be liberated from the epistemological and material hegemony of the symbiotic colony. This is the kind of structural violence that TallBear's approach seeks to redress. Historicizing philosophies of biology in such ways demonstrates the different possibilities for and consequences of theorizing relation. It also helps us to recognize the limits even of theories of diversity such as James's, which remains situated in a Euro-American scientific tradition.[55]

Rather than looking for answers to crises of modernity in the life forms to be found in Indigenous blood, there is much more to be gained from holding open space for Indigenous forms of relational world-building. To a deracinated phenomenology that cannot admit its own colonialist inheritances, TallBear says, "We cannot allow the old science/religion, and all of those other human/animal, life/not-life divides—upheld by the institutions that govern our lives, the land, and the lives of nonhumans who have been savaged by Western analytic frameworks, animacy hierarchies, and the institutions they produce—to continue to pervert understanding of our lifeways."[56]

Luyet's experimental philosophy was a product of his affiliations with a science forged from Christian (and in many cases, specifically Catholic) origins, not a rejection of the religious in favor of the secular. It worked to establish and formalize as "real" what would otherwise have remained speculative and open to analysis as such. What interests me most is not to sift out what was religious versus what was secular in his approach or to pinpoint his personal religious beliefs. Rather, it is to recognize the broader ways in which Luyet's scientific agenda was fundamentally constrained by the colonizing power relations inherent in Christianity and its history of persecution of Indigenous modes of existence.

What I have tried to show here is that Luyet himself was working within the constraints of a life/not-life boundary that was *both* a religious and a secular inheritance.[57] While it might seem as though his Catholicism would pose problems for a research agenda focused on the manipulation of life, I have attempted an exploration of thought that does not take their opposition for granted. I have tried to show how intrinsic to Luyet's reformulations of the cell and science itself were untapped possibilities for less violent ways of imagining relations between humans and nonhumans. That these possibilities have remained latent says as much about how practices of speculation work in shaping the meanings of life as it does about the power of experiment to define the real. The stories Luyet told about the past had profound consequences for the questions that got asked in his present, including the way he sought to circumscribe speculation itself.

NOTES

1. Joanna Radin. *Life on Ice: A History of New Uses for Cold Blood* (Chicago: University of Chicago Press, 2017).
2. The extractive dimensions of "salvage" as an animating force in anthropology are discussed in Jacob Gruber, "Ethnographic Salvage and the Shaping of Anthropology," *American Anthropologist* 72, no. 6 (1970): 1289–1299.
3. Adele Clarke and Joan H. Fujimura, *The Right Tools for the Job: At Work in Twentieth-Century Life Sciences* (Princeton, NJ: Princeton University Press, 1992).
4. Basile Luyet, *Biodynamica* 1, no. 1 (1934): frontmatter.
5. Hannah Landecker, *Culturing Life: How Cells Became Technologies* (Cambridge, MA: Harvard University Press, 2007).
6. From the vantage point of the present, his eugenic views have proven more damaging to his legacy. Andrés Horacio Reggiani, *God's Eugenicist: Alexis Carrel and the Sociobiology of Decline*, vol. 6 (New York: Berghahn, 2007).
7. D. Keilin, "The Leeuwenhoek Lecture: The Problem of Anabiosis or Latent Life: History and Current Concept," *Proceedings of the Royal Society of London, Series B, Biological Sciences* 150, no. 939 (1959): 149–191. More recently, anthropologists who have taken life as their object have demonstrated the present-day manifestations of this age-old question in the contemporary life sciences. Stefan Helmreich has considered how limits themselves have come to define the biosciences and in doing so has articulated a distinction between life forms—embodied bits of vitality—and forms of life—ways of thinking and acting. Stefan Helmreich, "What Was Life? Answers from Three Limit Biologies," *Critical Inquiry* 37, no. 4 (2011): 671–696. Sophia Roosth, who shares my interest in latency, has demonstrated the persistence of interest in questions about inanimacy or, as she puts it,

what it means for life to be "not itself." Sophia Roosth, "Life, Not Itself: Inanimacy and the Limits of Biology," *Grey Room* 57 (2014): 56–81.

8. Donna Jeanne Haraway, *Modest-Witness@Second-Millennium.Femaleman-Meets-Oncomouse: Feminism and Technoscience* (New York: Routledge, 1997), 10.

9. For a related approach to the entanglement of Christianity with evolutionist thought, see Terence Keel, *Divine Variations: How Christian Thought Became Racial Science* (Stanford, CA: Stanford University Press, 2018).

10. Caroline Walker Bynum, *Wonderful Blood: Theology and Practice in Late Medieval Northern Germany and Beyond* (Philadelphia: University of Pennsylvania Press, 2007). The Catholic Church was constantly struggling to control its subjects' relationships to nonhuman forms of animacy through efforts to define a life/not-life binary. Bynum shows how the church worked with those beliefs in order to leverage its own authority. This meant allowing for the possibility that statues could bleed or that blood relics could renew themselves. In maintaining this kind of ambiguous relationship to life, the Catholic Church was able to engage in strategic uses in times of uncertainty.

11. Jenny Reardon and Kim TallBear, "'Your DNA Is Our History,'" *Current Anthropology* 53, no. S5 (2012): 233–245.

12. Mel Chen, *Animacies. Biopolitics, Racial Mattering, and Queer Affect* (Durham, NC: Duke University Press, 2012).

13. Kim TallBear, "Beyond the Life/Not Life Binary: A Feminist-Indigenous Reading of Cryopreservation, Interspecies Thinking and the New Materialisms," in *Cryopolitics: Frozen Life in a Melting World*, ed. Joanna Radin and Emma Kowal (Cambridge, MA: MIT Press, 2017), 187.

14. TallBear, "Beyond the Life/Not Life Binary," 193.

15. Alana Harris, ed., *The Schism of '68: Catholicism, Contraception and Humanae Vitae in Europe, 1945–1975* (Cham, Switzerland: Springer, 2018).

16. Eve Tuck and K. Wayne Yang, "Decolonization Is Not a Metaphor," *Decolonization: Indigeneity, Education and Society* 1, no. 1 (2012): 1–40. Tuck and Yang warn against invocations of "decolonization" that actually function to perpetuate colonial formations. I am suggesting that efforts to bracket speculation or to devalue it as merely metaphorical or fantastical function to efface its enduring power to shape the real.

17. On the liberatory potentials of speculation, see, for example, Ruha Benjamin, "Racial Fictions, Biological Facts: Expanding the Sociological Imagination Through Speculative Methods," *Catalyst: Feminism, Theory, Technoscience* 2, no. 2 (2016): 1–28; H. Love, "How the Other Half Thinks: An Introduction to the Volume," in *Imagining Queer Methods*, ed. Amin Ghaziani and Matt Brim (New York: New York University Press, 2019), 28. Elsewhere I have also pointed out the strategic uses of speculation to confound science's claim to the real: Joanna Radin, "The Speculative Present: How Michael Crichton Colonized the Future of Science and Technology," *Osiris* 34, no. 1 (2019): 297–315.

18. Beatrice Pellegrini Saparelli, Thomas Antoniette, and Jacques Dubochet, *Basile Luyet: Un Vie Pour La Science, 1897–1974* (Sion, Switzerland: Editions de Musees Cantonaux du Valais, 1997).

19. Notes in this section, unless otherwise specified, come from Radin, *Life on Ice*.

20. John H. Evans, *Contested Reproduction: Genetic Technologies, Religion, and Public Debate* (Chicago: University of Chicago Press, 2010).

21. Alexis Carrel, "Latent Life of Arteries," *Journal of Experimental Medicine* 12, no. 4 (July 23, 1910), 460–486.
22. For accounts that situate Luyet in a genealogy with Spallanzani, see A. S. Parkes, "The Freezing of Living Cells." *Scientific American* 194, no. 6 (1956): 105–118; and Bronwyn Parry, "A Bull Market? Devices of Qualification and Singularisation in the International Marketing of US Sperm," *Bodies Across Borders: The Global Circulation of Body Parts, Medical Tourists and Professionals*, ed. Bronwyn Parry, Beth Greenhough, Tim Brown, and Isabel Dyck (London: Routledge, 2015): 53–72.
23. Untitled document in the unprocessed Archives du Père Basline Luyet, Carton No 1, D 400.1, Florimont, Geneva. In the fall of 2018, I had the chance to visit Luyet's newly recovered archives outside Geneva at Florimont, the Swiss school where he had taught biology for nearly half a century before moving to the United States. The school was founded by Francis de Sales in 1905. Espousing Roman Catholic beliefs, though accepting all students of all religions, it persists in a leafy suburb of Geneva called Lancey. In its basement archives is a modest collection of materials from a period of Luyet's time teaching at the Jesuit St. Louis University in the 1950s and early 1960s. It remains unclear how these materials found their way back to Florimont from St. Louis.
24. See a notable example written by a physicist: Erwin Schrödinger, *What Is Life? The Physical Aspect of the Living Cell, with Mind and Matter and Autobiographical Sketches* (Cambridge: Cambridge University Press, 1967). The influence of physics on the then nascent science of molecular biology has been discussed at length in the history of biology. A few of the more significant examples are Pnina Abir-Am, "The Discourse of Physical Power and Biological Knowledge in the 1930s: A Reappraisal of the Rockefeller Foundation's 'Policy' in Molecular Biology," *Social Studies of Science* 12 (1982): 341–382; Lily E. Kay, *The Molecular Vision of Life: Caltech, the Rockefeller Foundation, and the Rise of the New Biology* (New York: Oxford University Press, 1993); and Nicolas Rasmussen, *Picture Control: The Electron Microscope and the Transformation of Biology in America, 1940–1960* (Stanford, CA: Stanford University Press, 1997.
25. Luyet, *Biodynamica* 1, no. 1 (1934): x.
26. On the relationship between Christianity and natural history, see Janet Browne, *The Secular Ark: Studies in the History of Biogeography* (New Haven, CT: Yale University Press, 1983).
27. Basile Luyet, "Working Hypotheses on the Nature of Life," *Biodynamica* 1, no. 1 (1934): 2.
28. Luyet, "Working Hypotheses," 3.
29. Historian Jan Sapp notes that in the late 1920s, such theories had begun to circulate but were not readily embraced—namely, because they implied that parts of the cell might have nonearthly origins, though Luyet did not explicitly state this possibility. E. B. Wilson, the leading cell theorist of the time, observed, "To many, such speculations may appear too fantastic for present mention in polite biological society; nevertheless it is within the range of possibility that they may some day call for more serious attention." E. B. Wilson, quoted in Jan Sapp, "Living Together: Symbiosis and Cytoplasmic Inheritance," in *Symbiosis as a Source of Evolutionary Innovation: Speciation and Morphogenesis*, ed. Lynn Margulis and Rene Fester (Cambridge, MA: MIT Press, 1991), 17.
30. Basile Luyet, "The Case Against the Cell Theory," *Science* 91, no. 2359 (March 15, 1940): 252–255.

31. Luyet, "The Case Against the Cell Theory," 252.
32. Luyet, "The Case Against the Cell Theory," 252. The nature of the argument here is strikingly similar to that formalized around the same time by the Jewish physician and theorist Ludwik Fleck in his *Genesis and Development of a Scientific Fact*. I have no evidence that Luyet was in contact with Fleck, nor does he cite him here. Nevertheless, Fleck's ideas had just begun to circulate throughout communities of life scientists at this moment. Ludwik Fleck, *Genesis and Development of a Scientific Fact*, ed. Thaddeus J. Trenn and Robert K. Merton, trans. Fred Bradley and Thaddeus J. Trenn (Chicago: University of Chicago Press, 1979).
33. Luyet, "The Case Against the Cell Theory," 253.
34. A more dominant metaphor at the time may have been the cell as chemical factory, a construct that focused on questions of autonomy as opposed to symbiosis. Andrew Reynolds, "The Theory of the Cell State and the Question of Cell Autonomy in Nineteenth and Early Twentieth-Century Biology," *Science in Context* 20, no. 1 (2007): 71–95.
35. Unprocessed Luyet archives, Florimont.
36. Edward Baring, *Converts to the Real: Catholicism and the Making of Continental Philosophy* (Cambridge, MA: Harvard University Press, 2019), 9.
37. I want to note here that this idea of situation gets taken up in important ways by Donna Haraway in her feminist critique of objectivity. Donna Haraway, "Situated Knowledges: The Science Question in Feminism and the Privilege of Partial Perspective," *Feminist Studies* 14, no. 3 (1988): 575–599. Indeed, it is her idea of situation that I use to establish my own narrative here. Also important is her Catholicism, which she has written about in various of her works. The significance of Catholicism in her thought, as well as that of Bruno Latour, another key figure in the material-semiotic approaches that characterize science studies, are latent—as it were—in this chapter. A fuller explication is forthcoming in collaboration with my colleague, the philosopher of religion Noreen Khawaja. Read her book *The Religion of Existence: Asceticism in Philosophy from Kierkegaard to Sartre* (Chicago: University of Chicago Press, 2016) for insight into the possibilities inherent in such a project.
38. Baring, *Converts to the Real*.
39. Piero Camporesi, *Juice of Life: The Symbolic and Magic Significance of Blood* (London: Continuum, 1995).
40. "Deep-Freeze," *Time*, April 28, 1952, 68.
41. Unprocessed Luyet archives, Florimont.
42. Hannah Arendt, *The Human Condition*, 2nd ed. (Chicago: University of Chicago Press, 1998).
43. Basile Luyet, "Human Encounters with Cold, from Early Primitive Reactions to Modern Experimental Modes of Approach," *Cryobiology* 1, no 1 (1964): 4.
44. The absence of tension between the experimental model and Catholic reasoning is less surprising than one might think—in Luyet's time, Joseph Maréchal (who also had a background in biology) would be an important theorist of this position, called transcendental Thomism.
45. Terence Keel, *Divine Variations*.
46. Luyet, "Human Encounters with Cold," 6.

47. Discussed in Radin, *Life on Ice*.
48. Luyet, "Human Encounters with Cold," 6.
49. Luyet, "Human Encounters with Cold," 7.
50. Luyet, "Human Encounters with Cold," 8.
51. On NIH blood collection numbers, see Radin, *Life on Ice*, 185. On mid-twentieth-century cosmological narratives, see Nasser Zakariya, *A Final Story: Science, Myth, and Beginnings* (Chicago: University of Chicago Press, 2017).
52. Bynum, *Wonderful Blood*.
53. William James, *The Varieties of Religious Experience*, vol. 15 (Cambridge, MA: Harvard University Press, 1985).
54. Jenny Reardon, *Race to the Finish: Identity and Governance in an Age of Genomics* (Princeton, NJ: Princeton University Press, 2009). On varieties of symbiosis, see Jan Sapp, *Evolution by Association: A History of Symbiosis* (New York: Oxford University Press, 1994).
55. A rejection of such ontological reconfigurations in favor of a reflexive theorization of colonialism is at stake in Zoe Todd, "An Indigenous Feminist's Take on the Ontological Turn: 'Ontology' Is Just Another Word for Colonialism (Urbane Adventurer: Amiskwaci)," An (Un)Certain Anthropology: Uma (In)Certa Antropologia, October 24, 2014, umaincertaantropologia.org/2014/10/26/an-indigenous-feminists-take-on-the-ontological-turn-ontology-is-just-another-word-for-colonialism-urbane-adventurer-amiskwaci/.
56. TallBear, "Beyond the Life/Not Life Binary," 194.
57. Talal Asad, *Formations of the Secular: Christianity, Islam, Modernity* (Stanford, CA: Stanford University Press, 2003). More to the point regarding efforts to freeze life, the anthropologist Abou Farman is interested in tracing what can happen at the limits of the secular without asking religion to pick up the slack. This does not mean that metaphysical concerns are not in play. Rather, his depiction of cryonicists inspires me to view apparent contradictions in the worldview of cryonicists as meaningful. Examining what paradoxes are negotiated or resolved in a given project of future making is a helpful method for attempting to think about the role of religion in the contemporary sciences of life. Abou Farman, *On Not Dying: Secular Immortality in the Age of Technoscience* (Minneapolis: University of Minnesota Press, 2020).

CHAPTER 14

MAROON SCIENCE

Knowledge, Secrecy, and Crime in Jamaica

KATHARINE GERBNER

In the late 1970s, the anthropologist Kenneth Bilby spoke to a Jamaican Maroon named Ba Uriah, one of the oldest and most respected Kromanti specialists in Moore Town. "Ba Uriah was cryptic today," Bilby wrote in his journal. "He denied knowledge of Maroon nations, and he advised me to leave the drums alone, as they involved Science." Science, explained Bilby, was "Kromanti power," and Ba Uriah felt that he "[couldn't] manage it." Bilby pushed back: "I said I was interested in learning how to play the drums, and in learning *about* science, but not in *using* it." Bilby's distinction between learning and knowing was quickly dismissed. "Ba Uriah replied (very logically) that if you learn about Maroon Science, you must *use* it, it simply *has* to be used."[1]

The Jamaican Maroons are descendants of runaway slaves who created separate societies in the interior of the island. They began to use the word *science* as a synonym, replacement, or corollary for the Afro-Caribbean word *obeah* at some point after 1760.[2] *Obeah*, an African-derived term that is specific to the Anglophone Caribbean, is notoriously difficult to define, but it is often described as a ritual practice of healing, harming, and divining.[3] Obeah was criminalized in 1760 following a major slave rebellion called Tacky's Revolt, and it remains a crime in modern Jamaica.[4] Today many maroons will offer a sharp corrective if a visitor calls their practices obeah.[5]

Obeah and *science* are widely recognized as connected terms throughout the Anglophone Caribbean, a fascinating correlation that is only beginning to gain attention in academic writing.[6] This chapter examines the historical relationship between obeah and science at four different moments. First, I use colonial archival sources to show how the criminalization of obeah was part of a broader effort to criminalize Black epistemologies in the British Atlantic world. Recognizing the initial impetus for criminalizing obeah, I suggest, helps to demonstrate the way that the shifting categories of religion and science were influenced by the problems of colonial governance and the demands of policing the enslaved population.

Second, I use the Maroon story to explore the complex relationship among freedom, science, and obeah. The Jamaican Maroons play an ambiguous role in the history of freedom: they offer a victorious history of resistance against slavery but also a complicated counternarrative to any story of racial solidarity. In 1739, the Jamaican Maroons signed a peace treaty with the British, in which they agreed to turn over runaway slaves and help the British militia defeat slave rebellions. The Maroons thus play an unusual role in narratives about both obeah and science given their alliance with the colonial government and their tense relationship with non-Maroons.

Third, I examine the popularization of the word *science* in the Anglophone Caribbean and among the Maroons. The shift from obeah to science, however, is very difficult to trace due to the problematic nature of the archival record. Most colonial records about obeah and science are related to criminal trials, and the first several references to obeah as science come from white colonials rather than Afro-Jamaicans. As a result, it is difficult to know whether the term was in use in the eighteenth century or whether it emerged later in the nineteenth century.

Among the Jamaican Maroons, it is even more difficult to know exactly when the term *science* was integrated as a replacement or analogue for obeah. While it may have been in use as early as the eighteenth century, it was not until the late 1930s that one early anthropological account referred to both obeah and science. Academic knowledge about Maroon science during this period, however, was not widespread due to a preoccupation with "primitive" African survivals in anthropological studies from the 1930s. In the final section, I interrogate early anthropological archives, asking what the language of science may have meant for the Jamaican Maroons during this period. This section also asks broader questions about what it means to use anthropological archives as evidence for understanding the shifting categories of science and religion.

Given the problematic nature of archival sources and the inability to know, in a historical sense, the full extent of the relationship between obeah and science, the conclusion offers some thoughts about the relationship between Maroon science and historical methodology. The Maroons, in particular, use the term *science* to tell their own history, but they do so in a way that emphasizes secrecy. I end by considering the implications of both secrecy and the colonial archive for historical knowledge and ask what it would mean to narrate colonial history through the lens of Maroon science rather than solely using archival documents.

Overall, the story of Maroon science presents several epistemological and methodological challenges for scholars of science and religion. For most academic scholars, Maroon science is not *really* science—instead, it is glossed as "spirit working" or "power." But what does it mean to *not* redefine the Maroon invocation of science in our writing? Doing so, I suggest, raises questions about how scholars should handle unknowability and ambivalence in our scholarship. Addressing Maroon science and its complicated colonial and postcolonial legacy should be an exercise in humility for academic scholars, who need to recognize their own methodological choices and epistemological priorities.

OBEAH AND THE SCIENCE OF CRIME

On December 13, 1760, the members of the Jamaican Assembly entered a large, columned building in St. Jago de la Vega, or Spanish Town. Once inside the impressive structure, they finalized a new law that sought to prevent future rebellions like the one that had consumed the island for much of the previous year. The law that they created, "The Act to remedy the Evils arising from irregular Assemblies of Slaves," was the first to criminalize obeah. It had major consequences for the category of religion and the meaning of science.

Before 1760, white colonials had described obeah as a form of witchcraft or superstition, but it had never been criminalized. The "Act to remedy the Evils arising from irregular Assemblies of Slaves," by contrast, deemed obeah a capital offense, punishable by execution or transportation. The specific language of the Act is significant:

> And in order to prevent the many Mischiefs that may hereafter arise from the wicked Art of Negroes going under the appellation of Obeah Men and Women, pretending to have Communication with the Devil and other evil spirits, whereby the weak and superstitious are deluded into a Belief of their having full Power to exempt them whilst under their Protection from any Evils ...

The act continues:

> any Negro or other Slave who shall pretend to any Supernatural Power, and be detected in making use of any Blood, Feathers, Parrots Beaks, Dogs Teeth,

Alligators Teeth, Broken Bottles, Grave Dirt, Rum, Egg-shells or any other Materials relative to the Practice of Obeah or Witchcraft in order to delude and impose on the Minds of others shall upon Conviction thereof before two Magistrates and three Freeholders suffer death or Transportation . . .[7]

One of the most interesting things about the criminalization of obeah was the way that it drew on British conceptions of witchcraft while also seeking to distance the practice of obeah from witchcraft. As Diana Paton has shown, the colonial legislators in Jamaica were in a bind when it came to criminalizing obeah. In 1736, English Parliament had passed a new Witchcraft Act, which represented an important shift in the way that witchcraft was viewed in a legal context. This act reflected the increasingly skeptical attitude of elites toward claims of witchcraft. Not only did it decriminalize witchcraft, but also it created a new crime: *pretending* to practice witchcraft. It also became illegal to accuse someone of witchcraft. In other words, the main goal of the Witchcraft Act of 1736 was to suppress the belief that witchcraft existed.[8] As a result, lawmakers in Jamaica could not criminalize obeah as witchcraft because it was against the law to believe that witchcraft was real. But the legislators clearly saw obeah as a threat.

Traces of the 1736 Witchcraft Act are evident in the 1760 Slave Code. The lawmakers demonstrate skepticism when they write that "Obeah Men and Women" were "pretending to have Communication with the Devil and other evil spirits" and that "the weak and superstitious are deluded into a Belief." It may also be due to the 1736 Witchcraft Act that Jamaican legislators emphasized the category of obeah rather than witchcraft in the first place. The term *witchcraft* appears only once, almost as an afterthought, when the law reads that anyone "detected in making use of any Blood, Feathers, Parrots Beaks, Dogs Teeth, Alligators Teeth, Broken Bottles, Grave Dirt, Rum, Egg-shells or any other Materials relative to the Practice of Obeah or Witchcraft . . . [shall] suffer death or Transportation."[9]

There are differences as well. While the 1736 Witchcraft Act forbade accusations of witchcraft, the 1760 Slave Code did not have a similar clause. Instead, lawmakers seemed legitimately concerned about the practice of obeah—even as they simultaneously stated their disbelief in its power.[10] They were most worried about the ability of obeah practitioners to unite and bind enslaved people together—to create a sense of unity and solidarity—because this could aid a rebellion.

The criminalization of obeah had significant consequences. After 1760, obeah references became increasingly common in colonial histories, newspapers, official reports, novels, and plays. Even as they rejected and mocked obeah as barbaric and ridiculous, colonial writers remained fascinated by the practice. Colonial officials wielded more power to define and regulate the practice of obeah, thereby marginalizing Afro-Caribbean epistemologies.[11] Eventually, the criminalization and denigration of obeah would lead to the adoption of *science* as a preferred term throughout the Anglophone Caribbean.

The criminalization of obeah shows how the categories of witchcraft and superstition were central to the construction of religion and science. Obeah, with its emphasis on healing, had many overlaps with European ideas about medicine and science. Meanwhile, the institutional and ritual-based aspects of obeah demonstrate its overlap with the category of religion. Despite these similarities, obeah was excluded from the categories of religion and science because it was associated with slave rebellion and colonial resistance.[12]

FREEDOM AND COLONIAL GOVERNANCE

While the Jamaican Maroons were well-known obeah practitioners, they rarely play a role in historical narratives about the criminalization of obeah. That is because the Maroons fought *against* the rebel slaves in Tacky's Revolt. Indeed, the Maroons play an uncomfortable role in the history of obeah and science. Before the criminalization of obeah in 1760, it was the Maroons who were most closely associated with the practice of obeah in Jamaica. Colonial records from the 1730s describe Maroon Obeah men and women, especially the obeah woman Nanny, who fought the British state with tremendous success. This section examines the history of the Maroons in colonial Jamaica, suggesting that scholars should attend not only to the criminalization of obeah but also to the timing of the 1760 act. Despite its association with Maroon resistance, lawmakers did *not* criminalize obeah during the Maroon Wars of the early eighteenth century. There are a number of explanations for this fact, including the possibility that colonial authorities had yet to recognize the significance of obeah or that the meaning of obeah had shifted over the course of the eighteenth century. What is undeniable, however, is that the Maroons presented a different type of challenge for colonial governance than did enslaved rebels. Considering this counterfactual thus helps to illuminate the relationship between governance strategies and the shifting categories of religion, crime, and science.

The history of the Maroons is deeply intertwined with both resistance and obeah. Beginning in the sixteenth century, enslaved men and women escaped Spanish and then British colonizers.[13] The Maroons, as they became known, lived in the interior of Jamaica, in regions that were difficult to access from the coast. Over the course of the seventeenth century, two distinct groups of Maroons flourished in different parts of the island. The Windward Maroons, led by the obeah woman named Nanny, settled in the Blue Mountains and the John Crow Mountains on the eastern side of the island. The Leeward Maroons settled in Cockpit Country, which offered protection from outsiders.[14] Between 1660 and 1739, as the British system of plantation slavery grew, so did the Maroon communities. Subjected to terrorizing treatment and the brutal labor regime of sugar cultivation, scores of enslaved people risked torture, dismemberment, and death by fleeing into the mountains. Most new Maroons were runaway slaves who escaped on their own or in groups, sometimes following a rebellion. Others were captured during Maroon raids.[15]

The Maroons created governments that harnessed the religious and political authority of obeah. Obeah, as Kenneth Bilby has emphasized, cannot be reduced to one unitary thing. Instead, he has argued that obeah should be understood as "spiritual power" that the Maroons used for a variety of purposes. Among other things, obeah was central to Maroon military strategy. Indeed, medicine, politics, and religion become indistinguishable when narrating the role of obeah in Maroon history. Maroons followed obeah men and women in making military decisions, and they used obeah to catch British bullets without injury.[16]

Nanny, the leader of the Windward Maroons, is widely celebrated throughout Jamaica as one of the fiercest fighters of the Maroon Wars.[17] Contemporary Windward Maroons refer to themselves as *Yoyo*, which means "a descendent of Grandy Nanny."[18] Even in the colonial archive, Nanny is recognized as a powerful and dangerous obeah woman. In one reference, she is described as "the rebels old obeah woman," showing the close relationship between rebellion and obeah.[19] In his *Memoirs*, Philip Thicknesse describes Nanny as an "old Hagg, who passed [a] sentence of death" upon a white man. He provides an elaborate account of her appearance, which included "a girdle round her waste, with . . . nine or ten different knives hanging in sheaths to it, many of which I have no doubt, had been plunged in human flesh and blood. . . . That horrid wretch, their *Obea woman* would demand their deaths."[20] Obeah emerges in these records as a powerful force and Nanny as a fearless leader who terrified colonial authorities.

Given that obeah was widely recognized as a powerful tool of Maroon resistance, it is notable that obeah was *not* criminalized during the Maroon Wars of the early eighteenth century. This may have been because colonial officials did not yet understand the importance of obeah or because they did not believe they had the capacity to police obeah among Maroon communities. Either way, the Maroons presented a very different type of challenge to governance than did enslaved rebels. As a result, instead of criminalizing obeah, the British sought to make peace after decades of warfare with the Maroons.

On March 1, 1739, Col. Cudjoe (Kojo) and Accompong, leaders of the Leeward Maroons, met with the British colonial administrator John Guthrie and committed to a treaty. The 1739 treaty (1) recognized the Maroons as free, (2) gave them a grant of land, and (3) allowed them to sell produce in the island's markets. In return, the Leeward Maroons were expected to defend the island against invasion and rebellion, return runaway slaves to the British, hunt down Maroons who did not agree to these terms, and allow a white man to live in their settlements. The maroon leaders Cudjoe and Accompong were granted life tenure and the authority to administer any punishment except death.[21] The treaty with Nanny's Windward Maroons was similar, although it added clauses requiring Maroon militias to be headed by whites in order to track runaway slaves.[22] Both oral histories and colonial records suggest that the Windward or Nanny Maroons were resistant to signing a treaty with the British and only did so once Cudjoe (Kojo) promised to hunt down Maroons who did not make peace with the British.

The treaty transformed Maroon politics. Both oral and archival histories show that the Maroons maintained a strict rule against killing whites, defended the British population against slave rebellions, and returned runaway slaves. From the British perspective, the treaty was a success. The Maroons were granted autonomy, the enslaved lost an ally, and the colonial militia gained expert warriors. The Maroon alliance meant that Maroon towns—which had formerly offered refuge for runaways—were now sites of danger and death for the enslaved.

The difference in the white colonial responses to Maroon Obeah and rebel obeah provides insight into the relationship among religion, governance, and policing. During the Maroon Wars, the British tried to control the Maroons by defeating them through warfare but failed. Eventually, they recognized the Maroon forces as a separate governmental entity. In this context, obeah was a fearsome threat, but there was never any reason to criminalize it because the British could not police the Maroon population. Instead, the British opted initially for war, and when that did not succeed, they offered diplomacy. Enslaved rebels, however, represented a different challenge. As enslaved people, they were under near-constant surveillance and subject to both informal and formal policing. As a result, it made more sense to criminalize obeah as a strategy to govern and control the enslaved population.

As this counterfactual suggests, concerns about *governance* deeply influenced the emerging definitions of religion, superstition, and crime. Before 1760, white colonials had described obeah as a form of witchcraft or superstition, but they had never criminalized the practice. After Tacky's Revolt, colonial attitudes toward obeah shifted. In the 1760 act, *policing* strategies formed the basis for the definition of obeah. The act listed material objects, thereby associating obeah with specific items like alligator teeth and grave dirt. This definition made it easier to search slave huts for obeah-related objects. The depiction of obeah in the 1760 act also *excluded* healing, worship, and the community-building aspects of obeah. In doing so, white slave-owners excluded obeah from the category of religion and redefined it not only as superstition but also as a crime. As this history demonstrates, we cannot understand emerging categories of religion and science without calling attention to the categories of crime and nonreligion/superstition, which were used as a strategy of governance to control the behavior and movement of enslaved and colonized peoples.

FROM OBEAH TO SCIENCE

The term *science* entered the vernacular of Afro-Jamaicans sometime after the criminalization of obeah, but the nature of archival sources makes it very difficult to say when this occurred. Dianne Stewart has suggested that science is used "as a euphemism" for obeah "because it does not call to mind the historical baggage of immorality which cannot be avoided with the term 'Obeah.'"[23] Undoubtedly, the criminalization of obeah contributed to the adoption of science, although it was not the only impetus for the shift. The expansion of science was also connected to the links between colonialism and European collecting in the eighteenth century, as well as the popularity of spiritualism and mesmerism in the nineteenth and early twentieth centuries. This section traces the emergence of science as a replacement or analog for obeah, arguing that the constructions of science and religion were intimately affected by efforts to police enslaved and free people of color and by the concept of "African-ness."

In the colonial record, the stand-alone term *science* first appears in connection to obeah in the work of Bryan Edwards, a proslavery planter-politician whose *History, Civil and Commercial, of the British Colonies in the West Indies* was published in 1793. Edwards, citing a report transmitted by the agent of Jamaica to the Lords of the Committee of Privy Council, wrote that "all professors of Obi are, and always were, native Africans, and none other; and they have brought the science with them from thence to Jamaica, where it is so universally practiced."[24] It is impossible to know whether Edwards actually heard the word *science* used or whether he ascribed the word to obeah himself. But it is notable that he made the connection between obeah and science after the 1760 act. Edwards also knew that the knowledge about obeah was carefully guarded. "A veil of mystery is studiously thrown over their incantations," he wrote, "and every precaution is taken to conceal them from the knowledge and discovery of the White people."[25]

Edwards connected science and obeah in another way as well. Citing the same source, he described the trial of an obeah man after Tacky's Revolt of 1760. Before the man's execution, white colonists devised "various experiments" on him and other obeah men using "electrical machines and magic lanterns." In Edwards's telling, one obeah man received some "very severe shocks," after which he "acknowledged that his master's *Obi* exceeded his own."[26] The experiments with electricity, which occurred within the context of an obeah trial, are a reminder that European scientific experimentation is closely linked to the context of law. It was after the obeah man's trial that these experiments were conducted. As this anecdote emphasizes, the contours of European "science" were developing within colonialism and slavery.

While obeah was rarely linked with the stand-alone term *science*, it was occasionally referred to as an "occult science."[27] In a 1787 obeah trial, an enslaved man named George was accused of "procuring a phial of strong poison to destroy the white people on Stanton estate." The journalist who wrote about the case referred to the obeah man as "an adept in the occult sciences."[28] Here, the reference to science again comes from a white observer rather than an Afro-Jamaican practitioner.

In the decades before Emancipation, obeah became increasingly associated with criminal activity, and colonial authorities continued to arrest and persecute people of African descent who practiced obeah. Persecutions were most common if they were connected with rebellion or if an enslaved person was harmed by obeah.[29] In both cases, obeah persecutions aimed to strengthen and consolidate colonial power, either by discouraging rebellion or by protecting personal property (in the form of enslaved people "hurt" by obeah).

As obeah persecutions expanded, the language of science did as well. Throughout the British Empire, naturalists "explored" the Caribbean, taking specimens and speaking with informants, many of whom were of African descent or Indigenous.[30] Many scholars have shown how science and empire were intimately connected.[31] Naturalists often relied on Black and Indigenous people—who were rarely named—for their "discoveries," while many slave traders and slave owners prided themselves on being part of Enlightenment science. Bryan Edwards, for example, was elected to the Royal Society the year after publishing his *History*, while slave traders played a crucial role in collecting and literally circulating knowledge and specimens from the colonies to the "metropole."[32]

After Emancipation, obeah became part of the "Mighty Experiment" to see whether African-descended people were "ready" for freedom. Obeah was frequently compared to voodoo in the "black republic" of Haiti. During this period, the legal definition of obeah shifted away from the more serious crime of witchcraft to the more banal accusation of fraud. Rather than considering it a capital offense, post-Emancipation obeah regulations mimicked the English vagrancy law of 1824, which sought to police fraudulent activity. Even as obeah was downgraded as a crime, however, it became more widely prosecuted.[33]

References to science expanded in the nineteenth century. In *The Gleaner*, a Kingston newspaper established in 1834, *science* is a broad term, used in reference to photography, phrenology, and physiognomy, as well as healing, chemistry, and pills.[34] Meanwhile, some scholars have hypothesized that the popularization of the term *science* in the Anglophone Caribbean was linked to the circulation of occult literature like *The Sixth and Seventh Books of Moses*, published by the Chicago-based de Laurence Company.[35]

Obeah trials from the early twentieth century frequently link occult sciences, magnetism, and obeah. In 1917, for example, John Bell was accused of practicing obeah in Smith's Village, Jamaica. A search at his home turned up several items, including "parcels of human hair, two packs [of] card, small phials of funny smelling liquid, gunpowder, sulphur and several other things" that were known to be "implements for practising obeah." The investigating officers also found "pamphlets dealing with magnetism and other occult sciences."[36] In this case, it is interesting to note that while the magnetism pamphlets were mentioned, they were not necessarily considered evidence for "practising obeah." Instead, the evidence for obeah was confined to the "parcels of human hair" and other nontextual objects that were considered to be "African."

For those who were accused of obeah, science was sometimes used as a defense. While not located in Jamaica, the 1921 trial of Anita Smith shows how science could be defined in contrast to obeah. Smith, who was accused of aiding and abetting the practice of obeah, had allegedly fallen into a trance and told her interlocutors to "take a bath on the Third Stage of Science."[37] In her defense, Smith claimed that she was not practicing obeah but was induced into a hypnotic trance using the techniques of Franz Mesmer. Her lawyer, reading from the *Encyclopaedia Britannica*, insisted that mesmerism did "not assume supernatural powers" and that "in France hypnotism and mesmerism were regarded as science."[38] The dynamic in the courtroom, which focused on the distinction between science and obeah as a difference in "supernatural" belief versus "natural" science, demonstrates how the context of law was instrumental for the articulation of the difference between superstition and science.

Smith's case also shows that there was considerably more at stake in the debate about obeah and science than whether something was natural or supernatural. Both the prosecution and the defense were fixated on whether Smith's possessions and actions were African or not. The prosecution was less interested in Smith's books or letters, which were not considered "obeahistic." Here, the racial underpinnings of science are especially clear. If Smith's actions were deemed too African, then she would be convicted of practicing obeah. If she could prove that she was practicing science, then she would likely be released. Material objects like grave dirt or hair, as well as any objects that were considered specifically African, were used as evidence of obeah. Science in this context was operating as a racialized formation defined by its association with European texts and materials.

SCIENCE, KNOWLEDGE, AND THE SEARCH FOR THE "PRIMITIVE"

By the early twentieth century, science was firmly connected to obeah, and some individuals who were accused of practicing obeah argued that they were practicing science. However, this did not mean that the term "science" was in use in the rural Maroon communities of Jamaica. When I first began my research for this chapter, I could find no evidence that the term "science" was used in the Maroon communities in the early twentieth century, though I suspected it had been. In my research, I examined all the published and digitally accessible early anthropological accounts of the Maroons, looking for references to "science."[39] I found none. It was not until I was able to travel to Jamaica to examine the field notes of one anthropologist, Archibald Cooper, that I found several references to "science" tucked into Cooper's scribbled notes about everyday life in Accompong Town in the late 1930s.[40] These field notes, which are currently held at the West Indies and Special Collections at the University of the West Indies Mona, were never published, and Cooper never completed his dissertation.[41]

In contrast to Cooper's field notes, none of the published anthropological accounts from the first half of the twentieth century mention science. Instead, they are focused on a search for African survivals. This section examines the production of academic knowledge about the Jamaican Maroons and suggests that the focus on African-ness in anthropology and the obsession with African-ness in obeah trials were not dissimilar. While anthropologists viewed African-associated practices as positive "primitive" survivals, prosecutors in obeah trials viewed them as criminal. The fact that "science" went unmentioned in published anthropological accounts is an example of how scholarly interest in a particular topic (i.e. African survivals) can lead to the silencing of other important terms and concepts, such as "science" and its relationship to obeah.

The first academic scholars to study the Jamaican Maroons were associated with Melville J. Herskovits, one of the founders of African American anthropology.[42] Herskovits himself visited Accompong Town for one night in 1934, where he met with Col. Rowe and took photographs of the Maroons.[43] A year later he helped Katherine Dunham, an African American student, arrange a month-long stay at Accompong Town.[44] Dunham published a short memoir of her visit to Accompong and wrote about her experiences to Herskovits.

Dunham's memoir, which is based on her journal and moves day by day through her visit, begins with disappointment. "It is unfortunate," she wrote on her tenth day among the maroons, "the meagre material culture that these people have retained in spite of their voluntary isolation."[45] Her focus on cultural survivals is reminiscent of that of Herskovits, who prized "primitive" cultural material with clear analogues in African cultures. She was also disappointed that Maroons purchased "all of the household luxuries and necessities" in town, but she was fascinated by the *abeng*, a horn that, she noted, was "still used in Accompong to call special meetings."[46] By the end of her month, she had also become interested in what she called *obi*. "Obi runs like a dark current under the inscrutable surface of Maroontown. They all know it, I am sure."[47] She wrote in her journal that she was partially initiated into the "cult of obi." Conversations about obi, she noted, took place in the greatest secrecy, giving "emphasis to the mystery of our secret."[48]

Dunham did not mention science in her letters or field notes. Neither did Zora Neale Hurston, who visited Accompong in 1936, a year after Dunham. Hurston, like Dunham, was in communication with Herskovits throughout her visit to Jamaica.[49] But unlike Dunham, Hurston had extensive experience doing ethnographic fieldwork in Florida and the Bahamas.[50] Hurston was not impressed by the Maroons. Writing to Herskovits nearly a year after her visit to Accompong Town, she concluded that "the Maroons are highly over rated. They are the showpiece of Jamaica like a Harlem night club." Yet despite her general disdain, Hurston was impressed by what she deemed the "primitive medicine" of the Maroons. In the same letter to Herskovits, she highlighted this point, writing in all capital letters "THAT IS IMPORTANT," before adding that it would "be worth a study," though she did not have time to conduct it properly.[51]

Hurston's observations about Maroon medicine are sprinkled throughout her writings. In chapter 3 of *Tell My Horse*, she wrote about the "medicine man" she met in Accompong and how she was a "spectator while he practiced his arts."[52] She learned "the terrors and benefits of Cow-itch and of that potent plant known as Madame Fate." She also accompanied the medicine man to the "God Wood," which she identified as a birch gum. "It is called 'God Wood' because it is the first tree that ever was made," she explained. "It is the original tree of good and evil. [The medicine man] had a covenant with that tree on the sunny side." Hurston visited the God Wood multiple times with the medicine man. "One day we went there to prevent the enemies of the medicine man from harming him," she wrote. He took a "strong nail and a hammer" and drove the nail into the tree "up to the head." Later he sent Hurston back to fetch the hammer, which he had dropped next to the tree.[53]

While the term *science* did not appear in the letters or publications of Herskovits, Dunham, or Hurston, it was in use at the time. The anthropologist Archibald Cooper, who spent a year living in Accompong Town between 1938 and 1939, noted that several maroons used the term "science man." Ba Will, one of to the Maroons who spoke frequently to Cooper, referred to a science man as someone with "a great will power so that his power is greater than the power of the spirits he calls upon."[54] At other points, "science" emerges in Cooper's field notes in direct connection with obeah. In one of Cooper's notes, titled "Obeah Men," Cooper noted that he "asked Sa Liz where Ba Will learned his Science," and

Sa Liz responded the "he didn't know, but that his grandfather knew Science."[55] At another point, Cooper's interlocutors distinguished between being an "obeah man," being a "science man," and "just profess[ing]."[56] As Cooper's notes indicate, the term "science" was clearly in regular use in Accompong Town during the 1930s, and its' semantic field was connected with, though not identical to, the term "obeah."

The conflict between Cooper's field notes and the early published anthropological accounts is significant. The term *science* was in conflict with the anthropological program of "recovery" in the 1930s. Herskovits, Dunham, and Hurston all aimed to write about, observe, and preserve "primitive" material that they could connect with African analogues. The rhetorical implications of science, with its association with modernity, fit awkwardly with the search for survivals.

The study of primitive societies was political in another way as well. As Hurston herself noted, the Jamaican government wanted to portray itself as modern. In a letter to Herskovits, she wrote that "the government is eager to give the impression that all primitive expression is done by the Maroons." Yet she found some non-Maroon groups to be more "interesting," including some "right in Kingston who do more than the Maroons."[57] As Hurston's observations suggest, while American anthropologists were invested in the primitive, colonial governments, like that of Jamaica (a British colony until 1962), were deeply invested in the category of the modern. Hurston's connection of urban Kingston with the primitive and her denial of Maroon society as interesting were a repudiation of the Jamaican government's stance.

As this brief overview demonstrates, the knowledge about Maroon science remained largely inaccessible to—or ignored by—academic scholars in the first half of the twentieth century. The search for African survivals led most early anthropologists to ignore or diminish the meaning of science within Maroon communities. The historian of science Londa Schiebinger has studied this phenomenon in the eighteenth-century Caribbean. She has shown how European naturalists sometimes "forgot" or failed to record important information about plants they brought back to Europe. Specifically, she has shown how European collectors left out information about plants that were used as abortifacients. She names this phenomenon *agnotology*, or the study of that which is forgotten.[58] Here, a similar process was occurring among anthropologists who were studying African survivals among the Jamaican Maroons.

THE SCIENCE OF HISTORY

Science had become a dominant term among Maroons by the time the anthropologist Kenneth Bilby arrived in the 1970s. In the oral histories that he recorded, science has many meanings. Frequently, science is a sacred object.[59] In other cases, science refers to spirit possession. The Maroon leader Nanny is often described as being "under Science," which means that she is possessed by a spirit. In the twentieth century, Nanny herself is one of the spirits who does the possessing. She and the other Old People, including Accompong and Cudjoe, are considered to be the most powerful spirits who can enter the bodies of living Maroons.

Maroon science can be called many things, including a ritual practice, a method of experimentation, an epistemology, a legal system, a religion, and "Kromanti power," as Bilby himself has described it. But it is also, undoubtedly, a way of telling history. Maroons use the language of science to narrate their fight with the British colonial government and their claims to freedom. Given the problematic nature of the colonial archive and the inability to know, in an archival sense, how Maroons adopted the term "science," I conclude by offering some thoughts about what the language of science could mean for narratives of history.

According to twentieth-century Maroons, science was a central component of the Maroon Wars. Nanny's fighters, for example, "have Science."[60] Crucially, science was also central to the 1739 peace treaties. While British colonial authorities viewed the treaties as a means to contain and regulate the Maroons, the Maroons continue to see the treaties as sacred charters that could not be broken, and these documents play a foundational role within Maroon society.[61] According to Sydney McDonald (1978), the treaty documents were composed using a feather from a particularly powerful bird, "a chickenhawk" inhabited by a *pakit*, or personal spirit. "That is a Science," he explained. It was the science of the writing utensil that made the treaty powerful.[62] Other oral histories suggest that blood was used as the ink on the treaty. "A blood and feather," says Hardie Stanford (1991). "Dem cut de feather, pick de feather lik nib, and write ina de blood."[63]

As a way of narrating the past, Maroon science raises important questions about historical methodology. If the colonial records from the Maroon Wars do not refer to science but contemporary Maroon histories of the treaties do, then should academic historians use the term *science* to narrate colonial history?

Does it matter if there is no archival evidence that the Maroons used the term *science* until the twentieth century? Whose records should be authoritative?

My own research about the Maroons began with a letter written by an eighteenth-century Moravian missionary named Zacharias George Caries. He was the first Protestant missionary to be stationed in Jamaica, and he met the Maroon leader Accompong in 1755, sixteen years after Accompong signed the peace treaty with the British colonial government. Caries, who did not frequently make sartorial observations, described his appearance in detail. Perhaps the most striking aspect of his appearance, however, was the "silver Medal" that hung from the chain around Accompong's neck. On one side "was King George ye 2nd's Picture"; on the other was "his Commission with this subscription *Captain Acampong*."[64]

While I discovered Accompong in the diary of an eighteenth-century German missionary, it is science that keeps him alive for contemporary Maroons. Accompong's science maintains him in the position of own master, as he is sometimes spotted riding through town on a horse. Sometimes he is wearing a black military jacket; other times it is white.[65] In many ways, the archival histories complement and reinforce the oral histories, and vice versa. But there are also divergences. Whose Accompong is more accurate?

The differences between the colonial archival record and Maroon oral histories reflect not just contrasting perspectives but also underlying differences in the epistemological grounding of colonial and Maroon knowledge. Colonial epistemology, like historical methodology, prioritizes the written record over other forms of knowledge production and remembrance. Maroon science is a reminder of their incompatibility.

The epistemological problems involved in telling Maroon/colonial history are especially significant when it comes to the memory of the 1739 peace treaties, which granted the Maroons "freedom" in exchange for allying with the colonial state against enslaved men and women. What does it mean, for example, that the treaties themselves—written documents that held legal power for the colonial British state but were later ignored by the postcolonial government—are identified as products of science by the Maroons, who insist on their continued legitimacy?[66] What can this fact tell us about the relationship among textuality, notions of the sacred, law, and science?

Among the questions, there are a few tentative answers. First, the sacred charter of 1739 provides a poignant reminder of the role of power in determining the authority of some written documents over others. Second, the integration of science into the Maroon story of freedom indicates that both science and freedom were created within the context of colonialism and slavery. The scientific experimentation in the obeah trials, including the torture of the obeah practitioners after Tacky's Revolt, shows how "modern" science could be used as a performance of power and colonial domination. The freedom of the Maroons, meanwhile, existed only through the continued enslavement of people of African descent. Science and freedom have complicated meanings today, but they were both forged, at least in part, as colonial tools of supremacy, created for the maintenance of control over others.

Still, Maroon science is not only about colonialism and slavery. In his conversation with Ba Uriah, which began this chapter, Bilby told Ba Uriah that he was interested in learning "how play the drums, and in learning *about* Science, but not in *using* it."[67] Here, two aspects of science are especially notable. First, science is "play[ing] the drums." While drums are not usually associated with modern science, historically drumming was considered one of the most dangerous practices within slave societies. Slave codes throughout the Atlantic world shared an obsession with preventing drumming and other forms of communication. Drums were used to bring people together and to communicate complex messages. Drums, in other words, represented—and continue to provide—a vital system of knowledge making and knowledge distribution.

Finally, Ba Uriah's emphasis on use and secrecy is also helpful in telling a different story about science. The encounter reveals a tension between Bilby's perception of knowledge and Ba Uriah's proposition that knowledge can never just exist—it must always be "used." In this way, Maroon science is both a knowledge system and a critique of the idea that any kind of science is universal or fixed. Moreover, Ba Uriah's reticence to teach Bilby science shows that secrecy is foundational to Maroon science. If modern science was forged through colonialism and slavery using a myth of transparency and universality, Maroon science provides an implicit critique by maintaining a code of secrecy.

For the critical study of science and religion, the story of Maroon science underscores the impossibility of coherence and universality. It is also a reminder that the categories of science and religion were formed in the crucible of colonialism and slavery and that their evolving significance was deeply influenced by the effort to criminalize and police enslaved and free people of color. Colonial lawmakers intentionally excluded African diasporic practices like obeah from the categories of both religion and science. Jamaican Maroons, meanwhile, developed their own epistemological priorities and historical narratives that identified Maroon science as a central feature of Maroon history. Historical methodologies can reveal the existence of these dynamics, but the problematic nature of the colonial archive makes it impossible to make definitive statements about the relationship between Maroon science and obeah, either now or in the past. Instead, academic historians should embrace a position of epistemological humility as they seek to uncover the legacy of the fraught histories of science and religion.

NOTES

Acknowledgments: I am deeply grateful to the many scholars who provided invaluable feedback on this chapter. The Critical Approaches to Science and Religion workshop in August 2019 was an incredibly stimulating experience that provided the impetus for writing this piece. I am especially grateful to Myrna Sheldon, Terence Keel, Eli Nelson, Brent Crosson, and Mona Oraby for their thoughtful suggestions. Kenneth Bilby generously read and commented on this chapter, and he also provided me with new material from his field notes that I was able to integrate into the chapter. Jerome Handler, Caitlin Rosenthal, Lesley Lavery, and Susanna Drake all provided guidance and support in the early stages of writing. Finally, I am grateful to the 2018–2019 Young Scholars in American Religion who offered feedback at our October 2019 meeting: Shari Rabin, Joseph Blankholm, Samira Mehta, Jamil Drake, Alexis Wells-Oghoghomeh, Chris Cantwell, Matthew Cressler, Sarah Dees, and Melissa Borja, as well as our mentors, Sylvester Johnson and Sally Promey.

1. Kenneth M. Bilby, *True-Born Maroons* (Gainesville: University of Florida Press, 2005), 7.
2. The epistemological relationship between the terms *obeah* and *science* is extremely complex, and the choice of one word over another depends largely on context and chronology. In his ethnographic research, Kenneth Bilby found that Maroons sometimes use *obeah* and *science* as synonyms, while at other times they reject the term *obeah*, with its negative connotations, in favor of the term *science*. I am deeply grateful to him for sharing his thoughts and suggestions on this topic with me.
3. Jerome Handler and Kenneth Bilby refer to obeah as part of a broader "medicinal complex" that was used for "socially beneficial goals such as healing, locating missing property, and protection against illness and other kinds of misfortune." Jerome S. Handler and Kenneth M. Bilby, "On the Early Use and Origin of the Term 'Obeah' in Barbados and the Anglophone Caribbean," *Slavery and Abolition* 22, no. 2 (August 2001): 87. Kamau Brathwaite has similarly emphasized the medical properties of Obeah, writing that "the principle of obeah is . . . like medical principles everywhere, the process of healing/ protection through seeking out the source or explanation of the cause." E. Kamau Brathwaite, "The African Presence in Caribbean Literature," *Daedalus* 103, no. 2 (1974), 74–75. Dianne Stewart has gone further, arguing that Obeah should be understood as "an institution entailing much more than expert knowledge of botanic therapeutic

properties." Dianne M Stewart, *Three Eyes for the Journey: African Dimensions of the Jamaican Religious Experience* (Oxford: Oxford University Press, 2004), 43. As these scholars indicate, there is a great deal at stake in how we understand and categorize Obeah. Diana Paton has meticulously documented the cultural politics of Obeah over the last three centuries, showing how it has been excluded from the categories of religion and science. Obeah remains a crime to this day in Jamaica. For recent approaches to understanding and interpreting Obeah, see especially Diana Paton and Maarit Forde, eds., *Obeah and Other Powers: The Politics of Caribbean Religion and Healing* (Durham, NC: Duke University Press, 2012); Jerome S. Handler and Kenneth M. Bilby, *Enacting Power: The Criminalization of Obeah in the Anglophone Caribbean, 1760–2011* (Mona, Jamaica: University of the West Indies Press, 2013); Kelly Wisecup, "Knowing Obeah," *Atlantic Studies* 10, no. 3 (September 2013): 406–425; Diana Paton, *The Cultural Politics of Obeah: Religion, Colonialism and Modernity in the Caribbean World* (Cambridge: Cambridge University Press, 2015); and J. Brent Crosson, "What Obeah Does Do: Healing, Harm, and the Limits of Religion," *Journal of Africana Religions* 3, no. 2 (April 17, 2015): 151–176.

4. There is currently a debate in Jamaica about whether the Obeah laws should be repealed. See, for example, "Editorial: Repeal Unconstitutional Obeah Law," *The Gleaner*, June 21, 2019, http://jamaica-gleaner.com/article/commentary/20190621/editorial-repeal-unconstitutional-obeah-law.

5. Again, context is important here. Overall, however, the negative connotations of Obeah have led to a preference for the term *Science*, perhaps especially in conversation with non-Maroons.

6. Stephan Palmié makes this connection in *Wizards and Scientists: Explorations in Afro-Cuban Modernity and Tradition* (Durham, NC: Duke University Press, 2002). J. Brent Crosson, whose work also appears in this volume, has written extensively on the topic in a series of articles based on his fieldwork in Trinidad. See especially J. Brent Crosson, "Oil, Obeah and Science," *Cosmologics*, July 2016, http://cosmologicsmagazine.com/brent-crosson-oil-obeah-and-science/; J. Brent Crosson, *Experiments with Power: Obeah and the Remaking of Religion in Trinidad* (Chicago: University of Chicago Press, 2020), 21–22.

7. "An Act to remedy the Evils arising from irregular Assemblies of Slaves . . ." (1760). *Acts of Assembly, Passed in the Island of Jamaica, from the Year 1681 to the Year 1769 Inclusive*, vol. 2 (Kingston, Jamaica, 1787), 26-31.

8. Paton, *The Cultural Politics of Obeah*, 40.

9. "An Act to remedy the Evils arising from irregular Assemblies of Slaves . . ." (1760). *Acts of Assembly, Passed in the Island of Jamaica, from the Year 1681 to the Year 1769 Inclusive*, vol. 2 (Kingston, Jamaica, 1787), 26–31. For a longer discussion of this, see Diana Paton, *The Cultural Politics of Obeah*, 40–41.

10. Paton, *The Cultural Politics of Obeah*, 41.

11. Diana Paton, "The Trials of Inspector Thomas: Policing and Ethnography in Jamaica," in *Obeah and Other Powers: The Politics of Caribbean Religion and Healing*, ed. Diana Paton and Maarit Forde (Durham. NC: Duke University Press, 2012); Stewart, *Three Eyes for the Journey*.

12. On a related point, focusing on secularism and religion, see Jason Ānanda Josephson-Storm, "The Superstition, Secularism, and Religion Trinary: Or Re-theorizing Secularism," *Method and Theory in the Study of Religion* 30, no. 1 (January 2, 2018): 1–20.

13. When the English captured the island in 1655, some Spanish maroons—known as *Varmahaly negroes*—fought against them, while others eventually aided the British in their effort to conquer the island. For the history of the seventeenth-century Jamaican Maroons, see especially Barbara Klamon Kopytoff, "The Maroons of Jamaica: An Ethnohistorical Study of Incomplete Polities, 1655–1905" (PhD diss., University of Pennsylvania, 1973), and "The Early Political Development of Jamaican Maroon Societies," *William and Mary Quarterly* 35, no. 2 (1978): 287–307; and Mavis C. Campbell, *The Maroons of Jamaica 1655–1796: A History of Resistance, Collaboration and Betrayal* (Granby, MA: Bergin & Garvey, 1988).
14. Kopytoff, "The Early Political Development of Jamaican Maroon Societies," 290.
15. Kopytoff, "The Early Political Development of Jamaican Maroon Societies," 293–294.
16. Kopytoff, "The Early Political Development of Jamaican Maroon Societies"; Bilby, *True-Born Maroons*.
17. Nanny is now a national hero, and her image appears on the Jamaican $500 bill. Bilby, *True-Born Maroons*, 193. See also Jenny Sharpe, *Ghosts of Slavery: A Literary Archaeology of Black Women's Lives* (Minneapolis: University of Minnesota Press, 2002), ch. 1.
18. Bilby, *True-Born Maroons*, 483.
19. Cited in Sharpe, *Ghosts of Slavery*, 25.
20. Philip Thicknesse, *Memoirs and Anecdotes of Philip Thicknesse, Late Lieutenant Governor of Land Guard Fort, and Unfortunately Father to George Touchet, Baron Audley* (Dublin: William Jones, 1790), 74–77.
21. The full text of the treaties is available in Kopytoff, "The Maroons of Jamaica," 366–383. See also Barbara Klamon Kopytoff, "Jamaican Maroon Political Organization: The Effects of the Treaties," *Social and Economic Studies* 25, no. 2 (1976): 90.
22. Kopytoff, "Jamaican Maroon Political Organization," 90. In Windward Maroon oral histories, reluctance to accept peace with whites is a persistent theme. See Bilby, *True-Born Maroons*, 261–273.
23. Here, Stewart is discussing oral histories about Nanny, the Windward Maroon leader. Stewart, *Three Eyes for the Journey*, 254–255, n. 98.
24. Bryan Edwards speculates that Edward Long, another Jamaican slave-owning historian, may have been the original source for these comments. Bryan Edwards, *The History, Civil and Commercial, of the British Colonies in the West Indies* (Dublin: Luke White, 1793), 2:82–84.
25. Edwards, *The History, Civil and Commercial*, 2:84–85.
26. Bryan Edwards, *The History, Civil and Commercial, of the British colonies in the West Indies*, vol. 2 (London: Printed for John Stockdale, 1793), 100.
27. For more on occult sciences at this time in Europe, see Jason A. Josephson-Storm, *The Myth of Disenchantment: Magic, Modernity, and the Birth of the Human Sciences* (Chicago: University of Chicago Press, 2017).
28. I am grateful to Diana Paton for bringing this record to my attention. *Morning Chronicle and London Advertiser*, August 1, 1788, https://www.caribbeanreligioustrials.org/Case/Details/609.
29. Paton, *The Cultural Politics of Obeah*, 115.
30. Judith A. Carney, "African Rice in the Columbian Exchange," *Journal of African History* 42, no. 3 (2001): 377–396; Londa Schiebinger, "Agnotology and Exotic Abortifacients:

The Cultural Production of Ignorance in the Eighteenth-Century Atlantic World," *Proceedings of the American Philosophical Society* 149, no. 3 (2005): 316–343; Londa Schiebinger, *Plants and Empire: Colonial Bioprospecting in the Atlantic World* (Cambridge, MA: Harvard University Press, 2007); James Delbourgo, *Collecting the World: Hans Sloane and the Origins of the British Museum* (Cambridge, MA: Belknap Press of Harvard University Press, 2017).

31. James Delbourgo and Nicholas Dew, *Science and Empire in the Atlantic World* (New York: Routledge, 2008).

32. Richard Sheridan, "Edwards, Bryan (1743–1800), Planter and Politician," in *Oxford Dictionary of National Biography*, 2008, https://doi.org/10.1093/ref:odnb/8531; Kathleen S. Murphy, "Collecting Slave Traders: James Petiver, Natural History, and the British Slave Trade," *William and Mary Quarterly* 70, no. 4 (2013): 637–670.

33. Diana Paton, "Obeah Acts: Producing and Policing the Boundaries of Religion in the Caribbean," *Small Axe: A Caribbean Journal of Criticism*, no. 28 (March 2009): 1–18. For an overview of Obeah laws in the British Caribbean, see Handler and Bilby, *Enacting Power*.

34. "Kingston Gleaner Newspaper Archives | Jan 26, 1866, p. 1," accessed July 16, 2019, https://gleaner.newspaperarchive.com/kingston-gleaner/1866-01-26/; "Kingston Gleaner Newspaper Archives | Apr 05, 1866, p. 3," accessed July 16, 2019, https://newspaperarchive.com/kingston-gleaner-apr-05-1866-p-3/; "Kingston Gleaner Newspaper Archives | Oct 23, 1868, p. 3," accessed July 16, 2019, https://newspaperarchive.com/kingston-gleaner-oct-23-1868-p-3/.

35. William Lauron DeLaurence, a white man born in Cleveland, Ohio, gained fame as a professional hypnotist and magnetic healer in the late nineteenth century. He published several books that became widely popular in the Anglophone Caribbean. Donald W. Hogg, "Magic and 'Science' in Jamaica," *Caribbean Studies* 1, no. 2 (1961): 1–5; Palmié, *Wizards and Scientists*, 207; W. F. Elkins, "William Lauron DeLaurence and Jamaican Folk Religion," *Folklore* 97, no. 2 (1986): 216. The anthropologist Donald Hogg, one of the first to write about Science and Obeah, suggested that Science was linked to "illegal occultist books smuggled into the island from the United States." Hogg, "Magic and 'Science' in Jamaica," 1. See also Paton, *The Cultural Politics of Obeah*, 232–237.

36. *The Gleaner*, May 19, 1917.

37. "Ex-constable Alleged Obeahist. Young Woman as 'Medium.' Startling Revelations in Court," *Port of Spain Gazette*, August 18, 1922, https://www.caribbeanreligioustrials.org/CaseSource/Details/1166.

38. "Obeah Case Continued. Mesmerism Not Obeah, Counsel Argues. Counsel Upholds Science," *Port of Spain Gazette*, August 26, 1922. For a more in-depth analysis of this case, see Paton, *The Cultural Politics of Obeah*, 238–239.

39. The majority of my research for this article was done during pandemics (both Zika and Covid), which significantly affected my ability to travel to archives in Jamaica for several years.

40. Archibald Cooper Papers, West Indies and Special Collections, University of the West Indies Mona. The boxes are not numbered, but several references to science can be found in the light gray box with a white label w/ a red border. See, for example, folder 1 labeled "Duppies" and p. 86 in the large blue folder, dated April 6, 1939.

41. Archibald Cooper's papers were donated to the University of the West Indies Mona campus by Cooper's brother-in-law, the anthropologist McKim Marriot in 1975. UWI

Mona librarian Kenneth Ingram acquired them and anthropologist Barbara Kopytoff helped to organize the papers. The anthropologist Kenneth Bilby has donated Barbara Kopytoff's field research, including a copy of Cooper's field notes, to the National Anthropological Archives at the Smithsonian, but they have not been processed and are thus inaccessible to scholars as for 2022 (when this chapter went to press). As a result, the UWI Mona West Indies & Special Collections remains the only place to consult Cooper's records. The WISC has identified the Cooper papers as a priority for digitization, and they will likely be available online in the near future. I am grateful to Kenneth Bilby for helping me to track down Cooper's papers, to Ms. Yulande Lindsay at the UWI Mona West Indies and Special Collections, and to James Robertson for his insights into the Archibald Cooper papers

42. Kenneth Bilby, preface to an edited volume of Barbara Kopytoff's writings on the Jamaican Maroons. The volume was never published, but Bilby's preface provides a very helpful overview of anthropological work about the Jamaican Maroons in the twentieth century. I am grateful to Kenneth Bilby for giving me permission to cite the preface and to Jerome Handler for sending me a copy.

43. Herskovits corresponded with Col. Rowe after his return, and Col. Rowe thanked him for the photographs he sent. The letters are located in Herskovits's file about his "Trip to Haiti (1933–1935)," Melville J. Herskovits (1895–1963) Papers, box 8, folder 22, Northwestern University Archives.

44. Dunham, who was pursuing an undergraduate degree in social anthropology at the University of Chicago, won a fellowship to study with Herskovits and performed fieldwork in Jamaica, Martinique, Trinidad, and Haiti. "Correspondence Between Katherine Dunham and Melville J. Herskovits," Melville J. Herskovits (1895–1963) Papers, box 7, folder 12, Northwestern University Archives; Dennis Wepman, "Dunham, Katherine (22 June 1909–21 May 2006)," in *American National Biography*, April 1, 2016.

45. Katherine Dunham, *Journey to Accompong* (New York: Henry Holt, 1946), 41.

46. Dunham, *Journey to Accompong*, 41, 53.

47. Dunham, *Journey to Accompong*, 149.

48. Dunham, *Journey to Accompong*, 149.

49. Hurston felt that Dunham was not prepared for fieldwork and derided her conclusions in letters to Herskovits.

50. Herskovits tried to dissuade Hurston from visiting Jamaica and Haiti and urged her to spend more time in the Bahamas, where he felt she had more of a "background." Melville J. Herskovits to Zora Neale Hurston, April 22, 1936, Melville J. Herskovits (1895–1963) Papers. box 9, folder 32, Northwestern University Archives.

51. Zora Neale Hurston to Melville J. Herskovits, April 6, 1937, Melville J. Herskovits (1895–1963) Papers, box 9, folder 32, Northwestern University Archives.

52. Zora Neale Hurston, *Tell My Horse: Voodoo and Life in Haiti and Jamaica* (New York: Harper Perennial, 2008), 27.

53. Hurston, *Tell My Horse*, 27.

54. "Ba Will on Spiritual Power," p. 215. Large blue folder, organized by Barbara Kopytoff. In the light gray box with a white label w/ a red border. Archibald Cooper, "Papers relating to the Maroons, 1938-1939." UWI Mona Library, West Indies and Special Collections.

55. "Obeah Men (inf. Sa Liz), June 10, p. 168. Large blue folder, organized by Barbara Kopytoff. In the light gray box with a white label w/ a red border. Archibald Cooper, "Papers relating to the Maroons, 1938-1939." UWI Mona Library, West Indies and Special Collections.
56. "Chas. Reid," April 6, 1939, p. 69. Large blue folder, organized by Barbara Kopytoff. In the light gray box with a white label w/ a red border. Archibald Cooper, "Papers relating to the Maroons, 1938-1939." UWI Mona Library, West Indies and Special Collections.
57. Zora Neale Hurston to Melville J. Herskovits, April 6, 1937. Box 9, Folder 32. Melville J. Herskovits (1895–1963) Papers. Northwestern University Archives.
58. Schiebinger, "Agnotology and Exotic Abortifacients" and *Plants and Empire*.
59. This is the case in one oral history about two brothers fleeing slavery. When one brother struggles to navigate through Back River during his escape, he is able to find his way through Science, in this case referring to his *jege*, which Bilby glosses as an "oracle bundle." His "Science [i.e., his *jege*] drop out of him pocket," explained Henry Shepherd (1991). At that moment, the river swiftly changed direction, creating a path for him. Bilby, *True-Born Maroons*, 108–109.
60. Caleb Anderson (1991), as transcribed in Bilby, *True-Born Maroons*.
61. Barbara Klamon Kopytoff, "Colonial Treaty as Sacred Charter of the Jamaican Maroons," *Ethnohistory* 26, no. 1 (1979): 45–64.
62. Sydney McDonald (September 4, 1978), as transcribed in Bilby, *True-Born Maroons*, 286–288.
63. Hardie Stanford (1991), as transcribed in Bilby, *True-Born Maroons*, 288.
64. Entry for March 27, 1755, *Diary of Brother Caries' Voyage to Jamaica and Jamaica Diary, Oct 1754–Dec 1755*, Moravian Church House, London (uncatalogued).
65. Kopytoff, "Religious Change Among the Jamaican Maroons," *Journal of Social History* 20, no. 3 (Spring 1987): 475–476, 484 n. 73.
66. After Emancipation, the colonial government repealed the treaties and stripped the Maroons of their previous privileges. For a reprinted copy of the 1844 act, see Kopytoff, "The Maroons of Jamaica," 384–391.
67. Bilby, *True-Born Maroons*, 7.

CHAPTER 15

OBEAH SIMPLIFIED? SCIENTISM, MAGIC, AND THE PROBLEM OF UNIVERSALS

J. BRENT CROSSON

MAGIC, SCIENCE, AND RELIGION: PARTIAL CONNECTIONS, PARTIAL INCOHERENCE

Science, according to the *Dictionary of the English/Creole of Trinidad and Tobago*, is the appeal to spiritual powers to achieve effects, ranging from attempts to influence the criminal justice system to efforts at resolving an affliction that biomedical doctors are unable to cure.[1] The term through which subaltern religious practices were criminalized as witchcraft or superstition in the Anglophone Caribbean—*obeah*—bears almost the same definition in this dictionary, and *obeah* and *science* have been synonyms for some time in the region. Some of the first British colonial accounts of obeah described it as a science that helped to motivate eighteenth-century slave rebellions—uprisings that led to obeah's initial criminalization.[2] Legally defined as any "assumption of supernatural power," obeah continues to be a crime in much of the Anglophone Caribbean, and science continues to be (among other things) a lexical equivalent for obeah in the region's English creoles.[3] This chapter starts from this seemingly strange lexical equivalence to defamiliarize some of the key "universals" through which Atlantic modernity was forged: magic, science, and religion.

As the Haitian anthropologist Michel-Rolph Trouillot argued, the West has been a project rather than a place, forged through the normative force of what he called "North Atlantic universals."[4] Science and religion are two important North Atlantic universals through which the normative force of this project has been enacted. Yet, as Jason Josephson-Storm argues, these two categories have been defined by additional terms that they both allegedly oppose:

superstition, magic, and *witchcraft*.⁵ These three words bear very different resonances, but they all point toward how the categories of religion and science have been defined by what they exclude.⁶ In popular discourse in Trinidad and the Anglophone Caribbean, the presumed universalisms of both religion (as prototypically Christian) and science (as the prototype for modern rationality) are often defined against obeah (as alleged superstition, atavistic magic, or harming witchcraft). This North Atlantic making of *religion* and *science* through a shared opposition to African-identified *magic* lays bare the moral-racial foundations of modernity.⁷

At the same time that science and African-identified superstition have opposed each other in this popular discourse about what it means to be modern, *science* and *obeah* are synonyms in the region. This equation of science and obeah diverges from any logic of opposition between categories of secular reason and religion or superstition. Obeah, in the region's English creoles, can be neither a romantic alternative to science nor its atavistic foe; it is science. Nor can this equation be reduced to an instrumental and legitimating move, through which science becomes a positive mask for those prone to negative accusations of African-identified obeah.⁸ Science, as we shall see, has been just as dangerous and morally ambivalent as obeah, and this is part of the basis for these two terms' lexical equivalence in the region. While this equivalence might seem to be a local perversion of North Atlantic universals, I argue that it has something more universal to say about the equation of magic and science in a broader picture of modernity. Instead of seeing magic and science as opposed, as others have assumed in their story of the West,⁹ how have these terms often doubled for each other in Western modernity? How has this doubling been informed by the ways that the boundaries of religion and race have been defined?

While doubling might signal uncanny semblance, it also signals disjuncture and difference. I argue that moving between obeah and science reveals divergent notions of the word *science* and its relation to a project of North Atlantic universalism. As projects of reform, North Atlantic universals can never be universal in practice. They are premised on the geo-racial assumption that the West can be universal, while the rest remain particular and in need of tutelage. This assumption has drawn the line between science and magic, science and ethno-science, or true religion and false superstition.¹⁰ Both science and religion (as North Atlantic universals) are premised on aspirations of universalism through transcendence of context. Transcending the determinations of tradition, kin, or gendered and racialized embodiment signals the liberation of modern man, even though such transcendence is (explicitly or implicitly) figured as the property of white males. Yet this provincial and impossible project of liberatory transcendence is only one particular way to think about the universal. Obeah/science opens up another way to theorize about universalization, foregrounding its dangers, limitations, and afflictions.

Obeah/science is thus a situated way of rethinking universals (and the question of universalization itself). To rethink such concerns, however, it remains necessary to delve into some particulars. Otherwise, we risk lapsing back into a dream of universal coherence that the words *magic*, *science*, and *religion* have often sought to convey in Western anthropological discourses. As the case studies of this chapter show, the situated translation of obeah, science, and magic around the Atlantic involved discontinuities, partial incoherence, and ontological divergences. In this project of situated (mis)translation, obeah/science was taken up by those typically called "Western esotericists" (even though these Westerners were usually interested in adopting practices and identities that were marked as non-Western).

Rather than assuming that Caribbean spiritual workers simply appropriated a legitimating discourse of "spiritual science" from these "Western esotericists" (whom they often read), I travel in the opposite direction to show how the translation of obeah/science into a Western discourse of "scientific magic" reveals different understandings of what scientific practice entails. Simplifying a more complicated situation, I contrast the scientism of Western esoteric and anthropological discourses of magic-as-science with the dangerous powers of obeah/science. Scientism, in brief, denotes an ideology of universalism, truth, rationality, and transcendence bound up with the reformatory project of the West.[11] It is actually impossible to fully disentangle scientism from the heterogeneous practice of the divergent endeavors that get called science, but it is a distinction that I think is essential to any conversation about the categories magic, science, and religion. I start with some rather brief ethnographic context on obeah before focusing on the ways that one of the best known "Western esotericists" (Aleister Crowley) and a little-known Scottish theosophist (who adopted several Afro-Caribbean pseudonyms) took up obeah (and scientism) within their own projects of scientific magic. I then conclude by looking at the ways that the cross-Atlantic travels of obeah redefine both the question of universals and the relationship among religion, science, and magic.

SCIENCE/OBEAH/SPIRITUAL WORK

For the past decade and a half, I have done research with spiritual workers in a region of southern Trinidad that I call Rio Moro.[12] *Spiritual work* is the more neutral term for practices of problem solving that are popularly known (and stigmatized) as obeah. Since obeah simultaneously refers to spiritual power and the potential to misuse such powers, it is both an empowering and a stigmatizing term in the Anglophone Caribbean. The region of Rio Moro itself has a strong relationship with these powers and stigmas of obeah. While Trinidad is popularly known as an industrialized urban society with a vibrant carnival and a high murder rate, this rural region has often been detached from national visions of modern development through its association with obeah.

When I arrived there in 2010 for a year and a half of residency and research, the national media had latched onto a series of "mass possessions" at Rio Moro's secondary school as a sensational example of what articles called the region's affinity for "African witchcraft" or "obeah." Starting in 2010 and continuing for two consecutive years in the weeks preceding All Souls' Day, between ten and thirty female pupils enacted these possessions, closing down the school on a number of occasions. These students yelled obscenities, spoke in men's voices, broke the straps of ambulance stretchers, and resisted police officers with a force that seemed to exceed their thirteen- and fourteen-year-old frames. Similar disturbances had, in fact, happened at other schools in the nation (and across the world), but the secondary school's location in Rio Moro, a rural area popularly marked as the national epicenter of obeah, ensured that these incidents became a media spectacle, pitting discourses of both scientific rationality and born-again Christianity against the spectral forces that supposedly existed at the limits of the island's petroleum-fueled modernity. Alongside neo-Pentecostal "spiritual warfare" directed against obeah, advocates of "science" insisted on a natural or medical explanation for the events to dispel the allegedly superstitious beliefs of the region. Proponents of religion and science, while diverging in their opinions on the reality of spiritual forces, both wanted to reform Rio Moro by severing it from the alleged atavisms of "obeah" and "witchcraft."

While Rio Moro's "demons" provoked a familiar opposition between "African witchcraft" and either science or (Christian) religion, residents of Rio Moro often had a different story to tell about the events at the school. On the one hand, my interlocutors were painfully aware of the stigmas of superstition that they bore in this debate about the limits of rationality and moral community in Trinidad, stigmas that had justified Rio Moro's exclusion from national programs of development and narratives of economic progress. On the other, they often had a story to tell that did not sit easily on either side of the apparent battle between "African superstition" and scientific rationality or Christian moral reform. It was a story encapsulated by the word *science*.

Thus, when I asked Ashok, the father of one of the afflicted pupils at Rio Moro Secondary, what had caused the disturbances at the school, he told me matter-of-factly "science." My own image of science at that time was far removed from spiritual harm, and I asked him how science could possibly cause the afflictions at Rio Moro Secondary. "It have both good and bad science," he said by way of explanation. Science could perform beneficial works *and* harm, Ashok avowed. Instead of dispelling the occult forces of superstition, science actually represented the ambivalence of these powers.

Like the spiritual workers I knew in Rio Moro, Ashok asserted that science could index morally ambivalent spiritual powers—a characterization that unsettled my own preconceptions of a science constituted by the divide between matter and spirit, nature and supernature, facts and fetishes, or the benign pursuit of knowledge and malevolent intentions. As I talked about science with my interlocutors over the course of the next years, my "epistemic disconcertment"[13] at the equation of obeah and science led me to ask what the events at Rio Moro Secondary could say about the omnipresent adversaries of secular modernity—science and religion—that structured projects of reform at the school. While often opposed in contemporary narratives, science and religion both attained coherence through their opposition to alleged obeah, revealing the ways that North Atlantic universals take shape over and against supposed magic or superstition across a wide variety of contexts.[14] What concerns me more in this chapter, however, is the ways that categories of magic and science have doubled for each other. Yet this doubling might happen in divergent ways, revealing the marked heterogeneity that the terms *science* and *religion* conceal.

Elsewhere I have discussed the complex issues of gender, national sovereignty, neo-Pentecostal spiritual warfare, and modernization theory that the possessions at Rio Moro Secondary invoked.[15] In this chapter, I cannot go into depth on these issues, nor can I detail the ethnographic complexities of the school possessions. These possessions, in the context of the argument here, are one example of how

OBEAH SIMPLIFIED? SCIENTISM, MAGIC, AND THE PROBLEM 353

science is equated with obeah and endowed with potentially afflicting or dangerous powers in Rio Moro. Much of my other work is about understanding how obeah's association with affliction and retributive justice, particularly in response to police brutality and the violence of law, helps us to rethink the assumptions underlying pervasive understandings of religion.[16] Here, my focus is on how the association of science with the potentially harming power of obeah unsettles the assumed unity of the terms *magic*, *science*, and *religion* as the "trinary"[17] that founds Western modernity.[18]

OBEAH, SCIENCE, AND SCIENTISM

While the allegedly universal truths of natural science are so often pitted against the supposedly relativistic and otherworldly beliefs of religion in contemporary narratives of the "battle" or "war" between science and religion, this is only one way of conceiving of this relationship in North Atlantic modernity.[19] At different points in time (including the contemporary New Age), the term *science* has been particularly important in describing realms that would probably, for many observers, be religious, spiritual, or supernatural. In the parliamentary reports on the first obeah trials that followed the largest slave rebellion of the eighteenth-century British Caribbean, obeah was also described as a "science." Leaders of this 1760 uprising in Jamaica (usually known as Tacky's Rebellion) were identified by colonial officials as obeah practitioners who had administered oaths that bound rebels to secrecy and manufactured powders that might bring rebels luck and protection in battle. After the capture of these leaders, colonial officials were intent on trumping the imputed power—the obi or obeah—of their captives through what colonial observers called "experiments."[20] Electricity was the current scientific/magical fascination, and authorities subjected the alleged obeah practitioners to electrical shocks while projecting images on their bodies with "magic lanterns" (a kind of early form of slide projection). According to the Jamaican planter and historian Bryan Edwards, one of the leaders of the slave rebellion remarked that "his master's obi exceeded his own" after receiving particularly powerful shocks in these experiments.[21]

Rather than making European science into the evolutionary opposite of African superstition, as so many narratives of North Atlantic modernity would have it, these electrical experiments forged an uncanny relation of similitude between

European technology and (European conceptions of) obeah. This uncanny resemblance was borne out in the descriptions of obeah in the reports to Parliament on the conditions of slavery in the eighteenth century: "The professors of Obi ... have brought the science with them from thence to Jamaica, where it is so universally practiced, that we believe there are few of the larger Estates possessing native Africans, which have not one or more of them."[22] While such a description could be written off as an archaic eighteenth-century use of the word *science* to mean any kind of knowledge, this equation of science with obeah has endured in various forms to the present day. From the start of obeah's centuries-long criminalization in the Caribbean, African-identified obeah doubled for science (even as these electrical experiments tried to prove that it was an inferior science).

This equation of obeah and science, however, is contradicted by a powerful and pervasive narrative about science. As this word began to take on the meanings with which we are familiar in the course of the nineteenth century, scholars would place African fetishism at the very bottom of developmental frameworks, with another newly coined term (*scientist*) superseding Christian monotheism as the pinnacle of modern reason.[23] The year of Tacky's Rebellion, 1760, was also the year of the publication of Charles de Brosses's *Du culte des dieux fétiches*, which coined the new ism—*fetishism*—to define the kind of not-religion that supposedly characterized the entire continent of Africa.[24] In the nineteenth century, this ism would often play the opposite of another new ism— *scientism*. In this play of foils, science was thus pitted (and defined) against alleged "Black magic," charged with a project of reform in nineteenth-century evolutionary discourses. As one widely republished 1892 newspaper editorial entitled "How to Kill Obeah" asserted, "The rise of natural science in Europe and the fall of [African] fetishism are only two ways of looking at the same thing."[25] The best medicines against the Caribbean fetishism of obeah, the author asserted, were the plain and simple "natural facts" of science. In this view, science was a truth-making discourse, demonstrating facts and discarding backward superstitions.

As a triumphal program of progress and reform that reveals facts, the word *science* has long exerted the force of a North Atlantic universal.[26] It is necessary, however, to make a tentative distinction between such popular, triumphal discourses and the actual practice of a heterogeneous array of sciences, which can use very different methods that are not necessarily concerned with visible facts. During my period of long-term field research, for example, many of my Trinidadian interlocutors in Rio Moro were employed as manual laborers on a seismic petroleum survey that quite clearly involved a less-than-transparent practice of science. The contours of subsoil formations under my interlocutors' feet were visible to no one, including the geologists and geophysicists who sat in offices in the capital or in a heavily guarded air-conditioned computer trailer at my field site. These scientists remained as invisible to people in Rio Moro as the geological formations that the survey was trying to "see." While these scientists were literally hidden from view, their actual practices of mapping subsoil formations were no less straightforward. In new onshore petroleum fields, the predictions of geologists are wrong 75 to 80 percent of the time (a failure rate that rises to 85 to 90 percent for new deep-water surveys), resulting in the drilling of dry holes.[27] Relying on the reflected shock waves produced by buried dynamite and air cannons, these scientists performed the highly mediated task of transducing sound waves into visual images. Depending on different months-long ordeals of computer processing or the interpretations of different geologists, these surveys could produce divergent representations of what lay beneath the ground from the same seismic data. But these representations were all dependent on the local laborers who circumnavigated mined marijuana fields to lay the onshore seismic cables in Rio Moro or on the tense company negotiations with the unions of the fishermen whose boats would provide transport for offshore surveys there.

As was quite clear to the residents of Rio Moro who worked on these surveys, science was a less-than-transparent pursuit that depended on transnational relations of capital, the highly mediated sensing of invisible phenomena, local knowledge of terrain, and negotiations with labor.[28] While so often invoked in the singular, science denoted a heterogeneous range of methods and epistemologies in Rio Moro, referring to the divergent practices of school psychologists, petroleum geologists, geophysicists, and spiritual workers. Yet none of these sciences was simply about the demonstration of readily perceptible facts.[29] Broadly speaking, there were thus two ways of talking about science that exerted power in Rio Moro. The first discourse saw science as a transparent practice that replaced occult forces with facts (this was the kind of science that promised to dispel superstition and reveal the natural causes of the disturbances at Rio Moro's school). The second discourse insisted that science was a heterogeneous and occult practice, which included (what got called) obeah.

Philosophers of science have made a somewhat analogous distinction between two ways of talking about science. On the one hand, sciences are heterogeneous practices of experimentation that seek to produce novel conditions in which generally invisible or hard-to-perceive forces (such as subsoil fluids or electrons) can take shape as phenomena that seem to speak and act for themselves (even as specialists speak for or manipulate them in highly mediated ways). The philosopher of science Isabelle Stengers calls this experimental encounter with unknown or hard-to-perceive forces the "adventure of sciences," contrasting this adventure with the great narrative of Science (with a capital "S") disenchanting non-Western or premodern worlds.[30] The ideology (or as the philosopher of science Imre Lakatos suggested, the theology) of Science is closely related to what other scholars have called scientism—a popular discourse about Science as a reformatory organ of truth.[31] This scientistic discourse, as Peter Gottschalk notes, was particularly forceful in colonial settings.[32] In a similar way, one might say, an idealized notion of Religion (with a capital "R"), defined as the universal essence of human morality founded in an immaterial disposition of belief, exerted a potent reformatory force in colonial efforts at criminalizing and regulating subaltern religious practices.[33] During the secondary school possessions, these notions of Science and Religion were the primary North Atlantic universals through which the (in)adequacy of Rio Moro and its students was articulated.

What often goes unnoticed in this scientistic narrative, however, is the way that this notion of Science has often been the double of magic in modern Western representations rather than its other. Ostensibly, this doubling has revolved around attempts to denigrate and/or legitimize practices typically considered magic in Western genealogies of religion. Indeed, a key trope in Western anthropologists' representations of non-Western cultures has been that their magic is like Western science, making it rational, instrumental, or materialistic (but ultimately inferior to modern science). Non-Western or premodern magic was, in James George Frazer's famous turns of phrase, "false" or "bastard" science—a kind of primitive precursor to the modern European scientific endeavor.[34]

These foundational notions in anthropology sparked the protracted and multidisciplinary rationality debate, which I delve into elsewhere.[35] Considering anthropology's entanglement with Western racial theory, it seems no coincidence that this debate centered on Western anthropologists' representations of African-identified "magic" or "witchcraft." Twentieth-century anthropologists insisted that these practices were rational or scientific to the extent that they sought to uncover the hidden forces responsible for empirical social phenomena. However, these anthropologists stated that these African practices failed to engage in the open-ended, empirical experimentation that was the hallmark of Western science, instead relying on the rote explanations of received tradition.[36] For others, African magic was also unlike Western science in that science could *predict* phenomena using the mechanistic laws of an impersonal nature, while magic could seek only to *influence* empirical reality in line with personal ends.[37] Ultimately failing to attain the transcendence of Western science, African magic was also unlike Religion by virtue of these attempts at manipulating the world in many accounts. The hidden forces of religion proper (i.e., gods) were ultimately devoted to the elaboration of a moral order, whereas African magic tried to instrumentally influence the order of the world.[38] Not moral enough to be Religion and too interested or subjective to be Science, racial tropes of African magic were the constitutive others of these universal Western domains.

In my other work, I have questioned the conflation of religion with moral order, which a wide variety of people have selectively employed to elaborate a racist foil of immoral "Black magic."[39] In this chapter, however, I want to critique the notion of Science that also made African magic not-religion (and a bastard double of Science). More recent accounts of experimentation stemming from advances in quantum physics have suggested that scientists cannot stand apart from a law-governed nature, simply predicting outcomes without influencing the world around them. Experimentation with subatomic phenomena (according to one widespread interpretation in quantum physics) has shown that every measurement necessarily involves a transfer of energy that is transformative.[40] Scientists, in other words, interfere in the world, rather than simply peering at phenomena from an objective distance.[41] Experimentation is thus an ethically fraught act, even if one experiments only with seemingly innocuous nonhuman particles (the atomic bomb being the ultimate realization of this fraughtness).[42] Feminist science studies scholars have also roundly critiqued the idea of the objective scientist, divorced from the interests of power or interference in the world, as a masculinist colonial fantasy.[43] In short, Western anthropologists have relied on *scientistic* tropes of objectivity or rationality to make "African magic" into a (less evolved) double of Science. The doubling of science and obeah in Rio Moro, however, suggested something different. Before delving into how obeah/science diverges from the ideology of scientism, it is necessary to explore another line of equivalency between magic and science in Western modernity, one in which obeah itself plays a role.

THE SCIENTISM OF OCCULT SCIENCE

Science has been used not just to render African-identified or non-Western practices as morally inferior to religion and evolutionarily inferior (yet analogous) to science. When wielded by devotees of "Western esotericism" to refer to their own practices of avowed magic, spiritualism, or spirituality, science has been used to confer legitimacy, universality, rationality, and (in many cases) superiority to dogmatic (Christian) religion. In contrast with the ostensibly legitimizing, yet condescending tone of early anthropological accounts of non-Western practices, movements that scholars have (mis)placed under the heading of "Western esotericism" have equated science and magic to argue that spirit channeling and ceremonial magic are progressive, rational, and/or evolutionarily superior practices.[44] Nineteenth-century spiritualists and spiritists claimed to provide a scientific basis for the afterlife or spiritual evolution, and leaders of the Theosophical Society enamored with the "mysteries of the East" claimed to have uncovered a universal "spiritual science" that was superior to most forms of organized religion.[45] Emerging from (and breaking with) the Order of the Golden Dawn, the best-known ceremonial magician of the twentieth century (Aleister Crowley) claimed to be elaborating a purely scientific form of magic, scoffing at the intellectually inferior and superstitious approaches of magicians that preceded him in Victorian England.[46]

These legitimizing invocations of occult or spiritual science were not limited to what has been narrowly conceived of as the West. In Brazil and Cuba, a spiritist language of science has helped to establish class boundaries, racial hierarchies, and claims to authoritative rationality.[47] In Iran, movements of spiritual seeking and occultism have embraced science to define their projects as attempts at "cleansing metaphysical knowledge of superstition" or "rationaliz[ing] the unseen."[48] Scholarly accounts of esoteric or metaphysical movements across the world would thus seem to agree that science is an authorizing discourse associated with rationality and opposed to superstition. Moreover, the first known text written by an avowed obeah practitioner—*Obeah Simplified, the True Wanga* (1895)—speaks about obeah in such legitimizing terms as a science. Echoing theosophical reformatory discourses of occult scientism, this author (known only by the pseudonym Myal Djumboh Cassecanarie in the text) claims to purify "true" obeah of superstition, material fetishes, and charlatanry. It would only seem logical that

Caribbean spiritual workers would follow this post-eighteenth-century global trend in defining their practices as rational, advanced, and nonsuperstitious by calling them science. However, as I will argue, these widespread authorizing discourses (or scholarly interpretations) of occult or spiritual science yield a profound misunderstanding of many subaltern conceptions of obeah/science in the Caribbean.

Obeah Simplified, the True Wanga, was published in 1895 by the offices of *The Mirror*, a relatively short-lived newspaper in Victorian Trinidad, and the text refers to obeah not as religion but as science. Subtitled *A Scientific Treatise*, the book promises in the opening pages to present "an introduction and aide-memoire to the science" of what the author calls Obeah-Wanga (*wanga* is a seldom-used synonym for *obeah* in Rio Moro today). The author bemoans the criminalization of obeah and avows that alongside the "politico-social education" much advanced in the schools of the day, parents "should take great care that the aesthetics and true inwardliness of Wanga are inculcated upon their children of both sexes."

The author of the book is listed as Dr. Myal Djumboh Cassecanarie, a "Professor of Pneumatics in the Tchanga-Wanga University," consulting "quimboiseur" to the ex-president of Haiti, and "Member of the Principal West Indian and West African Scientific Societies." The author's name and credentials present an overtly transisland and transregional African identity. *Myal* is a term of Jamaican provenance for an African religious practice, *quimbois* is a rough French synonym for obeah still used in Martinique and Guadeloupe, and *Cassecanarie* is probably derived from the Haitian Vodou festival Casse Canarie.[49] The author was most likely active in transnational Theosophical societies, for a version of *Obeah Simplified* appeared in serial form a few years earlier, in 1891, in the journal of the International Theosophical Society (*The Theosophist*), founded by Westerners but headquartered in southern India, with the author listed as "Miad Hoyora Kora-Hon, F.T.S. [Fellow of the Theosophical Society]." From 1884, Kora-Hon was listed as a regular contributor to *The Theosophist*, writing articles on the "mystic lore" of the "kolarian tribes" of eastern India or on alchemy in Ireland. In this serial form, it becomes clear from the author's self-presentation that he is a foreigner in the West Indies who owned land and spent extended periods of time there. Christopher Josiffe speculates from the author's presentation as both insider and outsider to European society that he may have been from India, as the Theosophical Society counted many South Asians as members and Trinidad was home to a large number of East Indian "immigrants"[50] (although most of them were indentured laborers, not relatively privileged travelers). Stephan Palmié labels Cassecanarie as Trinidadian and groups him with other esotericists of West Indian and African descent, such as the "first Rasta" Leonard Howell, presumably because Cassecanarie identifies as an avowed obeah practitioner.[51] However, after painstaking research, Alexander Rocklin has found that the author was a Scottish member of the recently founded Theosophical Society who owned a plantation in Tobago (Trinidad's neighboring island).[52]

This author's use of *science* to describe occult or spiritual practices and his sarcastic hostility toward Christian religion are both characteristic of the Theosophical Society's membership. The cofounder of the Theosophical Society Henry Steel Olcott was a central early proponent of "scientific Buddhism," pioneering what has been a particularly persistent campaign to align Buddhism with the tenets of liberal Euro-American society.[53] The Theosophical Society espoused the belief that there was a universal occult science that formed the basis for all religious traditions and that Buddhism represented the purest extant emanation of such science. Although Olcott drew more on orientalist scholarship than on ancient texts and often saw contemporary popular Buddhism as a degraded practice, he converted to Buddhism and joined Sri Lankan religious leaders in attempting to reform contemporary Buddhist practice on the island into a "Protestant" or "scientific" Buddhism.[54] Far from an idiosyncratic invention, the use of *science* to describe obeah in the late nineteenth century was part of a much wider trend, in which European natural scientists, Sri Lankan Buddhist reformers, Hindu reformers (i.e., Arya Samajis), theosophists, and spiritualist mediums were all using *science* to describe and authenticate their practices as rational (and, as was often the case, as opposed to both superstition and dogmatic Christianity).[55]

A key prop of such discourses on occult or spiritual science was often the rejection of supernaturalism in favor of materialist or empirical explanations. As Victorian physicists exposed a whole array of new occult forces—from x-rays to electrons—and as psychical researchers attempted to experimentally prove (or disprove) the existence of spirits in laboratories, so, too, Theosophists insisted on natural, empirical forces that formed the basis for a kind of universal spiritual science.[56] Even when Olcott went on a Buddhist faith-healing tour of Sri Lanka to compete with Catholic miracle workers who were attracting converts, he insisted that he was using natural forces (which for him included auras and animal magnetism) to enact his healing.[57] Similarly, in *Obeah Simplified*, the author gives "scientific" explanations for seemingly miraculous accounts of the power of Obeah-Wanga (i.e., obeah "spells"), by which people leave their skin and fly, disappear altogether, or protect their gardens from thieves by means of rods that turn into serpents. The author acknowledges that such tales must appear supernatural fancy but insists that they are "natural" occurrences:

> The most thorough paced sceptic when he has finished this work may laugh at its contents, and *say* its statements are impossibilities, but he will never be able to conclude to his perfect mental satisfaction that such things have never happened, or that they can never again happen, and the only conclusion he *can* count upon with perfect certainty and cocksuredness, is—*that he does not know*. . . . The extra pious will probably regret that such things were published. . . . To students who are able to think of what they have read, and to subject their conclusions to analysis, it is only necessary to say, that no matter how ludicrous, how horrible, or improbable a picture may be the powers which are the cause of the acts depicted are neither good nor evil in themselves for they cannot but be natural.[58]

For the Theosophist author, the natural power in question was that of "the natural magical power of sound energized by the concentrated will," or what Cassecanarie calls "the true Wanga" or "the spell."[59] For spiritual workers in Rio Moro, *wanga* (as noted above) is a seldom-used synonym for *obeah*, and I have never heard it used in Trinidad outside of Rio Moro. In Haiti, *wanga* is a common word for a ritually prepared charm, bottle, or packet that congeals spiritual protection and power. For Cassecanarie, however, wanga is a "sound formula," and he proclaims that anyone who uses material objects—"bottles, rags, or other rubbish"—as a part of their practice is not a true obeah practitioner, but a "rascal" or "an impostor."[60] In the final chapter, Cassecanarie (with shades of De Brosses's polemics on fetishism) provides anecdotes about "archaic" European witchcraft to show that these outmoded practices used objects—wax figurines or dolls—that rendered them more akin to "fetishism" than to the science of "the true Obeah-Wanga."

Cassecanarie closes his "scientific treatise" by stating that his purpose has been to educate readers about proper obeah—or what he calls the "true Wanga"—and if his work leads the public to discern between impostors who use material objects and those true practitioners who rely primarily on words or spells, then he will consider that "the purpose of this pamphlet will have been accomplished."[61]

The science, then, is the natural power of sound formula, and other material practices or objects are purely superfluous (or superstitious). This notion would help to inspire the attempts of the best-known twentieth-century occultist—Aleister Crowley—to scientize magic. He incorporated Wanga and Obeah from his readings of Cassecanarie's text to provide one inspiration for his scientific "magick" in *The Book of Law* and his other writings on scientific sound formulas (spells) and concentrations of will.[62] Anti-Christian polemics aside, Cassecanarie's distinction between true and false obeah veered curiously close to post-Reformation distinctions between good and bad religion, with the former emphasizing words over material practices. Certainly, the idea that the true Wanga reproduced this crypto-Protestant bias on "good religion" would probably make Cassecanarie or Crowley turn in his grave.

As should be clear, modern discourses on occult science or scientific spirituality, such as those propagated by Theosophists, were using science to self-consciously work against Victorian notions of (Christian) religion. The notion of science employed, however, was indebted to narratives of scientism and, despite vehement anti-Christian rhetoric, to crypto-Protestant or post-Enlightenment notions of true religion. While actual spiritual workers have used material objects heavily in their practices, Cassecanarie's move from matter to word enacts a peculiarly Protestant purification of obeah. Similarly, in Henry Steel Olcott and Anagārika Dharmapāla's attempts to reform Sri Lankan Buddhism, *scientific* and *Protestant* were used more or less interchangeably to distinguish the right kind of Buddhism from alleged popular misapprehensions and superstitions.[63] Despite their invectives against closed-minded Christianity and their enthusiasm for occult science, Theosophist discourses on science were intertwined with attempts to reform non-Christian religions, which unwittingly recapitulated the very foundation of Protestant reform in antisuperstition invectives that condemned the use of material objects in favor of the word.[64] Like many of the responses to the possessions at Rio Moro's secondary school, these invectives promised to replace "hocus-pocus" with facts readily accessible to reasoning minds. While Crowley juxtaposed "the facts of [his own] occult science" with the "unscholarly hotch-potch" of the Theosophical Society cofounder Madame Blavatsky, his aim of a truly universal science, uniting the facts of psychology, physics, and "magick," was entirely in line with Theosophical discourse.[65] It makes a lot of sense (or a lot of science) that Crowley would have used Cassecanarie's Theosophical true Wanga as a key inspiration for his universal magick.

These projects of occult science were heavily indebted to notions of science as an authoritative North Atlantic universal that revealed transparent facts and universal formulas. *Obeah Simplified, the True Wanga* purported in its subtitle to be "a scientific but plain treatise" that revealed the natural truths of "what [obeah] really is"—one instance of the universal art of sound formulas. My interlocutors in Rio Moro, however, saw science not only as a program of transparent or naturalistic understanding but also as a morally ambivalent practice that was necessarily occult (in the sense of never fully revealed). Rather than representing the exposed truth of Afro-Caribbean spiritual work, Cassecanarie's elaboration of the true Wanga is a Theosophical polemic that is at odds with my own interlocutors' conceptions of science and obeah.[66]

While it is tempting to simply ascribe my interlocutors' notions of science to a mimicry of popular Victorian invocations of spiritual science that circulated via mail-order esoteric book catalogs in the Caribbean (texts that were, in some cases, also being read by many European Theosophists), it is necessary to look more closely at what practitioners of African-identified religions in Trinidad mean by science. Scholars have interpreted *science* in Caribbean spiritual work to refer primarily to imported, European "book magic," separable from spiritual workers' "local" or "African" practices.[67] Apart from denying spiritual workers their own agency, creativity, and intellectual capacities, these Eurocentric discourses of spiritual science misrepresent what *science* meant for my subaltern interlocutors.

As my interlocutors repeatedly made clear, these esoteric books, whose original provenance they often traced back to Egypt and Ethiopia rather than western Europe, were only a part of the story.[68] Effective science required embodied knowledge and unquantifiable capacities, and these books provided only recipes. According to idealized visions of replicability in science, experimental recipes should produce more or less unwavering results. For my interlocutors, however, science was dangerous precisely because it was not disembodied and universally replicable. In the hands of a bad cook, such recipes could go horribly awry. To use the metaphor employed by one of my interlocutors, efficacious science was like "sweet hand"—a colloquial phrase that referred to the embodied, idiosyncratic know-how that made one a good cook. Like the phrase "green thumb," sweet hand could include rote knowledge gleaned from books on cooking or gardening, but these phrases also expressed an embodied capacity that surpassed (or preceded) codified procedures. Rather than a simple importation of North Atlantic discourses of scientism (whether occult or natural), my interlocutors spoke of a science that began where rote formulas and transparent explanations were not completely sufficient for lived contexts of power. Science did not signal transparent facts or rational advancement, and it was not simply a legitimizing discourse. It was an ethically fraught process of accessing hard-to-perceive powers that was both potent and dangerous. This was why science, rather than an antisuperstition project, could be the cause of the afflictions at the secondary school for my interlocutors in Rio Moro.

CONCLUSION: MAGIC, SCIENCE, AND RELIGION (AGAIN)

The mass possessions at Rio Moro Secondary provided ample evidence of modernity as a project of reform in which either science or (Christian) religion promised to dispel superstition and inaugurate a new age of transparency.[69] These discourses of science and religion both took obeah as their object of reform, showing how science and religion are often defined as antisuperstition projects in Western modernity. Yet science was not simply opposed to alleged superstition in Western genealogies. Science frequently doubled for practices that had been classified as superstitions in colonial settings, and this doubling has a part to play in the transnational history of the relationship among the realms denoted as science, religion, and magic.

One strong branch of this Western genealogy of partially equating science and magic is found in anthropological studies of non-Western cultures. These studies have allegedly questioned the biases of Western modernity by revaluing non-Western magic as rationality, technology, or natural science.[70] These liberal projects of inclusion have bestowed the label *rational* or *scientific* on the activities of non-Western traditions, provided that these activities involved "objective investigation" of "what the West would refer to as the natural world."[71] These liberal projects of inclusion have attempted to recognize *certain aspects* of non-Western practices as science, but none of them has fully questioned the very notion of scientific rationality (and nature) on which these projects ultimately rest. Tacking in a different, yet resonant direction, movements (mis)labeled with the umbrella term "Western esotericism" have often called practices of magic scientific as a basis for claims to universality, authority, and/or rationality. Ceremonial magicians spawned from the Order of the Golden Dawn, Theosophists, and Spiritualists have often been intent on touting their practices as science since the latter half of the nineteenth century. As I have tried to show, however, these anthropological and esoteric invocations of science-magic have often been intensely scientistic, bound to a certain reformatory discourse of science (and, albeit implicitly, true religion).

This chapter has been an attempt to grapple with the equation of obeah with science made by "Western esotericists," drawing out their divergences from my Anglophone Caribbean interlocutors' ethics of obeah-science. Despite important differences of race, class, and legal status, those identified as obeah practitioners and those who have come to be known as "Western esotericists" overlapped in a number of ways. They often read many of the same books and spoke about a science that included spiritual forces. Nevertheless, they often (but not always) diverged on ontological and epistemological assumptions about science. Western esoteric descriptions of obeah (and spiritual science more broadly) were frequently imbued with the ideology of scientism. Science conferred legitimacy, universality, rationality, and (at least in many manifestations) superiority to dogmatic religion. For my interlocutors in Trinidad, in contrast, science was not necessarily a legitimating discourse; it was a powerful, often inscrutable, and potentially afflicting force equated with a highly stigmatized term (*obeah*).

Recent work by philosopher-scientists and anthropologists of science suggest that Caribbean notions of obeah-science might even offer a more faithful description of scientific practices and ethics. As Bruno Latour reminds us, much of what we know as science is not about concrete matters of fact, transparent to the gaze of a human observer. Scientific practice, he insists, addresses the invisible (or, to rephrase him, the hard-to-perceive). In accessing the far-off galaxy or the infinitely small world of subatomic phenomena, science builds incredibly extended and fragile chains of mediation that move between radically different mediums.[72] Radio waves and electrical charge differentials become visual images through complex concatenations of human and nonhuman forces. This does not mean that these scientific mediations are false or inaccurate. It does, however, imply that scientific experimentation is a relational, often fragile, and ethically fraught exercise in accessing the imperceptible. Perhaps my interlocutors in Trinidad were more attentive to this notion of science than Theosophists, occultists, or British colonial officials bent on using science to prove the superiority of their practices. While diverging radically in the kinds of phenomena they address, Caribbean spiritual work might legitimately share more with the fraught "adventure" of natural sciences than with the notions of Religion and Science that have structured narratives of North Atlantic modernity.

This divergence from scientism, however, does not redeem either *science* or *obeah*, for it underscores the ethical ambivalence of these two terms and their potential danger. Elsewhere I have shown how spiritual workers in Rio Moro enacted partial translations between various Christian, Hindu, Muslim, or esoteric practices and powers to treat a religiously diverse clientele. For spiritual workers, *science* was a key term in speaking about the material resonances between practices from different traditions. Because of its materiality, science could encompass precisely those material practices that North Atlantic notions of true religion (and, as we saw, scientific magic) deemphasized in favor of word and an interior disposition of belief or will. Yet because material practices could be codified (as a series of instructions and ritual designs in grimoires, for example), this allowed a potentially universal dissemination of esoteric science. At Rio Moro Secondary, my interlocutors told me, such accessibility was exacerbated by the internet, where free copies of most grimoires were available. Because the school was the only place it was then possible to access the internet, it made sense that an outbreak of science would happen there. In this way, the aspirationally context-transcendent universality of science was an inherent part of its danger.

It is also important to note that this assertion of the danger of scientific universalism is not necessarily morally superior or more palatable for Western liberal sensibilities. Antivaccine activists arguably mobilize such characterizations of a dangerous science in the contemporary United States to rail against the scientific conspiracy of universal vaccination programs and Big Pharma. In these contexts, scientistic discourses equating scientific authorities with truth may become palatable responses for readers of this chapter (assuming they are not conspiracy theorists who view science as a dangerous, threatening practice). This observation is intended not to conflate privileged antivaxxers at the center of a global empire with Caribbean practitioners of obeah/science but to refute any tendency to reconstruct a homogenizing moral divide that vindicates science (and obeah) by separating them from scientism. Rather, the equation of obeah and science could insist on something else: that any practice or discourse of science depends on particular and embodied conditions of power. Instead of a stable difference or similarity among magic, science, and religion, such conditions of power often animate partial connections, uncanny doublings, unexpected alliances, and contradictions hidden within single words. This is the kind of universalism that obeah/science marshals—the omnipresent, yet situated predicament of experimental powers rather than the transcendence that universal definitions promise.

NOTES

1. Lise Winer, ed. *Dictionary of the English/Creole of Trinidad and Tobago: On Historical Principles* (Montreal: McGill-Queen's University Press, 2009).
2. See *Report of the Lords of the Committee of the Council Appointed for the Consideration of All Matters Relating to Trade and Foreign Plantation, Part 3: Treatment of Slaves in the West Indies, and All Circumstances Relating Thereto, Digested Under Certain Heads, Part III* (London: Parliament, 1789), 117, https://babel.hathitrust.org/cgi/pt?id=mdp.39015084394389;view=1up;seq=4. See also Bryan Edwards, *The History, Civil and Commercial of the British Colonies of the West Indies*, vol. 2 (1793–1801; repr., New York: Arno Press, 1972).
3. Obeah has been decriminalized in Anguilla (1980), Barbados (1998), Trinidad and Tobago (2000), and St. Lucia (2004) but remains illegal in much of the region. Recent calls for the decriminalization of obeah in Jamaica and Antigua and Barbuda have met with considerable opposition, which argues that obeah is sinful, anti-Christian, and potentially damaging to national welfare. See J. Brent Crosson, "The Impossibility of Liberal Secularism: Political Violence, Spirituality, and Not-Religion," *Method and Theory in the Study of Religion* 30, no. 1 (2018): 35–55. In 2013, when Jamaica removed flogging with a whip as a punishment for obeah in order to sign the U.N. Convention Against Torture, it left the criminal status of obeah untouched. See Diana Paton, *The Cultural Politics of Obeah* (Cambridge: Cambridge University Press, 2015). In Jamaica, however, there are ongoing efforts toward decriminalization, although they have achieved no lasting juridical success as of the writing of this chapter. See Crosson, "The Impossibility of Liberal Secularism."
4. Michel-Rolph Trouillot, "North Atlantic Universals: Analytical Fictions, 1942–1945," *South Atlantic Quarterly* 101, no. 4 (2002): 839–858.
5. Jason Ānanda Josephson-Storm, "The Superstition, Secularism, and Religion Trinary: Or Re-theorizing Secularism," *Method and Theory in the Study of Religion* 30, no. 1 (2018): 1–20, and *The Myth of Disenchantment: Magic, Modernity, and the Birth of the Human Sciences* (Chicago: University of Chicago Press, 2017).
6. These oppositions, as Josephson-Storm notes in "The Superstition, Secularism, and Religion Trinary," might be either negative or positive (with magic, for example, often representing a romantic alternative to the dogma of religion or the rationality of science). For Josephson-Storm, *superstition* refers to a negative opposition to religion and secularism (with the latter term including science). *Magic*, in contrast, refers to a positive opposition, in which magic is a romantic alternative. This is often the case. However, as I show in this chapter, magic has often functioned as a negative term for not-religion or as the marker of an evolutionarily and morally inferior precursor to religion, particularly in colonial settings. One might also distinguish between witchcraft, which often involves an acknowledgment of a practice with real effects, and superstition, increasingly used in

the nineteenth century and onward to call witchcraft, magic, or obeah deluded beliefs without reality. However, witchcraft is often inseparable from the negative resonances of superstition (or the romantic ones of magic) when used in practice.

7. J. Brent Crosson, *Experiments with Power: Obeah and the Remaking of Religion in Trinidad* (Chicago: University of Chicago Press, 2020).
8. There are, however, a few cases of science used in anti-obeah trials as part of a defense. Far more common were defenses that represented the practices of those accused of obeah as accepted Christian practices or as analogous to such practices.
9. See Josephson-Storm, *The Myth of Disenchantment*.
10. See Crosson, *Experiments with Power*.
11. See, for example, Peter Gottschalk, *Religion, Science, and Empire: Classifying Hinduism and Islam in British India* (New York: Oxford University Press, 2012).
12. See Crosson, *Experiments with Power*.
13. Helen Verran, *Science and an African Logic* (Chicago: University of Chicago Press, 2001).
14. For example, Crosson, "The Impossibility of Liberal Secularism"; and Josephson-Storm, "The Superstition, Secularism, and Religion Trinary."
15. Crosson, *Experiments with Power*; J. Brent Crosson, "Catching Power: Problems with Possession, Sovereignty, and African Religions in Trinidad," in "What Possessed You? Spirits, Property, and Political Sovereignty at the Limits of 'Possession,'" ed. J. Brent Crosson, special issue, *Ethnos* 84, no. 4 (2017): 546–556, https://doi.org/10.1080/00141844.2017.1401704.
16. Crosson, *Experiments with Power*, "The Impossibility of Liberal Secularism," and "Catching Power"; J. Brent Crosson, "Oil, Obeah, and Science," *Cosmologics*, 2016, http://cosmologicsmagazine.com/brent-crosson-oil-obeah-and-science/, and "What Obeah Does Do: Healing, Harming, and the Boundaries of Religion," *Journal of Africana Religions* 3, no. 2 (2015): 151–176.
17. Josephson-Storm, "The Superstition, Secularism, and Religion Trinary."
18. On this trinary, see Josephson-Storm, *The Myth of Disenchantment*. There is considerable variation in how the terms *magic*, *sorcery*, and *witchcraft* are used. E. E. Evans-Pritchard made hard distinctions among these three terms in *Witchcraft, Oracles and Magic Among the Azande* (Oxford: Oxford University Press, 1937). However, in popular and scholarly usage, these distinctions are hardly ever stable. What gets called obeah in the Caribbean encompasses, in different contexts, the functions that Evans-Pritchard separately assigned to magic, witchcraft, and sorcery. I use magic as an imperfect umbrella category to talk about these functions and their foundational separation from religion by scholars. See, for example, Emile Durkheim, *The Elementary Forms of Religious Life* (1912; repr., London: Allen and Unwin, 1964); and Sir James George Frazer, *The Golden Bough: A Study in Magic and Religion* (New York: Macmillan, 1922).
19. See, for example, Josephson-Storm, *The Myth of Disenchantment*; and Ronald Numbers, "Introduction," in *Galileo Goes to Jail and Other Myths About Science and Religion*, ed. Ronald L. Numbers (Cambridge, MA: Harvard University Press, 2010), 1–7.
20. *Report of the Lords*, 117. See also Edwards, *The History, Civil and Commercial of the British Colonies of the West Indies*.
21. *Report of the Lords*, 117.
22. *Report of the Lords*, 117.

23. The neologism *scientist* was coined in 1833 by the British polymath William Whewell (see Numbers, "Introduction," x). *Natural philosopher* was the term commonly used to refer to the European men that we now call scientists prior to the popularization of Whewell's neologism over the course of the nineteenth century. For Eurocentric evolutionary hierarchies of "magic," "religion," and "science," see G. W. F. Hegel, *The Philosophy of History* (1837; repr. [trans. J. Sibree], Mineola, NY: Dover, 1956); Auguste Comte, *The Positive Philosophy of Auguste Comte*, trans. Harriet Martineau (London: J. Chapman, 1853); Edward Burnett Tylor, *Primitive Culture: Researches Into the Development of Mythology, Philosophy, Religion, Art, and Custom* (London: J. Murray, 1871); Frazer, *The Golden Bough*.
24. Charles de Brosses, *Du culte des dieux fétiches, ou Parallèle de l'ancienne religion de l'Egypte avec la religion actuelle de Nigritie* (England: Westmead, Farnborough, Hants, 1970 [1760]).
25. This editorial, while reprinted in newspapers across the Caribbean, appears to have been originally published on December 14, 1891, in *The Gleaner* (Kingston, Jamaica) by an anonymous author.
26. Trouillot, "North Atlantic Universals."
27. Gina Weszkalnys, "Geology, Potentiality, Speculation: On the Indeterminacy of First Oil," *Cultural Anthropology* 30, no. 4 (2015): 611–639, https://doi.org/10.14506/ca30.4.08.
28. See Crosson, "Oil, Obeah, and Science."
29. See Latour for a refutation of the notion that science is concerned with visible facts. Instead, he argues that science is concerned with what is invisible—forces that are too far away, too small, or too slow to be perceived directly. Bruno Latour, *On the Modern Cult of the Factish Gods* (Durham, NC: Duke University Press, 2010).
30. Isabelle Stengers, "Reclaiming Animism," *e-flux*, no. 36 (2012), https://www.e-flux.com/journal/36/61245/reclaiming-animism/. It is worth noting that "adventure of sciences," while positive in Stengers' usage, could easily conjure the violence of scientific experimentation in colonial settings and the representation of the colony as frontier laboratory.
31. Lakatos speculates that in the wake of the destabilization of Catholic theology as the universal explanation for truth during the Reformation, science took up this theological burden in western Europe. See Imre Lakatos, "Science and Pseudoscience," In *The Methodology of Scientific Research Programmes*, volume 1, ed. John Worrell and Gregory Currie (Cambridge: Cambridge University Press, 1978), 1–7.
32. Gottschalk, *Religion, Science, and Empire*.
33. See, for example, Talal Asad, *Genealogies of Religion: Discipline and Reasons of Power in Christianity and Islam* (Baltimore: Johns Hopkins University Press,1993); Crosson, *Experiments with Power*; and Robert Orsi, "Snakes Alive: Religious Studies Between Heaven and Earth," in *Between Heaven and Earth: The Religious Worlds People Make and the Scholars Who Study Them* (Princeton, NJ: Princeton University Press, 2006), 177–239.
34. Frazer, *The Golden Bough*. See also Tylor, *Primitive Culture*.
35. Crosson, *Experiments with Power*.
36. For example, Evans-Pritchard, *Witchcraft, Oracles and Magic*; and Robin Horton, "African Traditional Thought and Western Science, Part II: The 'Closed' and 'Open' Predicaments," *Africa* 37, no. 2 (1967): 155–187. See also D. W. Hogg, "Magic and 'Science' in Jamaica," *Caribbean Studies* 1 (1961): 1–5.
37. John Beattie, "Ritual and Social Change," *Man* 1, no. 1 (1966): 65.

38. Lucy Mair, *African Societies* (Cambridge: Cambridge University Press, 1974); Paton, *The Cultural Politics of Obeah*, 215.
39. Crosson, *Experiments with Power*.
40. Karen Barad, *Meeting the Universe Halfway: Quantum Physics and the Entanglement of Matter and Meaning* (Durham, NC: Duke University Press, 2007).
41. Ian Hacking, *Representing and Intervening: Introductory Topics in the Philosophy of Natural Science* (New York: Cambridge University Press, 1983).
42. Karen Barad, "No Small Matter: Mushroom Clouds, Ecologies of Nothingness, and Strange Topologies of Spacetimemattering," in *Arts of Living on a Damaged Planet: Ghosts of the Anthropocene*, ed. Anna Lowenhaupt Tsing, Heather Anne Swanson, Elaine Gan, and Nils Bubandt (Minneapolis: University of Minnesota Press, 2017), 103–120.
43. For example, Donna Haraway, "Situated Knowledges: The Science Question in Feminism and the Privilege of Partial Perspective," in *Simians, Cyborgs and Women: The Reinvention of Nature* (New York: Routledge, 1991); and Sandra Harding, *Sciences from Below: Feminisms, Postcolonialities, and Modernities* (Durham, NC: Duke University Press, 2008).
44. Throughout this chapter, I put "Western esotericism" (as well as "Western esotericists") in quotes to acknowledge the limitations of this term. The Western-ness of Western esotericism has to be located within the European genealogies of Egyptophilia, orientalism, and primitivism that have made divergent kinds of "Western esotericism" possible. Furthermore, accounts of "Western esotericism" have usually failed to take into account the extensive Black Atlantic uses of and experimentations with many of the very same texts and practices that have been named as central to Western esotericism (grimoires and Freemasonry being two examples).
45. David L. McMahan, "Modernity and the Early Discourse of Scientific Buddhism," *Journal of the American Academy of Religion* 72, no. 4 (2004): 897–933; Richard Noakes, *Physics and Psychics: The Occult and the Sciences in Modern Britain* (Cambridge: Cambridge University Press, 2019); Jeremy Stolow, "Salvation by Electricity," in *Religion: Beyond a Concept*, ed. Hent de Vries (New York: Fordham University Press, 2008), 668–687, and "The Spiritual Nervous System," in *Deus in Machina: Religion, Technology and the Things in Between*, ed. Jeremy Stolow (New York: Fordham University Press, 2013), 83–116.
46. Egil Asprem, "Magic Naturalized? Negotiating Science and Occult Experience in Aleister Crowley's Scientific Illuminism," *Aries* 8, no. 2 (2008): 139–165.
47. Roger Bastide, "Le spiritisme au Brésil," *Archives de sociologie des religions* 24 (1967): 3–16; Diana Espírito Santo, *Developing the Dead: Mediumship and Selfhood in Cuban Espiritismo* (Gainesville: University Press of Florida, 2015), 97–154; David J. Hess, *Spirits and Scientists: Ideology, Spiritism and Brazilian Culture* (State College: Penn State University Press, 1991).
48. Alireza Doostdar, *The Iranian Metaphysicals: Explorations in Science, Islam, and the Uncanny* (Princeton, NJ: Princeton University Press, 2018), 4–5.
49. On *quimbois*, see Margarite Fernández Olmos and Lizabeth Paravisini-Gebert, eds., *Sacred Possessions: Vodou, Santería, Obeah, and the Caribbean*. New Brunswick, NJ: Rutgers University Press, 1997), and *Creole Religions of the Caribbean: An Introduction from Vodou and Santería to Obeah and Espiritismo*, 2nd ed. (New York: New York University Press, 2011).
50. Christopher Josiffe, "Aleister Crowley, Marie de Miramar and the True Wanga," *Abraxas* 1, no. 4 (2013): 29–42.

51. Palmié 2002: 208.
52. This discovery will appear as a chapter on Cassecanarie in Alexander Rocklin, *Becoming Hindu: The Imposture of Religion and the Power of India in the Atlantic World* (forthcoming).
53. See McMahan, "Modernity and the Early Discourse of Scientific Buddhism"; and Donald S. Lopez Jr., *Buddhism and Science: A Guide for the Perplexed* (Chicago: University of Chicago Press, 2008).
54. McMahan, "Modernity and the Early Discourse of Scientific Buddhism."
55. See also Stolow, "Salvation by Electricity" and "The Spiritual Nervous System"; Cassie S. Adcock, *The Limits of Tolerance: Indian Secularism and the Politics of Religious Freedom* (New York: Oxford University Press, 2014), 43; "Science in Vedas," accessed July 10, 2019, http://www.aryasamaj.org/newsite/node/2358; and Swami Vivekananda, *Addresses at the Parliament of Religions* (Calcutta: Ramakrishna Mission, 1893).
56. See Noakes, *Physics and Psychics*.
57. See McMahan, "Modernity and the Early Discourse of Scientific Buddhism."
58. Myal Djumboh Cassecanarie, *Obeah Simplified, the True Wanga: A Scientific Treatise* (Port of Spain, Trinidad: The Mirror, 1895), 27.
59. Cassecanarie, *Obeah Simplified*, 74–75.
60. Cassecanarie, *Obeah Simplified*, 75.
61. Cassecanarie, *Obeah Simplified*, 75.
62. See Josiffe, "Aleister Crowley, Marie de Miramar and the True Wanga."
63. McMahon, "Modernity and the Early Discourse of Scientific Buddhism."
64. See, for example, Orsi, "Snakes Alive."
65. Crowley, quoted in Asprem, "Magic Naturalized?," 150.
66. In addition, many of the alleged ethnographic details in Cassecanarie's text are drawn or adapted from an earlier (and less complimentary) monograph by the British colonial official Hesketh Bell, *Obeah: Witchcraft in the West Indies* (London: S. Low, Marston, 1893 [1889]).
67. Littlewood and Hogg differentiate traditional obeah (as an African and rural practice) from "high science" (as a book-based, urban, and European practice). Roland Littlewood, *Pathology and Identity: The Work of Mother Earth in Trinidad* (Cambridge: Cambridge University Press, 1993); Hogg, "Magic and 'Science' in Jamaica." I found this distinction to be entirely untenable (and potentially racist); rural spiritual workers (like Western hermeticists) conceived of these texts as deriving from northeastern Africa, and they used them widely as another African source of inspiration.
68. The most popular grimoires in Rio Moro were *The Sixth and Seventh Books of Moses*, *The Petit Albert* ("Titalbay"), *The Greater Key of Solomon*, and *The Lesser Key of Solomon* (though many other grimoires were also studied, including various South Asian–identified works referred to as *Indrajal*). Grimoires have long been associated with Black Atlantic religion. *The Petit Albert*, for example, has recently been republished as the supposed "Spellbook" of the "voodoo queen of New Orleans," Marie Laveau (*The Spellbook of Marie Laveau: The Petit Albert*, trans. Talia Felix [Keighley, UK: Hadean Press, 2013]). The book currently marketed as the "obeah Bible" is an 1898 volume written (or compiled) by the largest purveyor of grimoires in the Caribbean, the Ohio-born hypnotist William DeLaurence, entitled *The Great Book of Magical Art, Hindu Magic and East Indian Occultism*. *The Sixth and Seventh Books of Moses*, a grimoire originally brought to the Americas by German immigrants (see Owen Davies, *Grimoires: A History of Magic Books* [Oxford: Oxford University

Press, 2010]), was cited by Zora Neale Hurston as one of the most important books for African American "hoodoo" practitioners. Zora Neale Hurston, *Moses, Man of the Mountain* (1939; repr., New York: Harper Perennial, 2009), vii–viii.

69. See also Nils Bubandt, *The Empty Seashell: Witchcraft and Doubt on an Indonesian Island* (Ithaca, NY: Cornell University Press, 2014).

70. On the rationality of African witchcraft, see Evans-Pritchard, *Witchcraft, Oracles and Magic*; on Indigenous traditional knowledge as scientific, see Sandra Harding, *Objectivity and Diversity: Another Logic of Scientific Research* (Chicago: University of Chicago Press, 2015); on magic as technology, see Bronislaw Malinowski, *Magic, Science and Religion* (1925; repr., Garden City, NY: Doubleday, 1954). For an alternative perspective, see John Wiredu, "How Not to Compare African Thought with Western Thought," in *African Philosophy*, ed. Richard A. Wright (Washington, DC: University Press of America, 1979).

71. Wiredu, "How Not to Compare African Thought," 137; Harding, *Objectivity and Diversity*, 90.

72. Latour, "Thou Shalt Not Freeze Frame," in *On the Modern Cult of the Factish Gods*.

WORKS CITED

Adcock, Cassie S. *The Limits of Tolerance: Indian Secularism and the Politics of Religious Freedom*. New York: Oxford University Press, 2014.

Asprem, Egil. "Magic Naturalized? Negotiating Science and Occult Experience in Aleister Crowley's Scientific Illuminism." *Aries* 8, no. 2 (2008): 139–165.

Bell, Hesketh. *Obeah: Witchcraft in the West Indies*. London: S. Low, Marston, 1893 [1889].

Brosses, Charles de. *Du culte des dieux fétiches, out Parallèle de l'ancienne religion de l'Egypte avec la religion actuelle de Nigritie*. London: Westmead, Farnborough, Hants, 1970 [1760].

Bubandt, Nils. *The Empty Seashell: Witchcraft and Doubt on an Indonesian Island*. Ithaca, NY: Cornell University Press, 2014.

Comte, Auguste. *The Positive Philosophy of Auguste Comte*, trans. Harriet Martineau. London: J. Chapman, 1853.

Crosson, J. Brent. "Catching Power: Problems with Possession, Sovereignty, and African Religions in Trinidad." In "What Possessed You? Spirits, Property, and Political Sovereignty at the Limits of 'Possession,'" ed. J. Brent Crosson. Special issue, *Ethos* 84, no. 4 (2017): 546–556. https://doi.org/10.1080/00141844.2017.1401704.

———. *Experiments with Power: Obeah and the Remaking of Religion in Trinidad*. Chicago: University of Chicago Press, 2020.

———. "The Impossibility of Liberal Secularism: Political Violence, Spirituality, and Not-Religion." *Method and Theory in the Study of Religion* 30, no. 1 (2018): 35–55.

———. "Oil, Obeah, and Science." *Cosmologics*, 2016. http://cosmologicsmagazine.com/brent-crosson-oil-obeah-and-science/.

———. "What Obeah Does Do: Healing, Harming, and the Boundaries of Religion." *Journal of Africana Religions* 3, no. 2 (2015): 151–176.

Davies, Owen. *Grimoires: A History of Magic Books*. Oxford: Oxford University Press, 2010.

Durkheim, Emile. *The Elementary Forms of Religious Life*. 1912. Reprint, London: Allen and Unwin, 1964.

Edwards, Bryan. *The History, Civil and Commercial of the British Colonies of the West Indies*, vol. 2. 1793–1801. Reprint, New York: Arno Press, 1972.
Evans-Pritchard, E. E. *Witchcraft, Oracles and Magic Among the Azande*. Oxford: Oxford University Press, 1937.
Frazer, James George, Sir. *The Golden Bough: A Study in Magic and Religion*. New York: Macmillan, 1922.
Gottschalk, Peter. *Religion, Science, and Empire: Classifying Hinduism and Islam in British India*. New York: Oxford University Press, 2012.
Harding, Sandra. *Objectivity and Diversity: Another Logic of Scientific Research*. Chicago: University of Chicago Press, 2015.
———. *Sciences from Below: Feminisms, Postcolonialities, and Modernities*. Durham, NC: Duke University Press, 2008.
Hegel, G. W. F. *The Phenomenology of Mind*, vol. 2. 1807. Reprint (trans. J. B. Baillie), London: Swan Sonnenschein, 1910.
———. *The Philosophy of History*. 1837. Reprint (trans. J. Sibree), Mineola, NY: Dover, 1956.
Hogg, D. W. "Magic and 'Science' in Jamaica." *Caribbean Studies* 1 (1961): 1–5.
Hurston, Zora Neale. *Moses, Man of the Mountain*. 1939. Reprint, New York: Harper Perennial, 2009.
Josephson-Storm, Jason Ananda. *The Myth of Disenchantment: Magic, Modernity, and the Birth of the Human Sciences*. Chicago: University of Chicago Press, 2017.
———. "The Superstition, Secularism, and Religion Trinary: Or Re-theorizing Secularism." *Method and Theory in the Study of Religion* 30, no. 1 (2018): 1–20.
Josiffe, Christopher. "Aleister Crowley, Marie de Miramar and the True Wanga." *Abraxas* 1, no. 4 (2013): 29–42.
Kant, Immanuel. *Religion Within the Limits of Reason Alone*. 1793. Reprint (trans. Theodore M. Greene and Hoyt H. Hudson), New York: Harper & Row, 1960.
Latour, Bruno. *On the Modern Cult of the Factish Gods*. Durham, NC: Duke University Press, 2010.
———. *We Have Never Been Modern*. Cambridge, MA: Harvard University Press, 1993.
Littlewood, Roland. *Pathology and Identity: The Work of Mother Earth in Trinidad*. Cambridge: Cambridge University Press, 1993.
Lopez, Donald S. Jr. *Buddhism and Science: A Guide for the Perplexed*. Chicago: University of Chicago Press, 2008.
Malinowski, Bronislaw. *Magic, Science and Religion*. 1925. Reprint, Garden City, NY: Doubleday, 1954.
McMahan, David L. "Modernity and the Early Discourse of Scientific Buddhism." *Journal of the American Academy of Religion* 72, no. 4 (2004): 897–933.
Noakes, Richard. *Physics and Psychics: The Occult and the Sciences in Modern Britain*. Cambridge: Cambridge University Press, 2019.
Numbers, Ronald L. "Introduction." In *Galileo Goes to Jail and Other Myths About Science and Religion*, ed. Ronald L. Numbers, 1–7. Cambridge, MA: Harvard University Press, 2010.
Olmos, Margarite Fernández, and Lizabeth Paravisini-Gebert, eds. *Creole Religions of the Caribbean: An Introduction from Vodou and Santería to Obeah and Espiritismo*. 2nd ed. New York: New York University Press 2011.
———. *Sacred Possessions: Vodou, Santería, Obeah, and the Caribbean*. New Brunswick, NJ: Rutgers University Press, 1997.

Orsi, Robert. "Snakes Alive: Religious Studies Between Heaven and Earth." In *Between Heaven and Earth: The Religious Worlds People Make and the Scholars Who Study Them*, 177–239. Princeton, NJ: Princeton University Press, 2006.
Report of the Lords of the Committee of the Council Appointed for the Consideration of All Matters Relating to Trade and Foreign Plantation, Part 3: Treatment of Slaves in the West Indies, and All circumstances Relating Thereto, Digested Under Certain Heads, Part III. London: Parliament, 1789. https://babel.hathitrust.org/cgi/pt?id=mdp.39015084394389;view=1up;seq=4.
Rocklin, Alexander. *Becoming Hindu: The Imposture of Religion and the Power of India in the Atlantic World*. Forthcoming.
"Science in Vedas." Accessed July 10, 2019. http://www.aryasamaj.org/newsite/node/2358.
Stengers, Isabelle. *The Invention of Modern Science*. Minneapolis: University of Minnesota Press, 2000.
——. "Reclaiming Animism." *e-flux*, no. 36 (2012). https://www.e-flux.com/journal/36/61245/reclaiming-animism/.
Stolow, Jeremy. "Salvation by Electricity." In *Religion: Beyond a Concept*, ed. Hent de Vries, 668–687. New York: Fordham University Press, 2008.
——. "The Spiritual Nervous System." In *Deus in Machina: Religion, Technology and the Things in Between*, ed. Jeremy Stolow, 83–116. New York: Fordham University Press, 2013.
Trouillot, Michel-Rolph. *Global Transformations: Anthropology and the Modern World*. New York: Palgrave Macmillan, 2003.
Tylor, Edward Burnett. *Primitive Culture: Researches Into the Development of Mythology, Philosophy, Religion, Art, and Custom*. London: J. Murray, 1871.
Vivekananda, Swami. *Addresses at the Parliament of Religions*. Calcutta: Ramakrishna Mission, 1893.
Weszkalnys, Gisa. "Geology, Potentiality, Speculation: On the Indeterminacy of First Oil." *Cultural Anthropology* 30, no. 4 (2015): 611–639. https://doi.org/10.14506/ca30.4.08.
Wiredu, John. "How Not to Compare African Thought with Western Thought." In *African Philosophy*, ed. Richard A. Wright. Washington, DC: University Press of America, 1979.

CONCLUSION

MYRNA PEREZ SHELDON, TERENCE KEEL, AND AHMED RAGAB

What work does a critical approach to science and religion inspire? This volume is meant to be a suggestive and inspirational collection of ideas rather than an exhaustive account of this new approach to the field. To conclude, we consider what is not fully realized in the collection and map suggestions for future students and scholars.

First, we urge a commitment to a global framework of analysis. Such a commitment would not simply fold in the study of more and various cultures, or detail more world religions, or recount the practice of science in places other than Euro-America. It would relentlessly confront the challenges of linguistic and conceptual discordance, thereby recognizing the study of science and religion itself as a colonial act. Euro-American sociological and anthropological studies of religion have been premised on hierarchies that rendered other cultures within the cognitive possibilities of Christianity. The very scaffolding of our imagination about secularity, faith, doctrine, piety, and myth making is dependent on the colonial aims of Western Christianity. And since Western science defined itself in conversation with Christianity, the definitions of scientific objectivity and rationality are themselves solutions to theological problems posed in Christian intellectual history. A global critical history of science and religion could move outside this violence with boldness and humility.

Several of the volume chapters emerge from the global focus of the contributors' research programs. Jason Ānanda Josephson Storm's piece draws on his 2012 book *The Invention of Religion in Japan*, which argues that the Japanese created religion during the nineteenth century in order to render their political and cultural life sensible to Western military powers.[1] Ahmed Ragab's chapter on the internet ecology of Muslim fatwas builds upon his reorientation of the history of science and technology away from Euro-America in a number of monographs.[2] Cassie Adcock's history of the sacred cow in her chapter expands on her 2013 book *The Limits of Tolerance: Indian Secularism and the Politics of Religious Freedom*, which was the first monograph to consider tolerance as a secular ideal.[3] And Katharine Gerbner's chapter in "Coherence" works from the deep immersion in the global archive from her 2018 book *Christian Slavery: Conversion in the Protestant Atlantic World*.[4] These writers have dedicated their careers to work outside of the received frames of Euro-American geographies and, in so doing, have demonstrated the necessity for a global frame for science and religion studies.

Second, we desire for the critical-historical study of science and religion to engage posthumanist and new materialist theory. Although these fields cannot be reduced to a single rubric, broadly they are efforts in the critical humanities to move beyond the linguistic turn to take up analyses of materiality and ontology. Posthumanism examines the boundaries blurred amidst the human, the nonhuman, and the more-than-human by the genomic and evolutionary sciences, as well as computers, cybernetics, and artificial intelligence. Whereas new materialism grapples with the ontological ramifications of questioning the traditional Western assumption that the physical world is inanimate, much feminist new materialism relies on readings of quantum mechanics or influences from continental philosophy for these radical ontologies.[5] Both posthumanism and new materialism inquire into science and religion in ways that are less concerned with critiquing sexism or racism or with valorizing the historical contributions of marginalized groups. Instead, they investigate how reformulations of basic dichotomies in Western thought destabilize the human. Thus, they disrupt identity categories such as race, sex, gender, class, and citizen. Generally, there is little contact between these bodies of thought and the historical and philosophical study of science and religion. This is undoubtedly due in large part to the reluctance of historians to participate in normative analyses in order to avoid the polarizing debates of popular depictions of science and religion. However, we maintain that by taking on these theories, a critical approach to science and religion can be emboldened to tackle material and ontological issues.

Many of the volume chapters directly engage with ontology through the lens of Indigenous theory. Terence Keel begins with the work of Sisseton Wahpeton Oyate scholar Kim TallBear in order to argue that the Western distinction between science and spirituality produces white supremacy. Keel's chapter builds on insights from his book *Divine Variations: How Christian Thought Became Racial*

Science, in which he contends that modern race science has reoccupied the theological problematics of Western Christianity.[6] Joanna Radin's chapter in "Coherence" also begins with TallBear's work, as it questions whether it is possible for Western scholarship to fully engage a "radically different understanding of life." Radin's chapter wrestles with the role of Christian resurrection theology in the Cold War–era efforts to freeze human blood, also the subject of her 2017 book *Life on Ice: A History of New Uses for Cold Blood*.[7] And Eli Nelson's chapter builds on his wider body of scholarship that not only refuses to translate Indigenous thought but that also critiques the colonial framing of new materialism itself.[8] Nelson's work highlights the colonial impulses of contemporary new materialism and posthumanism, which have often disregarded long-standing mutualisms between spirit and science, as well as life and nonlife in Indigenous thought.[9] We are encouraged that a critical approach to science and religion can prompt posthumanism and new materialism toward a self-reflection on their own colonial epistemologies and toward building a more rigorous account of race.[10]

Finally, we aim for a critical approach to science and religion to be responsible to communities outside of the academy. It is our contention that scholarship in this vein will make it possible for analyses of these categories to have a more tangible relationship to work for justice in the world. Our critique of the existing field is largely aimed at its preoccupation with overly abstract philosophical and cosmological questions. However, we recognize that critical theories are their own forms of specialized knowledge, with attendant jargon and expertise. And so, we continually push ourselves and our scholarship outside of university spaces. Importantly, the intellectual seed for this volume began with a five-year digital magazine project of the Harvard Divinity School, *Cosmologics: A Magazine of Science and Religion*.[11] That project also helped inspire the founding of the Center for Black, Brown, and Queer Studies, an independent center whose mission is to "create an alternative space that is committed to addressing inequities" by raising awareness through the lenses of critical race, postcolonial and Indigenous, and LGBTQ studies.[12] Keel's work on race, religion, and science has now manifested in an interdisciplinary lab that studies police violence in Los Angeles.[13] These beginnings push our collective task away from the university toward the political urgencies of our communities.

However, we are conscious that there is always more to do and that it is difficult to move beyond the habits of the academy and its hierarchies. In this, we fight against the intuition that our scholarly task is solely to characterize—or even to critique. We reject the notion that if we show what is wrong with the existing world, this will be enough to make a new one. And so the contributors of this volume write critical history that is future oriented, that is unafraid not only to describe white supremacy but also to suggest how it might be undone, that writes the history of abortion politics but also argues how we should view the morality of abortion, and that recognizes that if we are not honest about the moral urgency of our analyses, we are lying to ourselves about who we are and why we think, write, and teach.

Ultimately, our hope is that the limitations, faults, and gaps in this collection will engage, enrage, and embolden future scholars. We hope that they will take our weaknesses as opportunities to build new strengths. Our work here is not an attempt at a final word but is instead a faithful invitation to dialogue because we do this scholarship not for its own sake or as fealty to an imagined standard of excellence but as a means to learn what it is to care for one another in the world.

NOTES

1. Jason Ānanda Josephson, *The Invention of Religion in Japan* (Chicago: University of Chicago Press, 2012).
2. Ahmed Ragab, *The Medieval Islamic Hospital: Medicine, Religion, and Charity* (Cambridge: Cambridge University Press, 2015); Ahmed Ragab, *Piety and Patienthood in Medieval Islam* (London: Routledge, 2018); Ahmed Ragab, *Medicine and Religion in the Life of an Ottoman Sheikh: Al-Damanhuri's "Clear Statement" on Anatomy* (London: Routledge, 2019).
3. Cassie Adcock, *The Limits of Tolerance: Indian Secularism and the Politics of Religious Freedom* (Oxford: Oxford University Press, 2013).
4. Katharine Gerbner, *Christian Slavery: Conversion and Race in the Protestant Atlantic World* (Philadelphia: University of Pennsylvania Press, 2018).
5. Karen Barad, "Quantum Entanglements and Hauntological Relations of Inheritance: Dis/Continuities, Spacetime Enfoldings, and Justice-to-Come," *Derrida Today* 3, no. 2 (2010): 240–268.
6. Terence Keel, *Divine Variations: How Christian Thought Became Racial Science.* (Stanford, CA: Stanford University Press, 2018).
7. Joanna Radin, *Life on Ice: A History of New Uses for Cold Blood* (Chicago: University of Chicago Press, 2017).
8. Eli Nelson, " 'Walking to the Future in the Steps of Our Ancestors': Haudenosaunee Traditional Ecological Knowledge and Queer Time in the Climate Change Era," *New Geographies* 09: Posthuman (2017): 133–138, and "Canoes in Space," *Ventricles* (podcast), season 1, episode 5, produced by Shireen Hamza, October 3, 2018, https://ventricles.simplecast.fm/dc580fc6.

9. Zoe Todd, "An Indigenous Feminist's Take on the Ontological Turn: 'Ontology' Is Just Another Word for Colonialism," *Journal of Historical Sociology* 29, no. 1 (2016): 4–22.
10. Jasbir K. Puar, "'I Would Rather Be a Cyborg Than a Goddess': Becoming-Intersectional in Assemblage Theory," *PhiloSOPHIA* 2, no. 1 (October 2, 2012): 49–66; Margaret Rhee, "In Search of My Robot: Race, Technology, and the Asian American Body," *Scholar and Feminist Online* 13, no. 3 (2016).
11. *Cosmologics*, www.cosmologicsmagazine.com.
12. Center for Black, Brown, and Queer Studies, www.bbqplus.org.
13. The BioCritical Studies Lab, https://www.terencekeel.com/bcs-lab.

LIST OF CONTRIBUTORS

EDITORS

Myrna Perez Sheldon is associate professor jointly appointed in Classics and Religious Studies and Women's, Gender, and Sexuality Studies, and executive director, Cutler Scholar's Program, Ohio University. Sheldon is a historian of evolutionary theory, a feminist and critical race theorist, and a scholar of religion. She earned her PhD from the History of Science Department at Harvard University; previously held a postdoctoral fellowship at the Center for the Study of Women, Gender, and Sexuality at Rice University; and was a research fellow at the Darwin Correspondence Project at Cambridge University and at the Harvard Divinity School. She has published on sexuality, race, Christianity, and evolutionary theory in *Signs: A Journal of Women and Society*, *American Quarterly*, *Historical Studies in the Natural Sciences*, *Studies in the History and Philosophy of the Life Sciences*, and elsewhere. She is the principal investigator of a three-year grant, "Critical Approaches to Science and Religion," from the Templeton World Charity Foundation and the president of the Board at the Center for Brown, Black, and Queer Studies.

Terence Keel is associate professor at the Institute for Society and Genetics and at the Department of African American Studies, University of California, Los Angeles. Keel is a community-engaged scholar working across Black studies, history, religion, and science studies. He has written widely about the history of racism and its connections to science, medicine, religion, and politics. He is the author of the award-winning book *Divine Variations*, which documents the intellectual legacy shared between modern scientific racism and Christian thought. He is the founding director of the BioCritical Studies Lab—a community-engaged interdisciplinary research space located in the Division of Life Sciences at UCLA—which is committed to using quantitative and qualitative methods to understand how discrimination and resilience are embodied in human life. He earned his PhD from the Committee on the Study of Religion at Harvard University.

Ahmed Ragab is associate professor of the history of medicine at Johns Hopkins University; chair of the Medicine, Science, and Humanities Program at Krieger School of Arts and Science, Johns Hopkins University; director of the Center for Black, Brown and Queer Studies. Ragab received his MD from Cairo University School of Medicine in 2005 and his PhD from the École Pratique des Hautes Études in Paris in 2010. His research focuses on the history of medicine, science, and religion and the development of cultures of science and cultures of religion in the Middle East and the Islamic world. He also studies and publishes on gender and sexuality in the medieval and early modern Middle East, postcolonial studies of science and religion, and other questions in the history of science and religion. He is the author of *The Medieval Islamic Hospital: Medicine, Religion and Charity* (2015), *Piety and Patienthood in Medieval Islam* (2018), and *Medicine and Religion in the Life of an Ottoman Sheikh* (2019).

CONTRIBUTORS

Cassie Adcock is associate professor of history and South Asian studies, Washington University in St. Louis. Adcock's research focus is the history of religion and political culture in India. She is the author of *The Limits of Tolerance: Indian Secularism and the Politics of Religious Freedom* (2013).

J. Brent Crosson is associate professor of religious studies and anthropology at the University of Texas at Austin. Crosson is the author of the book *Experiments with Power: Obeah and the Remaking of Religion in Trinidad* (2020), which won the 2021 Clifford Geertz Award from the American Anthropological Association and was short-listed for the 2021 Albert J. Raboteau Prize from the *Journal of Africana Religions*. He has published other pieces in journals such as *Small Axe*, *Anthropological Quarterly*, *Method and Theory in the Study of Religion*, and *Ethnos*.

Katharine Gerbner is associate professor of history at the University of Minnesota. Gerbner's research explores the religious dimensions of race, authority, and freedom in the early modern Atlantic world. She is the author of *Christian Slavery: Conversion and Race in the Protestant Atlantic World* (2018).

Joseph L. Graves Jr. is professor of biological sciences in the Biology Department, North Carolina A&T State University. Graves is an evolutionary biologist whose current work includes the genomics of adaptation and the application of this research to the evolutionary theory of aging and microbial responses to novel nanomaterials.

Kathryn Lofton is Lex Hixon Professor of Religious Studies and American Studies and professor of history and divinity, Yale University. Lofton is a historian of religion who has written extensively about capitalism, celebrity, sexuality, and the concept of the secular. She is the author of *Oprah: The Gospel of an Icon* (2011) and *Consuming Religion* (2017).

Erika Lorraine Milam is Charles C. and Emily R. Gillispie Professor in the History of Science, Princeton University. Milam specializes in the cultural and intellectual history of field-based life sciences. She is the author of *Looking for a Few Good Males: Female Choice in Evolutionary Biology* (2010) and *Creatures of Cain: The Hunt for Human Nature in Cold War America* (2019).

Eli Nelson is assistant professor of American studies, Williams College. Nelson specializes in the history of Native science in settler colonial and postcolonial contexts.

Osagie K. Obasogie is the Haas Distinguished Chair and Professor of Law at the University of California, Berkeley School of Law, with a joint appointment in the Joint Medical Program and School of Public Health. Obasogie's scholarly interests include constitutional law, bioethics, sociology of law and medicine, and reproductive and genetic technologies. He is the author of *Blinded by Sight: Seeing Race Through the Eyes of the Blind* (2013) and *Beyond Bioethics: Toward a New Biopolitics* (2018).

Joanna Radin is associate professor of history of medicine and a core member of the Program for History of Science and Medicine at Yale University. Radin's research and teaching focus on the history of biomedical futures, which considers ways of accessing liberatory possibilities for the sciences of life and medicine latent within our present. She is the author of *Life on Ice: A History of New Uses for Cold Blood* (2017), the first history of the low-temperature biobank, and coeditor, with Emma Kowal, of *Cryopolitics: Frozen Life in a Melting World* (2017).

Suman Seth is Marie Underhill Noll Professor of the History of Science and chair, Department of Science and Technology Studies, Cornell University. Seth works on the history of medicine, race, and colonialism; the physical sciences (particularly quantum theory); and gender and science. He is the author of *Difference and Disease: Medicine, Race, and the Eighteenth-Century British Empire* (2018) and *Crafting the Quantum: Arnold Sommerfeld and the Practice of Theory, 1890–1926* (2010).

Jason Ānanda Josephson Storm is a historian and philosopher of the human sciences. Storm is currently professor of religion and chair of Science and Technology Studies at Williams College. He received his PhD from Stanford University and his MA from Harvard University, and he has held visiting positions at Princeton University, École Française d'Extrême-Orient, and Universität Leipzig in Germany. He is the author of the award-winning *The Invention of Religion in Japan* (2012), *The Myth of Disenchantment: Magic, Modernity and the Birth of the Human Sciences* (2017), and *Metamodernism: The Future of Theory* (2021).

Tisa Wenger is professor of American religious history, Yale Divinity School. Wenger is a historian of American religion with research and teaching interests in the cultural politics of religious freedom and the intersections of race, religion, and empire in U.S. history. She is the author of *We Have a Religion: The 1920s Pueblo Indian Dance Controversy and American Religious Freedom* (2009) and *Religious Freedom: The Contested History of an American Ideal* (2017) and the coeditor, with Sylvester Johnson, of *Religion and U.S. Empire: Critical New Histories* (2022).

INDEX

AAAS. *See* American Association for the Advancement of Science
abolition, 15–46
abortion, 378; biopolitics and, 62–63, 70–71, 73–75, 79, 81; Black women and, 77–78; Catholicism on, 66–67, 307; Christian nationalism and, 19, 62–69, 79–81; eugenics and, 62, 66–67, 70, 72–73, 81; feminist theology of, 63–64, 80–82; fetal life and, 75–77; Indiana state legislature on, 61–63, 70, 81; in Japan, 72–73; legalization campaigns, 71, 73; for nonbinary persons and transgender men, 82n11; population control and, 18–19, 71–72; reproductive justice activism, 77–79; sacred and, 64, 69–70; Supreme Court on, 61–62, 70, 74, 76, 82n1; trust and, 77–82
Abortion of the Human Race, The (film), 64
ABS. *See* American Breeders' Service
accidental settlement hypothesis, 289–290
Accompong (Maroon leader), 331, 339–340
active nihilism, 58n23
activism, scholarship and, 17–18
Adam (biblical figure), 256–257, 259; on *Newsweek* cover, 267, 268; in polygenism, 229, 233–234, 248, 260. *See also* pre-Adamism
Adam's Ancestors (Livingstone), 230
adaptation, 316
Adcock, Cassie, 376

adventure, of sciences, 356
Adventures of Roderick Random, The (Smollett), 239
Africa: Atkins on, 236–241, 252n28; human origin in, 205, 264–265, 267, 268, 269, 270, 272–273
African magic, 356–357
African-ness, 336
Africans: Long and humanity of, 245–248, 253n60; polygenism and, 232, 237–238, 248, 252n28
Agamben, Giorgio, 298n68
Agassiz, Louis, 258–259, 261–263
agency, nonhuman, 139, 283
agnotology, 338
agriculture. *See* scientific agriculture
Algic Researches (Schoolcraft, H.), 118
Algic Society, 113–114
"Algonac, A Chippewa Lament at Hearing the Reveille at the Post of St. Mary's" (anonymously published), 105
Algorithms of Oppression (Noble), 56n1
alienation, 44
American Association for the Advancement of Science (AAAS), 262, 271–272
American Breeders' Service (ABS), 313
American Civil Liberties Union, 271
American Foundation for the Study of Genetics, 313
American Indians. *See* Native.
anchoring processes, 148–149, 151–153

Anglophone Caribbean. *See* Caribbean, Anglophone
Anglophone societies: conflict thesis in, 1; racism of, 10; Western identity in, 7–8
Animacies (Chen), 295n11
animacy, 306–307, 321n10. *See also* inanimacy
animism: Kānaka Maoli, 278, 282–284, 288–289, 292–295; new, 282–284; objectivism and, 278, 283, 289, 292, 294, 295n10, 297n37; postmodernism and, 278, 283
Anishinaabe, 112; Christianity among, 114–115, 117–118, 122; on Copper Rock, 109–110; Ojibwe, 105–107, 109–110, 113–115, 117–119; H. Schoolcraft on, 109–110, 120
Answers in Genesis website, 256–257
anthropology, 53–54; African-ness in, 336; archives from, 326; Jamaican Maroons in, 338; on magic and science, 356–357, 364; settler colonialism and, 43; white supremacy in physical, 265–266
anti-Blackness, European immigrants facing, 11
anti-Jewish beliefs. *See* anti-Semitism
antiorthodoxy, performance of, 231–232, 239, 241, 243, 248–249
antipoaching. *See* poaching
anti-Semitism, 9, 19, 235–236
antivaccine activists, 366
Anzaldua, Gloria, 102
appearance, of websites, 188–189
apps, Islamic, 192–193
Apted, Michael, 219
archives, 193, 326, 340
Arendt, Hannah, 314
ART. *See* assisted reproductive technology
Asad, Talal, 177n6
assisted reproductive technology (ART), 2–3
Atkins, John: on Africa, 236–241, 252n28; *The Navy Surgeon*, 237–241, 244; on polygenism, 232, 235–241, 244, 249–250, 252n28; on slavery, 238; on spontaneous generation, 240–241; *A Treatise on the Following Chirurgical Subjects*, 236–237; *A Voyage to Guinea, Brasil, and the West-Indies*, 237–238

Atlantic slave trade, 3
Attenborough, David, 220
Atwood, Margaret, 220
Australopithecus africanus, 265
Ayala, Francisco, 269

Bachman, Charles, 261
Bachman, John, 262, 264
backstage, in social media, 195–196
Bacon, Francis, 287
al-Balāgh Cultural Society, 182, 187
Barad, Karen, 76
Baring, Edward, 312
Bashford, Alison, 73
basmala, 189, 191
Baudrillard, Jean, 190
Ba Uriah, 325, 341
Beauchamp, Tom L., 92
Bell, John, 334
Belmont Report, The, 92
Benezet, Anthony, 232, 249
Benjamin, Ruha, 56n1
Bennett, Jane, 139–140
Bering land bridge migration theory, 272, 297n50
Bible Defense of Slavery (Priest), 259
biblical literalism, 256–257, 259
biblical monogenism, 249, 256–257, 259
Big Science, 132, 155n17
Bilby, Kenneth: Ba Uriah and, 325, 341; on Jamaican Maroons, 325, 330, 339, 341, 347n59; on obeah, 330, 342nn2–3; on science, 325, 339, 341, 342n2, 347n59
Bingham, Abel, 114–115
Biodynamica (journal), 309–310, 315
bioethics: eugenics and, 90–91, 94–95, 97; Holocaust and, 19, 87–91, 96–97; medical ethics and, 87–88, 90, 97; Nuremberg Code, 91; originalism and, 19, 88, 95–97; principlism in, 92–94
biopolitics: abortion and, 62–63, 70–71, 73–75, 79, 81; of nation-state, 70–71, 73–74; racial, 62–63
Birth Control and Love (Guttmacher), 72

Birth of Bioethics, The (Jonsen), 87, 91
black drop effect, 296n32
Black freedom designs, 52
Black Lives Matter, 78
Black Skin, White Masks (Fanon), 44
Black witchcraft, 3–4
Black women, abortion and, 77–78
Bland, Sandra, 27
Blavatsky, Helena, 140–141, 362
blood-freezing technology, 278–279, 317, 377; Indigenous peoples and, 304, 318–319; Luyet and, 305, 313–314, 318
Blumenbach, Johann Friedrich, 249, 259
border, 102, 104
border thinking, 293
botany, 263
boundary, 101 102, 104
Boutwell, William, 116
braided knowledges, 283–284
Brakke, David, 35
Brathwaite, Kamau, 342n3
Brazil, 358
breeding, 167–168, 313
Bright God of the Electric Telegraph and Telephone. *See* Denden-Myōjin
Brinkmann, Svend, 30
Britain: Jamaican Maroons and, 326, 329–331, 339–340, 344n13, 347n66; slavery and, 245–247, 249, 253n60, 326, 329–331; treaties with Jamaican Maroons, 331, 339–340, 347n66; witchcraft in, 328
British colonialism, in India: colonial improvement, 166–168; cow protection under, 160, 163–164, 167–168; eugenics in, 169–170; Hinduism and, 103, 163, 168; Malthusian perspective in, 164–166, 168–170; RCAI, 161–166, 168–169, 171–173, 178n28; scientific agriculture and, 161–169
Broca, Paul, 264
Brosses, Charles de, 354
Brown, Wendy, 52, 54, 59n34
Bruno, Giordano, 229–231, 260
Bryan, William Jennings, 271
Bryceson, Derek, 211

Buddhism: Edison promoting, 141; in Japan, 127–128, 132–136; scientific, 360; Shinto and, 132–136; Theosophical Society and, 360, 362
Buell, Denise, 48
Buffon, Comte de, 244
Burch, Ruth, 58n23
Butler, Judith, 25, 77
Bynum, Caroline Walker, 321n10
Byrd, Jodi, 298n68

Cahiers de valasains de folklore (journal), 308
Cain (biblical figure), 259
Caldwell, Charles, 260
California, Southern, 68
Camus, Albert, 49–50, 54
Canaan (biblical region), 256, 259
Cann, Rebecca, 267
canoe. *See* wa'a kaulua
Caplan, Arthur, 87, 97
cardinal directions, 287–288
Caribbean, Anglophone: grimoires in, 366, 371n68; obeah criminalized in, 329, 348, 354, 359, 367n3; obeah decriminalized in, 367n3; science in, 326, 329, 348, 352, 357–359, 363, 365–366; spiritual workers in, 279–280, 350–351, 358–359, 361, 363, 365–366, 371n67; universals and obeah in, 349–350, 352, 366–367. *See also* Jamaica; Trinidad
Caries, Zacharias George, 340
Carrel, Alexis, 305, 308–309
Carson, Rachel, 221
Cass, Lewis, 107–108, 110–111, 115
Cassecanarie, Myal Djumboh, 358–363, 371n66
categorical universals, 278, 280
Catholicism: on abortion, 66–67, 307; eugenics and, 66; evolution and, 256, 258; experimental model and, 323n44; Haraway on, 323n37; on life/not-life binary, 321n10; of Luyet, 278–279, 305–306, 308, 312, 320; phenomenology and, 312; Reformation and, 369n31; H. Schoolcraft on, 108–109; suspended animation research and, 309; on transubstantiation, 306, 318

cell, 323n34; Luyet on, 305, 310–312, 318, 320, 322n29; as symbiotic colony, 305, 310–312, 318
Center for Black, Brown, and Queer Studies, 377
Chang, David A., 287–288
Charlottesville, Virginia, Unite the Right rally (2017), 50
Chen, Mel, 295n11, 306
Chidester, David, 121
Childress, James F., 92
chimpanzee, 207, 210–213
Chimpanzees of Gombe, The (Goodall), 211
Chinese medical ethics, 24
Chippewa. *See* Ojibwe
cholera vaccine, 181
Christianity: of Anishinaabe, 114–115, 117–118, 122; biblical literalism, 256–257, 259; colonialism and, 46–48, 55, 106, 110, 122, 319, 375; creationism and, 67–69; Epistle to the Romans, 233–234; eugenics and, 271–272; evolution and, 256, 258, 271; forgery in New Testament, 34–35; gay rights opposition and, 68; Haraway on science and, 305; as modern science precursor, 6, 8; occult science and, 362; patriarchal, 62–64, 66; racism and, 9, 15n20, 41–42, 46, 50, 53–54, 59n28, 110, 271–272; H. Schoolcraft on, 108–110, 112–113, 116, 118, 121–122; scientific racism and, 41–42, 46, 50, 53–54; secular and, 49–50; in settler colonialism, 106, 110, 122; white supremacy and, 50, 59n28, 110. *See also* Catholicism; evangelicals; Protestantism
Christian nationalism: abortion and, 19, 62–69, 79–81; on fetus, 64; white, 19, 62–64, 67–68, 79–80
Christian nihilism, 54; Nietzsche on, 41, 46–50, 58n23; scientific racism and, 41–42, 50, 53
Christian Right. *See* Christian nationalism
Christian Slavery (Gerbner), 376
Chun, Wendy, 194–195
Civil Rights movement, 29, 67

climate change, 278, 283
coalescent analysis, 267, 269
Coetzee, J. M., 223n4
Cohen, Adam, 62
coherence, 277–278
cold, 315–317
Cold War, 304–305
colonial designs, 41–44, 46, 55–56
colonialism, 8, 279–280; Christianity and, 46–48, 55, 106, 110, 122, 319, 375; epistemologies and, 340; Hawai'i and, 292, 294; Indigenous settlement, colonial theories of, 297n50; modernism and postmodernism in, 278, 283; translation and, 277. *See also* decolonial theory; postcolonial theory
colonialism, British. *See* Britain; British colonialism, in India
colonialism, settler. *See* settler colonialism
Columbia (space shuttle), 291
commensurability, 277–278
commensuration, principlist, 92–94
Committee on Race and Eugenics, AAAS, 271–272
common sense, 283
common value system, 36
community fire, 316
complexity thesis, 106; conflict thesis and, 1–2, 4, 144–145; process social ontology and, 145–147, 152
Concerned Women for America (CWA), 68
confidence game, 26–27
conflict thesis, 6–7, 150; complexity thesis and, 1–2, 4, 144–145; polygenism and, 250
Congo, Democratic Republic of, 215
Conlin, Edward Grant, 271
conservation, 204, 213, 217, 219–221
conservatives, on environmentalism, 221
constitutional interpretation: living constitutionalists on, 95; originalism in, 88, 95–96
conversion, 213, 216, 225n37
Cook, James: in Hawai'i, 129, 143, 284, 288–289, 296n16; Obeyesekere and

Sahlins on, 129, 138, 143; ship of, 285, 287–288; Terra Nullius and, 286–287, 297n34; *wa'a kaulua* and, 284–286, 288, 296n16, 296n26
cooler, 26–27, 36–27
Coon, Carleton, 266
Cooper, Archibald, 336–338, 345n41
Cooper, Thomas, 260–261
Copper Rock, 109–110
Cosmologics (digital magazine project), 377
Council of Senior Islamic Scholars, 186
cow: eugenics and, 103, 169–170; in Hinduism, 103, 159–162, 168; Malthusian perspective on, 164–166, 168–169; Mayo on, 171–173; RCAI on, 161–166, 168–169, 171–173, 178n28; scientific agriculture and, 161–169
cow protection, 161; in colonial India, 160, 163–164, 167–168; Hindu nationalism and, 159, 173–176
Cow Question in India, The, 164
Crania Aegyptica (Morton), 261
Crania Americana (Morton), 261
creation care movement, 228n94
creationism, 8, 67–69, 267
Crichton, Michael, 221
crime, 332–333
critical approaches, as term, 5, 11, 12, 103, 145, 375, 377
critical Indigenous studies, 294, 299n71
critical race theory (CRT), 5, 8–10
Crowley, Aleister, 350, 358, 362
CRT. *See* critical race theory
cryobiology, 305, 309, 313–315, 317
Cryobiology (journal), 314–315
cryonicists, 324n57
Cuba, 358
Cudjoe (Kojo) (Maroon leader), 331
CWA. *See* Concerned Women for America
"Cyborg Manifesto, A" (Haraway), 81, 102

Dangerous Pregnancies (Reagan), 71
Dar al-Ifta website, 182, 184–186
Darrow, Clarence, 271
Dart, Raymond, 265

Darwin, Charles, 258–259, 262–265, 271
Davenport, Charles, 271
Davis, Angela Y., 37
Davis, David Brion, 242
Dawtas of the Moon, 3
Declaration of Independence, 261
decolonial theory, 5, 7, 293–294, 298n68, 299n70, 299n72
decolonization, 321n16
deep listening, 80
degeneracy, theory of, 259
DeGeneres, Ellen, 220–221, 227n79
deification, 102, 129–130, 142–144, 154
DeLaurence, William Lauron, 345n35, 371n68
Deloria, Vine, Jr., 299n70
demarcation problem (in science), 101
Democracy in Black (Glaude), 45
Democratic Republic of the Congo, 215
Denden-gu Shrine, 127–128, 138, 143
Denden-Myōjin (Bright God of the Electric Telegraph and Telephone), 127–128
Derrida, Jacques, 193
descent with modification, 265
designs: Black freedom, 52; colonial, 41–44, 46, 55–56; of neoliberalism, 52; of nonwhite others, 42, 44–45, 55; in racism, 45, 52, 56n1
Dharmapāla, Anagārika, 362
diabetes, 16n26
Dian Fossey Gorilla Fund, 220
difference, narratives of, 205
Digit (gorilla), 216–217, 220
Digit Fund, 217
disciplinary formations, 12–14
discipline, language and, 277
discrimination, design and, 56n1
District of Columbia v. Heller, 95–96
Divine Variations (Keel), 376–377
DNA, 318
Dobbs v. Jackson Women's Health Organization, 61, 82n1
Dobzhansky, Theodosius, 266
Dōshō, 128
double-hulled canoe. *See wa'a kaulua*

drag queen, 29
Dred Scott decision, 261
Drescher, Seymour, 233
drumming, 341
Du culte des dieux fétiches (Brosses), 354
Duke University Genomics, Race, and Identity Center, 257
Dunham, Katherine, 336–337, 346n44, 346nn49–50
Durkheim, Émile, 34, 70
Duster, Troy, 94
dynamic-nominalist process, 148–149

ecology, online, 182, 189–190, 192, 195
ecstasy, 212–213
Edenic narrative, on nature, 221–222
Edison, Thomas: Big Science and, 132, 155n17; deification of, 102, 129–130, 142–144; diary of, 139–140; Japan and, 102, 128–129, 131–132, 134–138, 143, 153; Japanese industry and, 141–143; lightbulb and, 102–103, 130–132, 137, 142, 153; Moore and, 131–132, 135–136, 156n32; shrines to, 128–129, 137–138, 143, 153; spiritualism and, 138–141, 144; Theosophical Society and, 140–141
Edwards, Bryan, 333–334, 344n24, 353
Egypt: Dar al-Ifta website, 182, 184–186; grand mufti, 181–182, 184–185, 196n1, 197n4
Egyptians, ancient, 261
Ehrman, Bart, 34–35
electrical experiments, obeah and, 353–354
Ellen Show, The (television program), 227n79
Elliott, Carl, 93
Else, Peter, 208
Emancipation, in Jamaica, 334, 347n66
Emancipation Proclamation (U. S.), 262
Emanuel African Methodist Episcopal Church, 18, 46, 51–52
Endeavor (ship), 285, 287–288
Enlightenment, 317; polygenism in, 204–205, 230–233, 248–250; racism in, 241–242
environmental grace, feminine, 209–210, 222
environmentalism, 221–222

Episcopal Church of America, 272
epistemic disconcertment, 352
epistemologies: animist, 282–284, 292, 294–295; border thinking, 293; Indigenous, 7, 282–284, 292, 294–295, 297n39; Jamaican Maroons and colonial, 340; Kānaka Maoli, 282–284, 292, 294–295, 297n39; spiritual-epistemological grace, 207; subaltern, 293–294
Epistle to the Romans (biblical book), 233–234
equivocal generation. *See* spontaneous generation
ergonic convergence, 148–149, 151, 153
Erving Goffman Archives, 38n11
esotericists, Western, 370n44; Crowley, 350, 358, 362; obeah and, 350, 362, 365; science and, 350, 358, 362, 364–365
Essay on Classification, An (Agassiz), 263
Essay on the Principle of Population (Malthus), 165
ethnology, 120, 122
eugenics, 180n65; AAAS on, 271–272; abortion and, 62, 66–67, 70, 72–73, 81; bioethics and, 90–91, 94–95, 97; cow and, 103, 169–170; evolution and, 271–272; Hindu nationalist, 174–176; India and, 103, 169–176; of Mayo, 170–173; transnational, 73
evangelical revivals, early 1830s, 114–115
evangelicals: on abortion, 64, 66–69; creation care movement, 228n94; creationism of, 67–69; gay rights opposed by, 68; racism and white, 258, 271
Evans, John, 92
Evans-Pritchard, E. E., 368n18
Eve (biblical figure), 229, 234, 256, 259–260, 267, 268. *See also* mitochondrial Eve hypothesis
evolution, 43, 57n11, 270; Catholics on, 256, 258; conservative Christians on, 256, 271; Darwin on, 258–259, 263–265, 271; descent with modification, 265; eugenics and, 271–272; in high school curriculum, 258; Lamarckian, 260;

Luyet and, 316; multiregional theory of human origin, 265–267, 269, 270; out-of-Africa replacement model, 267, *268*, 269, 270, 272–273; racism and, 264–266, 271–273; in *Scopes* trial, 271; U. S. surveys on, 255–256

Examination of Professor Agassiz's Sketch of the Natural Provinces of the Animal World and Their Relation to the Different Types of Man, An (Bachman, J.), 262

exception, state of, 294, 298n68

exclusive humanism, 59n28

existentialism, 312

experiment: in Catholicism, 323n44; Luyet on, 306–307, 309–310, 315, 318–319; obeah and electrical experiments, 353–354; in quantum physics, 357

Falwell, Jerry, 67

famine, 169

fanaticism, religious, opposition to, 50, 59n28

Fanon, Frantz, 44, 298n68

Farewell, The (film), 24

Farman, Abou, 324n57

Fatwa-Online, 186, 190

fatwas, 103, 183, 196; on Dar al-Ifta website, 182, 184, 186; by Egyptian grand mufti, 181–182, 197n4; on IslamOnline, 187–190; on Islamweb, 182, 188–189, 192; online ecology of, 182, 189–190, 195; Permanent Committee for Islamic Research and Fatwa, 185–186

Feldman, Noah, 58n26

feminine environmental grace, 209–210, 222

femininity, white, 204, 207, 209

feminist science studies, 4–6, 357

feminist theology, of abortion, 63–64, 80–82

feminist theory, 5–6; on fetal life, 76–77; mestiza feminism, 102

feminization, of religion, 6

Ferry, William, 112

fetishism, 354, 361

fetus, 61, 64, 75–77

Finney, Ben, 289–290, 297n49

fiqh al-aqaliyyāt (minoritarian jurisprudence), 187

Fire Thunder, Cecilia, 79

Fleck, Ludwik, 323n32

flood, biblical, 234, 256, 259, 264

Florimont school, 322n23

Floyd, George, 11

Ford, Christine Blasey, 35–36

forgery, 34–35

Forms of Talk (Goffman, E.), 31

Förster-Nietzsche, Elisabeth, 46

Fossey, Dian: antipoaching activities of, 215, 217–219, 227n80; biography and film on, 219–220, 227n79; DeGeneres on, 220–221; feminine knowledge and, 204, 209, 222; *Gorillas in the Mist*, 215–216, 218–219, 226n58, 226n63; grace and, 204, 207, 209–210, 222; Karisoke Research Centre, 215–216; murder of, 218–219, 227n83; Redmond and, 217, 219–220, 226n63

fossils, 264–265, 267

Foucault, Michel, 70

Fouts, Roger, 210

Fox, William Darwin, 262

fractionation, 313

Frame Analysis (Goffman, E.), 30, 33

Francis of Assisi (saint), 211, 225n20

fraud, 34

Frazer, James George, 356

Freedman's Inquiry Commission, 262–263

freedom, 326, 339–341

freedom, of religion, 3–4

French, Anishinaabe nations and, 107

frontstage, in social media, 195–196

Fujioka Ichisuke, 142

fundamentalist movements, 8

Galdikas, Biruté, 210

Galton, Francis, 168–169

Gandhi, M. K., 170

Garlow, James, 68–69

gay rights, opposition to, 68

Gender Advertisements (Goffman, E.), 25

Gender Trouble (Butler), 25

Genesis (biblical book), 229, 231, 234, 260, 267
Genesis and Development of a Scientific Fact (Fleck), 323n32
genetics, 11, 305; Answers in Genesis website and, 256–257; mitochondrial Eve hypothesis, 267, 268, 269; multiregional theory and, 266–267, 269, 270; nihilism and, 41–42; public on race and, 257–258, 270–273; settler colonialism and, 43; SIGMA, 16n26
genocide, 294
Genomics, Race, and Identity Center (GRID), 257, 272–273
genomic sequencing, 267, 269
geocentric solar system models, 269
geography: Kānaka Maoli, 287–288, 294; objectivist, 287
geopolitics, 293
Georgia state legislature, 78
Gerbner, Katharine, 376
Gibbes, Robert W., 262
Gillispie, Charles C., 42, 56n5
Glaude, Eddie, 45
Gliddon, George, 230
global frame, for science and religion studies, 375–376
Goffman, Alice, 39n21
Goffman, Erving, 22, 32, 39n21; on common value system, 36; Erving Goffman Archives, 38n11; *Forms of Talk*, 31; *Frame Analysis*, 30, 33; *Gender Advertisements*, 25; on lying, 18, 27–28, 36–37; on marginalized people, 28; obituary photographs of, 24–26; "On Cooling the Mark Out," 26–27; on performance, 28–29, 31, 35; *The Presentation of Self in Everyday Life*, 33; on ritual, 28, 34; self-help and, 29–30; on smiling, 24; *Stigma*, 33
golden ratio, 313
Gombe National Park, Tanzania, 210–214
Goodall, Jane: *The Chimpanzees of Gombe*, 211; conversion of, 213, 216, 225n37; ecstasy of, 212–213; feminine knowledge and, 204, 209, 222; grace and, 204, 207, 209–210, 222; *In the Shadow of Man*, 210–211; on moral evolution, 214, 225n28; public image of, 214, 222, 224n13, 226n40; *Reason for Hope*, 211–214; spirituality of, 210–214, 221, 225n20, 225n28
gorilla: conservation for, 217, 219–221; DeGeneres and, 220; Digit, 216–217, 220; Fossey and, 207, 215–221, 226n58, 226n63, 227nn79–80; Redmond and, 217, 226n63; threats to, 215–217, 219–220
Gorillas in the Mist (Fossey), 215–216, 218–219, 226n58, 226n63
Gottschalk, Peter, 356
Gould, Stephen Jay, 101, 210
grace: feminine environmental, 209–210, 222; secular, 204, 223n4; spiritual-epistemological, 207
Grafton, Anthony, 225n37
grand mufti, Egyptian, 181–182, 184–185, 196n1, 197n4
Gray, Asa, 263
Greeks, ancient, 15n20, 316–317
GRID. *See* Genomics, Race, and Identity Center
grimoires, 366, 370n44, 371n68
gross knowledge, 297n39
Guinea coast, Africa, 236–238, 241
Guttmacher, Alan, 72

habits, in new media, 194–195
Hachiman, 136, 155n21; Buddhism and, 132–135; Iwashimizu Hachimangū Shrine and, 128–129, 134, 137
Haeckel, Ernst, 260
Haiti, 334, 359, 361
Ham (biblical figure), 229, 259, 271
Handler, Jerome, 342n3
Haraway, Donna, 207, 323n37; on Christianity and science, 305; "A Cyborg Manifesto," 81, 102; *Primate Visions*, 222
Harrison, Ross Granville, 308
Harrison, William Henry, 120
Harrub, Brad, 267
Harvard Divinity School, 377
Harvard University, 262–263

Harvey, Graham, 297n37
Harvey, Sean, 111
Haudenosaunee, 284
Hawai'i: Cook in, 129, 143, 284, 288–289, 296n16; epistemologies in, 282–284, 292, 294–295, 297n39; KAHEA, 281; Kānaka Maoli activists in, 2–4, 281–283; Mauna Kea, 2–4, 281–284, 286, 289, 293–295; TMT for, 2, 281–283, 286, 289
Hayes, Harold, 219
Helmreich, Stefan, 320n7
Henneberg, Christine, 75–77
heresy, polygenism as, 205, 229
Herodotus, 249
Herskovits, Melville J., 336–337, 346n50, 346nn43–44
Heschel, Susannah, 47–48
heterosexual marriage, IVF in, 2–3
Heyerdahl, Thor, 289–290, 297n49
high school curriculum, 258
Hinde, Robert, 211
Hinduism: in British colonial discourse, 103, 163, 168; cow in, 103, 159–162, 168
Hindu nationalism: cow protection and, 159, 173–176; eugenics and, 174–176
Hindu Sangathan (Shraddhanand Sanyasi), 175
Hirabayashi Kinnosuke, 128, 143
historical methodology, 339–340
History, Civil and Commercial, of the British Colonies in the West Indies (Edwards), 333–334
History of Jamaica, The (Long), 232, 242–243, 245, 253n60
History of Science Society (HSS), 42–44, 46
Hochschild, Arlie Russell, 25
Hogan, Bernie, 195–196
Hogg, Donald W., 345n35, 371n67
Hōkūle'a, 289, 294–295, 296n22, 298n59, 299n70; as animist technology, 284, 292; Finney and, 290; Kane and, 290–292
Holocaust: bioethics and, 19, 87–91, 96–97; Nuremberg trials, 88–91, 96–97
Homo erectus, 265–266, 269, 270
hooks, bell, 277

Howard, Mary, 120
Howe, Samuel Gridley, 263
Howell, Leonard, 360
HSS. *See* History of Science Society
Human Condition, The (Arendt), 314
Human Destiny (Lecomte du Noüy), 225n28
humanity, of Africans, 245–248, 253n60
humanoid fossils, 264–265, 267
human origin: in Africa, 205, 264–265, 267, 268, 269, 270, 272–273; Darwin on, 258–259, 264–265; humanoid fossils in, 264–265; mitochondrial Eve hypothesis, 267, 268, 269; multiregional theory of, 265–267, 269, 270; narratives of, 204–205; out-of-Africa replacement model, 267, 268, 269, 270, 272–273; public on, 255–258
Hume, David, 231, 236, 249
Hurston, Zora Neale, 35, 337–338, 346nn49–50, 371n68
Huxley, Thomas H., 264
hyperlinks, 189–190
hyperreality, 190, 192, 196

ICR. *See* Institute for Creation Research
identity, Western, 7–10
Ige, David, 281
Iida Takesato, 137
immigration, 71, 173
impression management, 28–29, 31
inanimacy, 320n7
incandescent lightbulb. *See* lightbulb
incoherence, 278, 283–284, 350
India: colonial improvement in, 166–168; cow protection in colonial, 160, 163–164, 167–168; eugenics and, 103, 169–176; Hindu nationalism and cow protection, 159, 173–176; International Theosophical Society in, 359; Mayo on, 170–173; population growth in colonial, 168–170; RCAI on, 161–166, 168–169, 171–173, 178n28; scientific agriculture and colonial, 161–169
India, British colonialism in. *See* British colonialism, in India

Indian agents, U. S., 110–111, 116–120
Indiana state legislature, 61–63, 70, 81
Indian Service, U. S., 116–117, 120–121
Indigenous Americans. *See* Native
Indigenous animism. *See* animism
Indigenous peoples, 299n70; blood-freezing technology and, 304, 318–319; colonial theories on settlement, 297n50; decolonial theory and, 294, 298n68, 299n72; religion and rights of, 298n65; sovereignty of, 292–293, 295; state of exception and, 294, 298n68; stubbornness and, 294–295
Indigenous theory, 5, 7, 376–377; critical Indigenous studies, 294, 299n71; decolonial theory and, 299n72
Indigenous worldviews, 43; epistemologies, 7, 282–284, 292, 294–295, 297n39; life/not-life binary and, 306–307; ontologies, 8; postmodernism and, 278, 283
Institute for Creation Research (ICR), 67–68
integration, 45–46
International Theosophical Society, 359
In the Shadow of Man (Goodall), 210–211
Invention of Religion in Japan, The (Storm), 375
in vitro fertilization (IVF), 2–4
Iran, 358
Isaac, Benjamin, 15n20
ISIS, 194
Islam, 194; apps in, 192–193; Council of Senior Islamic Scholars, 186; fatwas by Egyptian grand mufti, 181–182, 197n4; feminist and queer scholars on, 6; Hindu nationalists on Muslims, 174–176; IVF and, 2–3; *salafi* and *wasati*, 188, 198n16; West encountering, 8
Islamic websites: appearance and usability of, 188–189; backstage and frontstage on, 195–196; fatwas on, 103, 182–192, 195–196; in online ecology, 182, 189–190, 192, 195; in securitized discourse, 193; as simulacra, 190–192; U. S. military and data from, 192
IslamOnline, 182, 191–192; fatwas on, 187–190; al-Qaraḍāwī and, 184, 187
Islamweb, 182, 184, 187–189, 191–192

Israel, 272
Itō Hirobumi, 136, 156n32
IVF. *See* in vitro fertilization
Iwadare Kunihiko, 142
Iwashimizu Hachimangū Shrine, 143; Edison and, 128–129, 134–135, 137–138, 141, 153; Hachiman and, 128–129, 134, 137

Jamaica: Emancipation in, 334, 347n66; obeah criminalized in, 325–329, 331–336, 341, 342n3, 354, 367n3; 1760 Slave Code, 327–328, 332; slavery in, 279, 325–333, 341, 353–354; Tacky's Revolt, 325, 329, 332–333, 341, 353–354
Jamaican Maroons, 344n17, 346n42; Africanness and, 336; in anthropology, 338; Bilby on, 325, 330, 339, 341, 347n59; British and, 326, 329–331, 339–340, 344n13, 347n66; British treaties with, 331, 339–340, 347n66; freedom and, 326, 339–341; historical methodology and, 339–340; Hurston on, 337–338; Kromanti of, 325, 339; Maroon Wars, 330–331, 339; medicine of, 330, 337; obeah and, 279, 325–326, 329–332, 335–338, 342, 342n2, 343n5; on science, 279, 325–327, 335–342, 342n2, 343n5, 347n59
James, William, 319
Jane Goodall Institute, 214
Japan, 154, 158n69, 375; abortion in, 72–73; Buddhism in, 127–128, 132–136; Denden-gu Shrine, 127–128, 138, 143; Edison and, 102, 131–132, 134–136; Edison and industry of, 141–143; Edison shrines in, 128–129, 137–138, 143, 153; Hachiman in, 132–137, 155n21; Iwashimizu Hachimangū Shrine, 128–129, 134–135, 137–138, 141, 143, 153; Meiji restoration, 135–136; Moore in, 131–132, 135–137, 156n32; Shinto in, 127–128, 132–137, 143, 155n23
Jefferson, Thomas, 260
Jeffries, John, 264
Jesus, 48
Jews, biblical, 234
Johnston, Charlotte, 114, 120

Johnston, George (Kahmentayha), 11, 105, 119; Christianity and, 117–118; as Indian subagent, 116–118; parents of, 106–107; H. Schoolcraft and, 116–118, 120, 122
Johnston, Jane (Obabaamwewe-giizhigokwe), 105–107; poetry of, 119; H. Schoolcraft and, 111–114, 119–120
Johnston, Susan. *See* Oshaguscodawaqua
Jonsen, Albert, 87, 91, 97
Josephson-Storm, Jason. *See* Storm, Jason Ānanda Josephson
Josiffe, Christopher, 359

KAHEA, 281–282
Kahmentayha. *See* Johnston, George
Kānaka Maoli: activists, 2–4, 281–283; animism of, 278, 282–284, 288–289, 292–295; epistemologies of, 282–284, 292, 294–295, 297n39; geography of, 287–288, 294; Hōkūle'a, 284, 289–292, 294–295, 296n22, 298n59, 299n70; Mauna Kea and, 2–4, 281–284, 286, 289, 293–295; TMT and, 2, 281–283, 286, 289; wa'a kaulua, 284–286, 288–292, 294–295, 296n16, 296n22, 296n26, 297n44, 298n59, 299n70
Kane, Herb, 290–292
Kant, Immanuel, 242
Karisoke Research Centre, 215–216
Katsura Kogorō. *See* Kido Takayoshi
Keel, Terence: on Christianity and racism, 15n20; *Divine Variations*, 376–377; on Nott, 230–231, 249–250; on SIGMA, 16n26
Keeler, Jacqueline, 79
Khimeini, Ruhallah al-, 197n4
Kidd, Colin, 236
Kido Takayoshi (Katsura Kogorō), 135–136
Kimmerer, Robin Wall, 283
Kipling, Rudyard, 160
Knox, Robert, 258
Koberlein, Brian, 282
Kohler, Robert, 225n20
Kojo (Cudjoe) (Maroon leader), 331
Koop, C. Everett, 65
Kopytoff, Barbara, 345n41, 346n42

Kora-Hon, Miad Hoyora, 359
Kortlandt, Adriana, 225n20
Kromanti, 325, 339
Kumer, Jerome M., 72

La Frontera/Borderlands (Anzaldua), 102
LaHaye, Beverly, 68
LaHaye, Timothy, 67–68
Lakatos, Imre, 356, 369n31
Lamarck, Jean Baptiste, 260
land bridge hypothesis, 272, 297n50
language, discipline and, 277
La Peyrère, Isaac, 57n11, 231, 233–234, 236, 248
latency, 279, 305, 309, 320n7
Latin Americans, SIGMA on, 16n26
Latour, Bruno, 323n37, 365, 369n29
Laveau, Marie, 371n68
Leakey, Louis S. B., 207, 212, 215
Lecomte du Noüy, Pierre, 225n28
Leeward Maroons, 330–331
life: latent, 309; life forms and forms of life, 320n7; life/not-life binary, 306–307, 319–320, 321n10; Luyet on, 305, 307, 310, 314–315, 320
life-denying orientations, 18, 46, 49–50, 55, 58n23
Life on Ice (Radin), 304, 318, 377
lightbulb: Edison and, 102–103, 130–132, 137, 142, 153; ergonic convergence and, 148–149, 153; Japanese industry and, 142; Swan and, 131
Limits of Tolerance, The (Adcock), 376
Lincoln, Abraham, 262
linkage disequilibrium, 269, 270
Lipsitz, George, 45
Listen, America! (Falwell), 67
Literary and Philosophical Society of Charleston, 261–262
Littlewood, Roland, 371n67
living constitutionalists, 95
Livingstone, David, 230
location data, 192
Lofton, Kathryn, 224n13, 226n40

Long, Edward, 241, 344n24; *The History of Jamaica*, 232, 242–243, 245, 253n60; on polygenism, 232–233, 242–244, 248; on slavery, 242, 245–249, 253n60

L. P. (author), 231, 234–236, 248–249

Luyet, Basil: archives of, 322n23; blood-freezing technology and, 305, 313–314, 318; Catholicism of, 278–279, 305–306, 308, 312, 320; on cell, 305, 310–312, 318, 320, 322n29; on cold, 315–317; on community fire, 316; cryobiology and, 305, 309, 314–315, 317; experimental philosophy of, 306–307, 309–310, 315, 318–319; on life, 305, 307, 310, 314–315, 320; phenomenology and, 312

Lyell, Charles, 258

lying: in academic life, 17–18; detection of, 23; forgery in New Testament, 34–35; E. Goffman on, 18, 27–28, 36–37; social science on, 23–24

Lynch, Gordon, 69–70

madhhabs (schools of law), 188, 192

madrasa (school), 197n4

magic, 280, 362, 367n6; African, 356–357; anthropology on science and, 356–357, 364; obeah and, 349–350, 354, 368n18; Western esotericists and, 358

magick, 362

magisteria, debates around, 277–278

magnetism, 334

Makimura Masanao, 135–137, 142–143

Malthus, Thomas, 164–166, 168–170

Managed Heart, The (Hochschild), 25

Mandrillus Sphinx (painting), 208

Marcel, Gabriel, 312

Maréchal, Joseph, 323n44

marginalized people, 28

mark, in confidence game, 26–27

Maroons, Jamaican. *See* Jamaican Maroons

Martin, Trayvon, 51

mass possession, 351–353, 362, 364

materialism: Atkins and, 244, 250; new, 139–140, 283, 376

Mauna Kea, 3–4; animism and, 282–284, 293–295; TMT, 2, 281–283, 286, 289

Mayo, Katherine, 170–173

McDonald, Sydney, 339

McKenny, Thomas, 116

McMurray, William, 114

medicine: of Jamaican Maroons, 330, 337; lying in, 24; medical ethics, 24, 87–88, 90, 97; obeah and, 342n3

Meeting the Universe Halfway (Barad), 76

Meiji restoration, 135–136

Men Before Adam (*Prae-Adamitae*) (La Peyrère), 234, 248

mesmerism, 332, 335

mestiza feminism, 102

Mexicans, SIGMA on, 16n26

Mexico border, migrants crossing, 11

Meyer, Manulani Aluli, 297n39

Mignolo, Walter, 293–294, 298n68, 299n70

migrants, 11

military, U. S., 192

Mills, Kenneth, 225n37

Ministry of Interior, Egyptian, 181

minoritarian jurisprudence (*fiqh al-aqaliyyāt*), 187

mitochondrial Eve hypothesis, 267, 268, 269

moderate (*wasati*), 188, 198n16

modernism, 278, 283

modernity, 364

Mohawk Interruptus (Simpson, A.), 292

monad, 260

monogenism, 236, 260, 262; biblical, 249, 256–257, 259; slavery and, 232–233, 248–250, 259, 271

Moore, William H., 131–132, 135–137, 156n32

morality: Goodall on moral evolution, 214, 225n28; Nietzsche on, 47–48, 57n17, 58n23; in public health policy, 18

Moral Majority, 67

Morton, Samuel, 261

Mother India (Mayo), 170–173

mountain gorilla. *See* gorilla

Mowat, Farley, 219–220

Mudd, Stuart, 72

INDEX 397

Mukharji, Projit, 283
multiregional theory, of human origin, 265–267, 269, 270
Museum of Comparative Zoology, Harvard University, 262
Muslims. *See* Islam
mustaftī, 182, 191
Myser, Cat, 93–94

Nanny (Maroon leader), 330–331, 339, 344n17
Narrative Journal of Travels (Schoolcraft, H.), 108–109
narratives: of human origin, 204–205; nature and Edenic, 221–222; rhetoric and, 203–204; secularization, 6
National Baptist Convention, 256
National Geographic magazine, 217
National Geographic Society, 207
National Institutes of Health, 317
nationalism, Christian. *See* Christian nationalism
nationalism, Hindu. *See* Hindu nationalism
nation-state, biopolitics of, 70–71, 73–74
Native: Christianity of, 114–115, 117–118, 122; land bridge hypothesis rejected by, 272; languages of, 111, 113–114; missions to, 112–113; polygenism and, 229. *See also* Anishinaabe; Oglala Lakota Nation; Ojibwe
Native Americans. *See* Native
native epistemologies. *See* Indigenous worldviews
Native Hawaiian. *See* Kānaka Maoli
Native science, 283–284, 292–294, 299n70
naturalism, 232–233, 236, 240
natural philosopher, 369n23
Navy Surgeon, The (Atkins), 237–241, 244
Nazism, 272
Neanderthal fossils, 264
NEC, 142–143
neoliberalism, 52, 54, 59n34
New Age religion, 140
new animism, 282–284

new materialism, 139–140, 283, 376–377
new media, 194–195
Newsweek (magazine), 267, 268
New Testament: Epistle to the Romans, 233–234; forgery in, 34–35; gospel of John, 48
Nietzsche, Friedrich: on nihilism, 41, 46–50, 55, 57n17, 58n23; *The Will to Power*, 41, 57n17, 58n23
nihilism, 18; active and passive, 58n23; Camus on, 49–50, 54; Christian, 41–42, 46–50, 53–54, 58n23; colonial designs and, 41–42, 46, 55–56; Nietzsche on, 41, 46–50, 55, 57n17, 58n23; scientific racism and, 41–42, 50, 53
Noah (biblical figure), 229, 234, 256, 259
Noble, Safiya, 56n1
noble savage, 304
nonbinary persons, abortion for, 82n11
nonwhite others, designs of, 42, 44–45, 55
North Atlantic universals, 348–350, 352, 354, 356, 363
Notes on Virginia (Jefferson), 260
Notre Dame (cathedral), 212
Nott, Josiah, 230–231, 249–250, 261–262
Nunberg, Geoffrey, 31
Nuremberg Code, 91
Nuremberg trials, of doctors, 88–91, 96–97

Obabaamwewe-giizhigokwe. *See* Johnston, Jane
obeah, 280; African-ness and, 336; Bilby on, 330, 342nn2–3; criminalization of, 325–329, 331–336, 341, 342n3, 348, 354, 359, 367n3; decriminalization of, 367n3; electrical experiments and, 353–354; in grimoires, 371n68; Jamaican Maroons and, 279, 325–326, 329–332, 335–338, 342, 342n2, 343n5; magic and, 349–350, 354, 368n18; medicine and, 342n3; superstition and, 332, 335, 348–349; trials for, 326, 333–336, 341, 353, 368n8; witchcraft and, 327–329, 348–349, 351, 367n6, 368n18

398   INDEX

obeah, as science, 329, 332, 348, 368n8, 371n67; Cassecanarie on, 358–363; Jamaican Maroons and, 279, 325–327, 335–342, 342n2, 343n5, 347n59; occult science and, 333–334, 345n35, 358, 360, 362–363, 365; scientism and, 350, 354, 365–366; spiritual workers and, 279–280, 350, 358–359, 363, 365–366; in Trinidad, 351–353, 355, 357–365; universals and, 349–350, 352, 366–367; Western esotericists and, 350, 362, 365
Obeah Simplified, the True Wanga (Cassecanarie), 358–363, 371n66
Obeyesekere, Gananath, 129, 138, 143
objectivism: animism and, 278, 283, 289, 292, 294, 295n10, 297n37; geography in, 287; new materialism on, 283; technologies of, 284, 288
occult science: Christianity and, 362; obeah and, 333–334, 345n35, 358, 360, 362–363, 365; scientism and, 358, 362–363, 365. *See also* esotericists, Western
Oglala Lakota Nation, 79
Ojibwe (Chippewa), 105–107, 113, 119; Christianity among, 114–115, 117–118; on Copper Rock, 109–110
Ōjin (legendary emperor), 134
Olcott, Henry Steel, 140–141, 360, 362
Omai, 285, 296n26
"On Cooling the Mark Out" (Goffman, E.), 26–27
online ecology, 182, 189–190, 192, 195
ontologies, 376; Indigenous, 8; of settler colonialism, 43–44, 46
Oprah (Lofton), 224n13, 226n40
orangutans, 243–244, 248
Order of the Golden Dawn, 358
Oreskes, Naomi, 228n94
orientalism, 8
originalism: bioethics and, 19, 88, 95–97; in constitutional interpretation, 88, 95–96
Origin of Species, The (Darwin), 263
Osborn, Henry Fairfield, 271

Oshaguscodawaqua (Susan Johnston), 106–107, 114
out-of-Africa replacement model, 267, 268, 269, 270, 272–273

Pacific Islanders, settlement of, 289–290, 297n49
Palestinian Arabs, 272
Palmié, Stephan, 360
Paracelsus, 229, 231, 260
passive nihilism, 58n23
Paton, Diana, 328, 342n3
patriarchy, Christian, 62–64, 66
Paul (saint), 213, 233–234
performance: E. Goffman on, 28–29, 31, 35; polygenism as antiorthodox, 231–232, 239, 241, 243, 248–249
performance studies, 25
Permanent Committee for Islamic Research and Fatwa, 185–186
personhood, of fetus, 61
perspectival geography, 287–288, 294
Petit Albert, The, 371n68
phenomenology, 312
Philomathean Society, University of Pennsylvania, 214
physical anthropology, 265–266
Piltdown Man, 264
Pinkola, Clarissa, 209
Planned Parenthood, 72
Planned Parenthood v. Casey, 74
Plato, 24
poaching, 215–219, 227n80
Poland, 11
police violence, studying, 377
polygenism: Agassiz on, 258–259, 261–263; Atkins on, 232, 235–241, 244, 249–250, 252n28; Bruno on, 229–231, 260; Darwin on, 259, 262–264; Hume on, 231, 236, 249; La Peyrère on, 57n11, 231, 233–234, 236, 248; Long on, 232–233, 242–244, 248; L. P. on, 231, 234–236, 248–249; Nott on, 230–231, 249–250, 261–262; Paracelsus on, 229, 231, 260; as performance of

INDEX 399

antiorthodoxy, 231–232, 239, 241, 243, 248–249; pre-Adamist, 260, 262, 271; racism and, 204–205, 230–233, 236, 242, 250–251, 260–263, 271; Voltaire on, 231, 235–236, 249. *See also* human origin polymorphisms, 318
Polynesian Voyaging Society, 284, 291–292
polyp, self-regenerating, 241
Popkin, Richard, 233–234
Popper, Karl, 101
population control, 18–19, 71–72
Population Crisis and the Use of World Resources, The (Mudd), 72
population growth, 168–170
possession, 351–353, 362, 364
Possessive Investment in Whiteness, The (Lipsitz), 45
postcolonial theory, 5–7, 298n68, 299n72
posthumanism, 376–377
postmodernism: animism and, 278, 283; nihilism and, 55
practical rationality, 129–130, 138, 143
Prae-Adamitae (Men Before Adam) (La Peyrère), 234, 248
pre-Adamism, 260, 262, 271
Prenatal Non-discrimination Act (PreNDA), 78
Prentice, Rockefeller, 313
Presentation of Self in Everyday Life, The (Goffman, E.), 33
Priest, Josiah, 259
Primate Visions (Haraway), 222
Principles of Biomedical Ethics (Beauchamp and Childress), 92
principlism, 92–94
privilege, 36
process social ontology: complexity thesis and, 145–147, 152; social kinds, 145, 147–149, 151–153
productive incoherence, 278, 283–284
Protestantism: eugenics and, 66; evolution and, 256; occult science and, 362; Reformation, 369n31
Providential Deism, 59n28

public health policy, 18
puériculture, 170
Putnam, Carleton, 266

Qaraḍāwī, al-shaykh al-, 184, 187
Qatar, 182, 184, 187–188
Qatari Ministry of Endowments and Religious Affairs, 182, 188
quantum physics, 357
queer theory, 5–6

race, 377; abortion and racial biopolitics, 62–63; Atkins on, 237–239, 244, 252n28; CRT, 5, 8–10; public on genetics and, 257–258, 270–273
Race After Technology (Benjamin), 56n1
racial segregation, 266
racism: Agassiz and, 262–263; Christianity and, 9, 15n20, 41–42, 46, 50, 53–54, 59n28, 110, 271–272; designs in, 45, 52, 56n1; Enlightenment, 241–242; evolution and, 264–266, 271–273; in high school curriculum, 258; in Israel, 272; of Long, 241–244, 248; in monogenism, 259–260, 271; nihilism and, 41–42, 50, 53; originalism and, 96; polygenism and, 204–205, 230–233, 236, 242, 250–251, 260–263, 271; scientific, 10–11, 18–19, 41–42, 46, 50, 52–54, 57n11; white evangelicals and, 258, 271. *See also* white supremacy
Radin, Joanna, 304, 318, 377
rationality, 129–130, 138, 143–144, 151, 356
Ray, John, 240–241
RCAI. *See* Royal Commission on Agriculture in India
Reagan, Leslie, 71
Reason for Hope (Goodall), 211–214
Rebel, The (Camus), 49
Redmond, Ian, 217, 219–220, 226n63
Reformation, 369n31
Reich, David, 269
Report (RCAI). *See* Royal Commission on Agriculture in India

reproductive justice activism, 77–79
Republican Party, 64
reverse habituation, 227n72
rhetoric, 203–204
Rio Moro, Trinidad, 354–356, 363; grimoires in, 366, 371n68; mass possession in, 351–353, 362, 364; spiritual workers in, 351, 361, 366
ritual, 28, 34
Rocklin, Alexander, 368
Rocks of Ages (Gould), 101
Roe v. Wade, 61, 74, 82n1
Romans, ancient, 15n20
Roof, Dylann, 18, 46, 50–54
Roost, Sophia, 320n7
Royal Commission on Agriculture in India (RCAI): Mayo compared with, 171–173; Report of, 161–166, 168–169, 171–173, 178n28
Royal Society, 284, 286–287, 334
rubella epidemic, 71–72
Rushdie, Salman, 197n4
Russia, 11
Rwanda. *See* Fossey, Dian

sacred: abortion and, 64, 69–70; Lynch on, 69–70
sacred cow. *See* cow
Sahai, Lala Hardev, 180n75
Sahlins, Marshall, 129, 138, 143
salafi (traditionalist), 198n16
salvage, scientific, 304–306, 318
Sapp, Jan, 322n29
Sartre, Jean-Paul, 312
Saudi Arabia, 185
Sault Ste. Marie, Michigan: evangelical revivals in, 114; Johnston family in, 102, 105–107, 111, 114, 117; H. Schoolcraft in, 107–108, 110–111
Scalia, Antonin, 95–96
Schaeffer, Francis, 64–65, 67, 69
Schaeffer, Frank, 64, 69
Schiebinger, Londa, 338
Schoen, Johanna, 73

scholarship: activism and, 17–18; as scripture, 17–18, 21–22
school (*madrasa*), 197n4
Schoolcraft, Henry Rowe, 105–107; *Algic Researches*, 118; Algic Society of, 113–114; on Christianity, 108–110, 112–113, 116, 118, 121–122; on Copper Rock, 109–110; ethnology and, 120, 122; expeditions under, 115–116; as Indian agent, 110–111, 116–120; on Indian languages, 111, 113–114; G. Johnston and, 116–118, 120, 122; J. Johnston and, 111–114, 119–120; *Narrative Journal of Travels*, 108–109
Schoolcraft, James, 112–113
Schoolcraft, William Henry, 111, 113
schools of law (*madhhabs*), 188, 192
Science (magazine), 271, 311
science, in Jamaica. *See* Jamaica; Jamaican Maroons
science, obeah as. *See* obeah, as science
science and technology studies (STS), 5, 13
scientific agriculture: breeding in, 167–168; in colonial India, 161–169
scientific Buddhism, 360
scientific claims, 151–152
scientific racism, 10–11, 18, 52, 57n11; bioethics responding to, 19; Christianity and, 41–42, 46, 50, 53–54; nihilism and, 41–42, 50, 53
Scientific Revolution, 149
scientific salvage, 304–306, 318
scientism, 356; feminist science studies and, 357; obeah and, 350, 354, 365–366; occult science and, 358, 362–363, 365
scientist, 354, 369n23
Scopes trial. *See Tennessee v. Scopes*
Scott, Dred, 261
scripture: scholarship as, 17–18, 21–22; science compared with, 21
Second Amendment, 95–96, 99n25
Second Great Awakening, 114
secular: Christianity and, 49–50; secularization narratives, 6
secular grace. *See* grace

securitized discourse, 193
segregation, racial, 266
seismic petroleum survey, 354–355
self-help, 29–30
service economy, 31–32
settler colonialism, 45, 55, 102; Christianity in, 106, 110, 122; Indigenous theory on, 299n72; ontologies of, 43–44, 46; science and religion in, 106, 121–122; TallBear on, 41, 43–44; Terra Nullius in, 286–287, 297n34
Shalin, Dmitri N., 38n11
Sharp, Andrew, 297n49
Sharpe, Granville, 248
Sheldon, Myrna Perez, 228n94
Shinto, 127–128, 143, 155n23; Buddhism and, 132–136; on science and technology, 136–137
Shoumatoff, Alex, 216
Shraddhanand Sanyasi, 175
SIGMA. *See* Slim Initiative in Genomic Medicine for the Americas
Simpson, Audra, 292, 294
Simpson, Leanne Betasamosake, 294
simulacra, 190–192
Simulacra and Simulation (Baudrillard), 190
Sisi, Abdel Fattah al-, 185
SisterSong, 77–78
situation, 312, 323n37
1619 Project, 10
Sixth and Seventh Books of Moses, The, 334, 371n68
Slave Code (1760) (Jamaica), 327–328, 332
slavery: Atkins on, 238; Atlantic slave trade, 3; Britain and, 245–247, 249, 253n60, 326, 329–331; evolution and, 264; in Jamaica, 279, 325–333, 341, 353–354; Jamaican Maroons and, 326, 329, 341; Long on, 242, 245–249, 253n60; monogenism and, 232–233, 248–250, 259, 271; polygenism and, 204–205, 230–233, 242, 248–250, 260–262, 271; in U. S., 96, 260–261
Slim Initiative in Genomic Medicine for the Americas (SIGMA), 16n26
smallpox vaccine, 181–182

smiling, 24–25, 27, 32
Smith, Anita, 335
Smith, Samuel Stanhope, 260
Smith, Wilfred Cantwell, 21–22
Smollett, Tobias, 239
Smyth, Thomas, 262
social kinds, 145, 147–149, 151–153
social media, 194–196
Somerset case, 245
sorcery, 368n18
South Carolina, 260–262
South Dakota, 79
Southern Baptist Convention, 256
sovereignty, 292–293, 295
Spallanzani, Lazzaro, 309
Spark Reproductive Justice Now, 79
speculation, 306–310, 320
speculative materiality, 278–279
spiritual-epistemological grace, 207
spiritualism, 138–141, 144, 358, 360. *See also* Theosophical Society
spiritual science, 358, 360, 362–363, 365
spiritual workers, 371n67; science and, 279–280, 350, 358–359, 363, 365–366; in Trinidad, 351, 361, 366
Spivak, Gayatri Chakravorty, 299n71
spontaneous generation, 240–241, 260, 309
Sri Lanka, 360, 362
Stand Firm (Brinkmann), 30
Stanford, Hardie, 339
Star of Gladness. *See Hōkūle'a*
state of exception, 294, 298n68
stem cell technologies, 308
Stengers, Isabelle, 356
sterilization, 72
Stewart, Dianne, 332, 342n3
Stigma (Goffman, E.), 33
Stoneking, Mark, 267
Storm, Jason Ānanda Josephson, 348–349, 367n6, 375
STS. *See* science and technology studies
stubbornness, 294–295
subaltern: epistemologies of, 293–294; speech of, 277; studies, 299n71

Sunni Islam. *See* Islam
superstition, 332, 335, 348–349, 364, 367n6
Supreme Court, U. S.: on abortion, 61–62, 70, 74, 76, 82n1; *District of Columbia v. Heller*, 95–96; Dred Scott decision, 261; *Planned Parenthood v. Casey*, 74; *Roe v. Wade*, 61, 74, 82n1; Scopes trial, 271
survivance, 294
suspended animation, 306, 309
Swan, Joseph, 131
symbiotic colony, cell as, 305, 310–312, 318

Tacky's Revolt (Jamaica), 325, 329, 332–333, 341, 353–354
Tahiti, 286, 292
TallBear, Kim, 41, 376–377; HSS address of, 42–44, 46; on life/not-life binary, 306–307, 319
Taney, Roger B., 261
Tanizaki Jun'ichiro, 153–154, 158n69
Tanzania. *See* Goodall, Jane
Taung boy, 265
Taylor, Charles, 59n28
Taylor, Telford, 88–91, 96–97
technological redlining, 56n1
technologies: animist, 284, 288–289, 292; objectivist, 284, 288
technology, Shinto on, 136–137
Tell My Horse (Hurston), 337
Tennessee v. Scopes (Scopes trial), 271
terrain, weaponization of, 11
Terra Nullius, 286–287, 297n34
terrorism, 193
thanatology, 89–90
Theosophical Society, 140–141, 358–360, 362–363
Theosophist (journal), 359
Thicknesse, Philip, 330
Thirty Meter Telescope (TMT), 2, 281–283, 286, 289
Thomas, Clarence, 62–63, 70
Thomism, transcendental, 323n44
Thompson, Bert, 267
Thompson, Nainoa, 290–291, 298n59

Thoughts on the Original Unity of the Human Race (Caldwell), 260
TMT. *See* Thirty Meter Telescope
Tobago, 360
tolerance, 376
Toshiba, 142–143
Tower of Babel, biblical, 256
traditionalist (*salafi*), 198n16
transcendental Thomism, 323n44
transgender men, abortion for, 82n11
translation, 277
transubstantiation, 306, 318
Treatise on the Following Chirurgical Subjects, A (Atkins), 236–237
Trembley, Abraham, 241
Trinidad, 349; grimoires in, 366, 371n68; mass possession in, 351–353, 362, 364; obeah in, 351–353, 355, 357–365; Rio Moro, 351–356, 361–363, 366, 371n68; science in, 351–355, 357–365; seismic petroleum survey in, 354–355
Trouillot, Michel-Rolph, 348
Trump, Donald, 68–69, 194
trust, 77–82
#TrustBlackWomen campaign, 77–78
Tuck, Eve, 321n16
Turkey, 2–3
Turner, Victor, 28
Types of Mankind (Gliddon and Nott), 230

Uganda, 215
Ukraine, Russian invasion of, 11
United States (U. S.): Black witchcraft in, 3–4; Christian nationalism in, 19, 62–69, 79–81; constitutional interpretation in, 88, 95–96; evolution surveys in, 255–256; Freedman's Inquiry Commission, 262–263; freedom of religion statutes, 3–4; Georgia state legislature, 78; Indian agents, 110–111, 116–120; Indiana state legislature, 61–63, 70, 81; Indian Service, 116–117, 120–121; Indigenous peoples and, 299n70; Mayo and immigration to, 173; military and Islamic data, 192; polygenism in, 260–262; public on

genetics and race, 257–258, 270–271; racial segregation in, 266; Second Amendment, 95–96, 99n25; slavery in, 96, 260–261; Supreme Court, 61–62, 70, 74, 76, 82n1, 95–96, 261, 271. *See also* Indians, American
Unite the Right rally (2017), 50
Unity of the Human Races Proved to Be the Doctrine of Scripture, Reason, and Science, The (Smyth), 262
universals, 278, 280; North Atlantic, 348–350, 352, 354, 356, 363; obeah and, 349–350, 352, 366–367; religion as, 348–349, 352, 356; science as, 348–349, 352–354, 363
University of Pennsylvania, 214
University of the West Indies Mona, 335, 345n41
updating, in new media, 194–195
Updating to Remain the Same (Chun), 194–195
usability, of websites, 188–189

vaccines: antivaccine activists, 366; fatwas on, 181–182
values, 17, 47, 49, 58n23
Veach, Lacy, 291, 298n59
Venus, transit of, 284, 286–287
Vesey, Denmark, 51–52
vitalism, 140
vitrification, 309
Viznor, Gerald, 294
Voltaire, 231, 235–236, 249
Voyage to Guinea, Brasil, and the West-Indies, A (Atkins), 237–238

wa'a kaulua (double-hulled canoe), 297n44; as animist technology, 284, 288, 292; Cook and, 284–286, 288, 296n16, 296n26; Hōkūle'a, 284, 289–292, 294–295, 296n22, 298n59, 299n70
Wallace, William, 130
wanga, 359, 361–362
wasati (moderate), 188, 198n16
Waubojeeg, 106–107
wayfaring, 278, 290–291
weaponization, of terrain, 11
Weaver, Sigourney, 219, 227n79

Weber, Bill, 220
websites, Islamic. *See* Islamic websites
Weidenreich, Franz, 265–266
Weingarten, Karen, 62
Well Versed (Garlow), 69
West: identity of, 7–10; ontological divisions of, 43–44; white supremacy in, 9–10
West, Cornel, 55
Western esotericists. *See* esotericists, Western
Whatever Happened to the Human Race? (film series), 64–65, 69
Whatever Happened to the Human Race? (Koop and Schaeffer, Francis), 65
Whewell, William, 369n23
white Christian nationalism. *See* Christian nationalism
White Coat, Black Hat (Elliott), 93
white evangelicals, racism and, 258, 271
white femininity, 204, 207, 209
whiteness, in bioethics, 93–94
white supremacy, 9, 44, 376, 378; abolition and integration in, 45–46; Christianity and, 50, 59n28, 110; evolution and, 272; nihilism and, 46; in physical anthropology, 265–266; scientific racism and, 10–11, 42, 50, 53–54
Wiegandt, Kai, 223n4
Williams, Daniel K., 67
will to power, 48
Will to Power, The (Nietzsche), 41, 57n17, 58n23
Wilson, Allan, 267
Wilson, E. B., 322n29
Wilson, E. O., 65
Windward Maroons, 330–331
Wisdom of God Manifested in the Works of the Creation, The (Ray), 240
witchcraft: Black, 3–4; in Britain, 328; obeah and, 327–329, 348–349, 351, 367n6, 368n18
Witchcraft, Oracles and Magic Among the Azande (Evans-Pritchard), 368n18
Witchcraft Act (1736) (Britain), 328
Witgen, Michael, 107
Wolpoff, Milford, 57n11, 266–267
Woman in the Mists (Mowat), 219–220

Women Who Run with the Wolves (Pinkola), 209
Woods, Rebecca, 178n36
World and All the Things Upon It, The (Chang), 287–288
World Church of the Creator, 272
Wynter, Sylvia, 45

Yang, K. Wayne, 321n16
Year of the Flood, The (Atwood), 220

Zaire, 215
Zimmerman, George, 51
zones of creation theory, 258, 261–263